Managing Open Innovation

Special acknowledgement

This research benefitted from financial support from the Research Foundation Flanders – FWO Vlaanderen (G014708N).

Managing Open Innovation

Connecting the Firm to External Knowledge

André Spithoven
Belgian Science Policy Office and Vrije Universiteit Brussel, Belgium

Peter Teirlinck
Hogeschool-Universiteit Brussel and Vrije Universiteit Brussel, Belgium

Dirk Frantzen
Vrije Universiteit Brussel, Belgium

Edward Elgar
Cheltenham, UK • Northampton, MA, USA

Published by
Edward Elgar Publishing Limited
The Lypiatts
15 Lansdown Road
Cheltenham
Glos GL50 2JA
UK

Edward Elgar Publishing, Inc.
William Pratt House
9 Dewey Court
Northampton
Massachusetts 01060
USA

A catalogue record for this book
is available from the British Library

Library of Congress Control Number: 2012935312

MIX
Paper from
responsible sources
FSC
www.fsc.org
FSC® C018575

ISBN 978 1 78100 020 5

Printed and bound by MPG Books Group, UK

Contents

Foreword

It is almost a decade since open innovation became a buzzword in the management literature on innovation. Open innovation emphasises the benefits to be had from searching beyond the company ramparts for new ideas, relevant cooperation partners and alternative ways of making money from innovation. Several tendencies have enhanced this process: universities have become mature players in the economic system and, in commercialising their scientific output, operate increasingly as entrepreneurial entities; the standard of education has risen significantly around the world, globalising the search radius for skilled labour; the greater availability of venture capital has stimulated start-ups and spin-offs because it allows ideas to be commercialised. These tendencies have led to an abundance of useful knowledge beyond the company walls. At the same time, it is increasingly common for firms with innovations in the pipeline to seek to commercialise them outside the company.

Though open innovation is quite a business-driven phenomenon, it would be short-sighted to ignore its importance to science, technology and innovation policy. The OECD Working Party on Innovation and Technology Policy (TIP) commenced a project on "Globalisation and open innovation" in 2006, drawing to some extent on earlier OECD work on the internationalisation of R&D. The Belgian Science Policy Office participated enthusiastically in both projects and this resulted in the bringing-together of two key dimensions of firm innovation strategy: the internationalisation of R&D and increasing openness in innovation. The main consequence of these broad tendencies is that national innovation policies can no longer be designed purely in the context of geographically-bound, knowledge-based activities or of vertically-integrated value chains in business. The main lesson that was learned from these projects was the need to synchronise public policies with today's open innovation management by creating favourable framework conditions.

Although much is written about open innovation it has not acquired the status of formal theory, because it is built on gradual improvements in our understanding of its constituent elements. From a small open economy perspective this book addresses four important issues deserving of further attention within the stream of open innovation literature: the inclusion of company heterogeneity so as to understand differences in company behaviour when it comes to open innovation; the relationship between public funding and a firm's innovation networking behaviour; the impact of R&D outsourcing on internal R&D activities; and the role of human resources in the process of deciding how to engage in open innovation practices. This book is a collection of empirical essays authored in recent years to address the need to better understand open innovation in these four areas. They are founded on official micro-level data on innovation and R&D, most of which was collected in Belgium through surveys by the Federal Cooperation Commission on Statistics (CFS/STAT). The essays in this volume can be considered as a source of inspiration and reflection on open innovation management at both corporate and policy-making level.

Dr Philippe Mettens
Chairman of the Board of Directors
Belgian Science Policy Office

General introduction

INTRODUCING OPEN INNOVATION

The process of innovation is dynamic and complex, and entrepreneurs have to deal with, and adapt to, the ever-changing environment in which they operate. Many textbooks, therefore, start by listing innovation models from every conceivable perspective (see, for example, Dodgson et al., 2008; Schilling, 2008; Tidd et al., 1997; Afuah, 1998; Trott, 1998; Rothwell, 1992), or by showing these models' sequential rise and fall (for example, Chesbrough, 2006a; Rothwell, 1992; Rothwell and Zegveld, 1985; Trott, 1998). Open innovation has been welcomed as the most recent member in this family of models. Authors rarely explain the precise demarcation or transition between the stages of the different innovation models. Chesbrough is one of the first, however, to undertake such an effort in describing the 'erosion' of the closed innovation model and transition to the open innovation model.

It is claimed that innovation is entering a new 'era' involving deliberate and intense knowledge exchanges between firms and other organisations (Chesbrough, 2003b). At the firm level, knowledge exchanges are used as a resource or as a strategy to generate additional revenue (Gassmann, 2006). To channel these exchanges and marry them to potential internal R&D efforts, the concept of the business model is revised and becomes central to the firm's innovation strategy (Chesbrough and Rosenbloom, 2002; Chesbrough, 2003a; Chesbrough, 2006b; Huston and Sakkab, 2007; Chesbrough and Schwartz, 2007). Thus far, the storylines on open innovation appeal to practitioners, and the majority of studies in this field read very well indeed. However, from an academic perspective the concepts are harder to grasp and there is a need for further empirical evidence based on large-scale datasets. By looking at several facets of open innovation that are relevant to innovation management and policy-making alike, and applying different measurement techniques, this book aims to bridge at least part of that gap.

This general introduction aims to position the ideas on open innovation in the literature. To prevent distortions from the specificities of national innovation systems the focus is mainly on Belgium. In the remainder of the introduction the main ideas on open innovation are highlighted by looking at the context in which they have taken root.

The concept of open innovation is attractive on two counts. On the one hand, it pretends to offer a new analytical framework to explain the process of innovation. On the other, it promotes a management toolbox containing practical guidelines on the organisation of innovation.

Like sports and fashion, open innovation is a topic on which practitioners and scholars are divided. It has attracted a huge amount of attention. It has received a great deal of hype and in a short space of time it has improved the academics' chances of getting published. Open innovation has given rise to special issues (for example, in R&D Management in 2006 and 2009; in Industry and Innovation in 2008 and in Technovation and Technology Analysis and Strategic Management in 2010) and takes centre stage in OECD studies (OECD, 2008c). The issue of open innovation is highly relevant, at least by contemporary standards. All of a sudden, it seems, open innovation has become the buzzphrase to explain everything there is to know in the field of innovation management and beyond. It is a textbook example of a successful crossover between management literature and economics.

But all this exposure has invited fierce criticism. Hardly surprising, given the diversity of the ideas and the fuzziness of its key concepts (Gann, 2005; Helfat, 2006). Though they haven't pronounced it dead, the critics frequently question the shelf-life of open innovation. However, although this criticism does contain elements of truth, much of it lacks a more satisfactory alternative to the majority of open innovation's central themes. In these cases, there is always the danger of throwing out the baby with the bath water.

The 'forces' or 'triggers' underlying open innovation have been discussed at length because there was a need to explain the changing innovative behaviour of enterprises. The genesis of open innovation is related closely to, even defined by, the model that precedes it: that is, closed innovation (Chesbrough, 2003a; Dahlander and Gann, 2007). The key facilitator for open innovation is the observation that knowledge is omnipresent in a firm's environment (Quinn, 2000): knowledge is distributed over various actors (von Hippel, 1988; Pisano, 1990; Lundvall, 1992; Rothaermel et al., 2006; Schilling, 2008); and knowledge is accessible through specialised markets (Arora et al., 2001; Howells, 1999; Chiesa et al., 2004; Cesaroni, 2004). The innovation process too is characterised by an increasingly distributed nature (Coombs et al., 2003; Acha and Cusmano, 2005; Howells et al., 2003).

Technical progress alters the knowledge requirements of firms (Pisano, 1990) and increases the complexity of the environment in which they operate. Therefore, because of this wide distribution of knowledge, firms need to tap into complementary knowledge and reach beyond their boundaries to extend their innovative activities.

A definition may well be in order before we go on to discuss the theoretical underpinnings of open innovation. Chesbrough summarises by saying that 'open innovation is the use of purposive inflows and outflows of knowledge to accelerate internal innovation, and expand the markets for external use of innovation, respectively' (Chesbrough, 2006a: 2). Though admittedly fuzzy, this definition is a good place from which to launch our discussion of the most salient theoretical underpinnings of open innovation.

Their originator formulates his ideas on open innovation according to six general principles (Chesbrough, 2003b) and, in his endeavour to demonstrate open innovation's validity as a new innovation management theory, he sums up eight elements by which it can be differentiated from the earlier, closed innovation model (Chesbrough, 2006a). Hence the – at least partial – explanation for the fuzziness of the concept: it aims to become an overarching 'theory' through which all the relevant facts can be explained, then morphs into a fully-fledged paradigm. As and when criticism is levelled, the theory adapts incrementally. The diversity of these theoretical 'building blocks' accounts for its multifaceted appearance and the rigorous task of selecting the relevant theories on which open innovation is grounded.

But what economic factors changed so dramatically that businesses were forced to adapt their innovation strategies? And how did these altered strategies influence the key characteristics of their innovation processes? How did this 'erode' the prevailing – that is, closed innovation – model? Chesbrough (2003a; 2006b) identified many factors. These can be reclassified into three categories: (i) supply side factors: a larger quantity and higher quality of skilled workers due to increased labour mobility; increased availability of intellectual property rights (IPRs); and the expansion of venture capital; (ii) technical factors: technical progress reducing the minimum scale of production to boost specialisation; the adoption of technical standards to facilitate an exchange of resources between firms; and new technologies affecting communication procedures; (iii) geographical factors: globalisation and internationalisation to extend a firm's commercialisation markets and the search environment for offshore partners and ideas. An increasing reliance on the knowledge present outside the boundaries of the firm has particular implications for those who operate in a small open economy context. Often, in this case, external knowledge can only be found outside the national borders (Spithoven and Teirlinck, 2006).

Chesbrough further highlights the open innovation model by couching his discourse on open innovation largely in monetary terms. Chesbrough's insights, though not explicit, imply a dynamic element that runs from closed to open innovation. Temporality, as implied by the shortening of the product cycle, should reduce the base in market turnover. Likewise, internal development costs rise and require longer depreciation periods, enlarging the base at the cost side. In addition, reducing development costs by partly externalising them does not reveal the origin of the leverage effect.

However, it is in this tying of the internal costs to the external and in the marketing of innovation through internal and external mechanisms, that the open business model acquires its meaning (Chesbrough, 2006b). Therefore, the business model is the internal mechanism for turning innovation to profit. It shows how R&D costs relate to profit from innovation. Profit, as emphasised by classical economists such as Ricardo and Marx, is the very essence of economics. This is no different for individual firms, for whom profit is the main target (Teece, 1986; Afuah, 1998; Chesbrough, 2003a; Laursen et al., 2007). Of course, firms aim to turn a profit and so have to compete. Innovation is one of the main ways to stay competitive. Wheelwright and Clark (1992) see three elements as critical here: international competition; fragmented markets; and rapid technical progress. Since profit is the main goal, and innovation is critical to it, any measurement of open innovation would need to take it into consideration.

Turnover results from dealings on the market and costs are incurred through buying on the market. Hence the stress that Chesbrough lays on the emerging knowledge and technology markets, a phenomenon already captured by many others (Howells, 1999; Arora et al., 2001; Chiesa et al., 2004). Chesbrough (2006b) also points to the existence of a greater number of innovation intermediaries. These elements are receiving more attention (Chiesa et al., 2004; Howells, 2006), yet such market relations are captured only by traded interdependencies, as Storper (1997) calls them. In several cases, untraded interdependencies are far more important. Since the ideas on open innovation are largely cast in monetary terms – that is, traded interdependencies – their shortcomings are similar to those associated with transaction cost economics (Storper, 1997). Nonetheless, the business models concept is also intended to capture value from untraded interdependencies. These do not, however, appear in Chesbrough's ideas on open innovation.

TYPES OF OPEN INNOVATION

Firms, like people, develop in different social and physical surroundings and so they see the world of open innovation differently (Nooteboom, 2001). The elements contained in the ideas on open innovation imply that there are different processes at work, and this leads to 'archetypes' (Gassmann and Enkel, 2004). Chesbrough (2003a) and Chesbrough and Crowther (2006) discuss these using the terms 'inbound' and 'outbound' open innovation. The former refers to the search strategy of finding external information sources to complement in-house R&D activities; the latter stresses external paths to markets in order to commercialise the innovation. Gassmann and Enkel (2004) add a third process – the 'coupled' process – in which strategic alliances unite both inbound and outbound open innovation (see also Jacobides and Billiger, 2006). The first three chapters concentrate on firm heterogeneity and highlight the coupled process by looking at incoming knowledge spillovers, research collaboration and strategic appropriability.

Obviously, even within these 'archetypes', open innovation means different things to different writers. In addition to inbound and outbound models, Dahlander and Gann (2007) see room for open models to which informal and formal network relations are central. Laursen and Salter (2006) distinguish between the variety of external sources that channel into the firm (so-called 'breadth') and the intensity by which they are used (referred to as 'depth'). These elements offer an indication of the appreciation of network relations (Spithoven and Teirlinck, 2005).

The OECD (2008c) refines the types of open innovation in two ways. First, the degree of openness is coupled to the strategic autonomy pursued by the firm and the time horizon corresponding to it. Some activities, like in-licensing or outsourcing R&D projects, imply that firms are able to access technology in a short space of time at the expense of losing some of their independency. This carries the potential threat of technological dependency. The traditional, vertically-integrated or 'closed' innovator can devote greater effort to time-consuming in-house development, in which case the firm's autonomy is protected against unwanted spillovers.

Second, in line with the general ideas connected to the literature on the outsourcing of R&D, the OECD (2008c) differentiates between (a) core, peripheral and step-out activities in the target market and (b) the technology used. Some activities, such as internal R&D, are core to the technologies used and are carried out to supply core markets; other activities, such as licensing, are peripheral to markets and technology. Finally, external corporate venturing and selling are step-out technologies offered in step-out markets. The OECD (2008c) recognises various hybrid forms in which some open innovation practices find a home. However, both of these OECD refinements (2008c) are convincingly argued derivatives of inbound and outbound types of open innovation. The refinements do, however, support Nooteboom's argument (2001) that firms differ in their open innovation strategies in response to different social and physical environments.

To date, the bulk of the literature on open innovation draws on case study material and is, therefore, criticised as reliant on 'anecdotal' evidence. The use of select cases may indeed point to the emergence of a new business practice. Chesbrough (2006a) acknowledges this lack of external validity. But case studies have the advantage of accommodating the richness and diversity of the concepts and dimensions; putting names to the faces of open innovation. Statistical yardsticks and econometric methods also have their drawbacks. Obviously, there is a greater need for integration and reconciliation between these two approaches to the identification of open innovation practices and their effects. This book relies heavily on statistical methods to discuss open innovation practices at some length, but does not fully capture the loss of the dynamic nature of innovation and there is a clear need for additional theory-building from case studies.

Previous paragraphs in this section have demonstrated that open innovation has not yet matured into fully-fledged theory. West et al. (2006) refer to open innovation as a 'set of practices' rather than as a theory. But this set of practices still contains blind spots and blank spaces to be filled by academics and practitioners.

It would be nothing short of a miracle if innovating ideas were to be flawless at the first attempt. Due to the many ideas that open innovation throws up, Chesbrough shoots wide of the target in several cases, but there is much work still to be done (West et al., 2006). This book addresses four areas for further development from both the theoretical and the empirical perspective. These areas include: addressing firm heterogeneity in innovation performance due to the relative fixation on firm-level analysis in general and large firms in particular; stressing the role of government in shaping corporate open innovation practices – especially in terms of networking; focusing on the implications for internal research and innovation activity of the growing tendency towards R&D outsourcing; and highlighting the role of internal R&D personnel in a firm's decision to engage in open innovation practices. These topics are relevant both to open innovation management in firms and to policy-making in the area of open innovation practice. They also form the backbone of this book.

PART ONE: FIRMS REACT DIFFERENTLY WHEN MANAGING OPEN INNOVATION

The first three chapters of this book consider the issue of heterogeneity in firm innovation behaviour and performance. One drawback with empirical analyses is that, whatever the firm population, they provide information on the mean behaviour of firms. However, management science is just as interested in the behaviour at the extremes – that is, in firms that perform exceptionally well, or exceptionally poorly. This is especially relevant if we consider that open innovators are most likely to be found among the more dynamic firms (Spithoven et al., 2010). Moreover, the fiction of a 'representative firm' may be highly misleading in view of the heterogeneity of innovative behaviour. This is of particular relevance, because open innovators are likely to be found among the more dynamic firms in particular.

Therefore, analyses must enable a systematic identification of the behavioural differences between poor and strong innovative performers, with regard to both innovation output and 'black box' content when it comes to describing the external knowledge environment in which the firm operates. The challenge is to incorporate these elements in a general framework with which to analyse the relation between innovative efforts and firm performance and stress the fact that each firm has its own specificities.

The first chapter uses a structural model to explain R&D intensity and innovative turnover and their impact on growth of turnover and employment. The empirical model is estimated using data on Belgium from the third wave Community Innovation Survey. Average regression estimates suggest that incoming knowledge spillovers, research cooperation, appropriability, workers' qualifications and export orientation all have a positive influence on R&D investment, and, through this, on innovative turnover and employment. (For a more elaborate argumentation along this lines, see Spithoven et al., 2010.) Quantile regressions are used to look into the behaviour of different types of firms and indicate that the average impact of incoming knowledge spillovers and research cooperation tends to increase quantile by quantile, whereas the opposite holds true for appropriability. These results corroborate the predictions of open innovation theory. They suggest that innovative firms devote an increasing portion of their R&D effort to the absorption of external knowledge, and that, among the strongest innovative performers in particular, we can find firms that are prepared to sell excess knowledge in disembodied form.

The second chapter focuses on the impact of incoming knowledge spillovers, research collaboration and strategic appropriability on research intensity and product innovative performance on a large sample of firms in

Belgium. Again, quantile regression is applied to differentiate between weak and strong innovators. Knowledge-related variables are shown to exert an indirect effect through research intensity on innovative turnover. The impact of knowledge-related variables on research intensity differs greatly from firm to firm. It would seem that this is especially the case as regards reliance on strategic appropriability.

Since the first two chapters focus on Belgium, the third seeks external validation. We estimate a model on a country-basis to explain research intensity and innovative turnover using harmonised, firm-level data for Belgium, Germany, Norway, Spain and Portugal. We also consider alternative measures of research intensity. Average estimates show that knowledge-related variables in particular, such as incoming knowledge spillovers, research cooperation and appropriability, help explain research efforts. In-house research intensity matters most in explaining innovative turnover in the more technologically advanced countries, whereas total research intensity, which also covers expenditure on advanced machinery, does so in the less advanced. Quantile regressions indicate that the joint relative impact of incoming knowledge spillovers and research cooperation tends to increase quantile by quantile in relation to that of appropriability, and that, if anything, the impact of research intensity on innovative turnover grows in the higher quantiles. This would suggest that dynamic innovators have a more open and efficient innovation process.

PART TWO: OPEN INNOVATION AND PUBLIC FUNDING PRACTICES

Public funding can be of benefit to the innovation process and the use of external knowledge in private firms. The main argument in favour of public intervention in R&D and innovation processes lies in the existence of market failures. Due to these failures unintended knowledge spillovers occur and the appropriability of benefits from the innovation hampers further activity. Enterprises are then assumed to avoid R&D investment through a lack of the necessary incentives. Therefore, support of innovation is a major ingredient of science and technology policy, which aims ultimately to enhance entrepreneurial performance and therefore stimulate economic growth.

The extent to which open innovation activities in firms benefit from public funding is less clear and is the subject of the second part of this book. The aim is to find a better understanding of the determinants and outcomes of research cooperation behaviour, with the focus on industry-science networking in particular. Special consideration is given to the potential influence that public funding exerts on research and innovation. Though the rationale for the public funding of innovation lies on firm theoretical ground, the empirical literature is ambiguous in its assessment of the benefits of public funding to R&D and innovative activity, and there is little consensus on the efficacy of R&D subsidies. This lack of consensus can be attributed to several factors. Among them we have a potential endogeneity problem, since public aid is not independent of other variables, and there is a risk of sample selection bias due to the existence of a distribution process, at the end of which a difference arises between the firms that benefit and those that do not.

Second, much of the existing literature on the effects of public funding revolves around the (mostly linear) relationship between input and output. The new ideas on open innovation, however, are well acquainted with the role of the business model in externalising innovation and making a profit from innovations not confined to the core business of the firm. In this respect, many public authorities have addressed the risk taken by enterprises to engage in innovation, or stimulated specific forms of innovative activity such as research cooperation, by providing public funding.

Chapter 4 draws on the results of the third-wave Community Innovation Survey to analyse the effects of public funding of innovation. These effects are predominantly viewed in terms of research intensity and innovative turnover. However, the way in which innovative networking at firm level indirectly influences the relationship between the funding and the effects remains a black box mystery. The analysis takes into account different sources of public funding which are regional, federal or European and uses quantile regression analysis to account for heterogeneous entrepreneurial performance. The main findings are that – in contrast with the absence of influence of funding by central government and EU funding other than the Framework Programmes – innovation funding by regional governments and by the EU Framework Programme has a positive direct impact on research intensity. Moreover, an indirect effect can also be found in the sense that this funding is positively related to and even targets involvement in research cooperation, which in turn positively affects research intensity. Therefore, funding policies can be assumed to contribute to the opening-up of the research process. In the relation between source of funding and research cooperation, both the type (industry-science versus inter-firm) and the geographical scope (domestic versus international) of cooperation matter. The indirect effect is strongest for weak innovators, indicating the highest need for public intervention to open up the research process for these innovators. In terms of innovative turnover, funding showed only an indirect influence through research intensity. The impact of the latter turned out to increase (quantile by quantile) with the strength of the innovator.

Chapter 5 analyses the impact of public funding for innovation on research cooperation by firms. This time, use is made of data provided by the third and fourth Community Innovation Surveys (CIS3 and CIS4). The chapter applies an improved econometric technique to measure the impact of different types of public funding on different types of research cooperation. The results show that there is a substantial impact of total public funding and that its statistical significance is highest in the case of cooperation with public research organisations. A division of the total funding shows that regional government funding in Belgium has, by far, the greatest impact. There is also a more modest influence, exerted by federal government funding, mainly on cooperation with business, whereas EU funding has an impact when combined with regional funding, and mainly on cooperation with public research organisations. These differences in impact can be explained by the organisation of science and technology policy in Belgium and by the distinct aims pursued by the respective public funding agencies.

Chapter 6 analyses the behavioural additionality of subsidies based on regional and EU framework program public funding awarded to business enterprises for the 'establishment' of research cooperation between industry and science. Account having been taken of their specificities, the science component distinguishes between universities, which are considered to be more (basic) research oriented, and public research centres, which tend to be oriented more towards applied research and the commercialisation of research findings. The main result, drawing on firm-level data provided by the OECD bi-annual business R&D surveys, is that funding by regional governments fosters the instalment of industry-science research cooperation. However, this positive effect is found only in the case of cooperation with public research centres and not with universities. This might be explained by the importance given to the prerequisite of commercialisation of research in the case of funding by regional governments. Public funding provided under the EU framework program is found not to have an impact on the instalment of industry-science cooperation, either with universities or with public research centres. This could be due the fact that EU funding is aimed at firms that are already cooperating and does not favour the set-up of new cooperation. However, the persistence over time of industry-science cooperation, as found in the analysis, highlights the importance of policymakers paying greater attention to measures that enhance the instalment of this kind of cooperation.

PART THREE: A CLOSER LOOK AT OPEN INNOVATION: R&D OUTSOURCING

Outsourcing activities are fuelling a heated debate among economists and the general public. Not that the purchase of goods and services from suppliers is anything new, but it is feared that the surge in outsourcing (off-shoring) displaces domestic employment. Due to the rising level of education among most of the populations of the developed world, the outsourcing of low-skilled labour has not yet set alarm bells ringing. However, with the increasing division of labour at firm level, the availability of specialist knowledge outside the firm's walls, the need to remain competitive through innovation and the swift development of emerging economies such as India and China, it is believed that R&D outsourcing could potentially affect highly-skilled employees and displace R&D jobs.

In the past two decades there has been a remarkable increase in the R&D outsourcing activities of firms in developed market economies. This is observed through the substantial increase in the number of R&D partnerships, through interactions between partners and through the significant growth of firm expenditure on external R&D. This has all taken place in a context of rapidly developing, elaborate R&D networks that combine different kinds of partnership, from straight licensing, consulting and arm's-length pure subcontracting, through more profound long-term types of subcontracting, to various forms of explicit R&D cooperation. These partnerships involve different types of partners, such as competing firms, firms connected by vertical supplier–customer relations, and public research organisations such as universities and public research centres. And they tend increasingly to become international in nature, involving partners from different countries.

The perception of the above-mentioned tendencies has led to a gradual emergence of new views on the purpose of R&D outsourcing and on its relation to in-house R&D and appropriation.

In Chapter 7 a cross-section regression analysis is presented of the impact of in-house R&D, appropriability and cooperation on the R&D outsourcing intensity of firms, making use of data from the third and fourth Community Innovation Surveys. The results indicate that in-house R&D has a strong positive impact on R&D outsourcing intensity, emphasising the importance of knowledge absorptive effort and, thereby, the complementary nature of in-house and external R&D. Strategic appropriability has a stronger positive impact on R&D outsourcing intensity than patenting, and does so mainly through complexity of design. This suggests that the required appropriation occurs in a more informal manner than through deliberate actions of legal protection or secrecy. Finally, formal cooperation has a significant positive impact on R&D outsourcing intensity, and does so mainly through cooperation with public research institutions, such as universities and research centres. This suggests that the firms in question tend to spend relatively more on external R&D. Estimates show that this can be partly explained by the fact that public funding authorities favour cooperative industry-science relations.

Chapter 8 assesses whether firms' strategies of R&D outsourcing determine changes in their internal R&D employment intensity (see for additional information on this Teirlinck et al., 2010). Four strategic decisions are investigated: to start, increase, decrease or stop outsourcing. It is found that internal R&D employment intensity decreases when firms decide to start or to increase R&D outsourcing. However, this finding conceals important differences according to the type and location of the contractor. Starting to outsource to research centres in the proper region seems to decrease internal R&D employment intensity, whereas the opposite is true in the case of outsourcing to firms and universities outside the proper region.

PART FOUR: THE ROLE OF HUMAN RESOURCES IN OPEN INNOVATION PRACTICES

Nowadays economies are characterised as knowledge intensive. This implies that knowledge – in addition to labour and capital – has become an important factor of production. This knowledge is said to be either codified or tacit. If it is codified, knowledge can be traded on a specialised market, provided it is standardised and fairly priced. But tacit knowledge is embodied in the minds of people, and is part and parcel of their skills. Insights from the 'distributed' or 'open' innovation literature emphasise the need for enterprises to access external knowledge in order to perform research and innovate. One way to do so is by means of research cooperation. A research partnership may improve learning efficiency in absorbing external knowledge, thereby increasing the impact of incoming spillovers on innovative performance. It may, at the same time, provide a mechanism to reduce outgoing spillovers, by helping to internalise them.

Since R&D and innovation are said to be ever more reliant on knowledge external to the firm, the complexity of the tasks involved in R&D and innovation has been examined, as, in particular, have the relations between internal R&D personnel qualification and occupation and the development and exchange of knowledge in research cooperation with the outside world. This type of research has not received much in the way of attention, which is all the more strange given that the topic of absorptive capacity (which, in our opinion, overly concentrates on the monetary element of R&D intensity) figures so prominently in the current literature.

Chapter 9 looks at the functions and the qualifications of all R&D personnel and relates these to external knowledge linkages (Spithoven and Teirlinck, 2010). Using the OECD business R&D survey, insights are offered into the way in which R&D outsourcing impacts on the organisation of the internal division of labour. The main conclusion is that the level of educational attainment is more important than on-the-job training if firms decide to engage in research cooperation. This implies that the skills necessary to external knowledge relation-building are largely to be found in people who already have a high degree as a threshold. Specifically, in terms of absorptive capacity, it is our finding that patent registration and innovative outputs also influence propensity to engage in knowledge development in R&D collaboration, but not in terms of engagement of knowledge exchange in R&D collaboration agreements.

Chapter 10 concentrates more specifically on the role of R&D experts and managers in knowledge development and exchange in research collaboration. In line with the resource-based view and the literature on organisational absorptive capacity it is demonstrated that the absorptive capacity embodied in the experts and managers who work for the firm lowers the propensity to engage in knowledge exchange in research cooperation and increases the propensity to engage in knowledge development in research cooperation.

PART ONE

Firms react differently when managing open innovation

1. Open innovation and the relation between innovative activities and firm performance

1.1 INTRODUCTION

The way firms manage innovation has changed profoundly in recent years (Chesbrough, 2003a; Chesbrough and Crowther, 2006). They used to invest heavily in specialised in-house R&D departments in order to generate innovation. This model has recently ceded ground to a more 'open' model, in which the research strategies of innovative firms increasingly rely on external knowledge. This encourages cooperation and implies a reorganisation of the research process, to allow for the internalisation of external ideas. Open innovation also implies a changing attitude to the appropriability of new ideas, since this may occur not only through product innovation, but through selling these ideas in disembodied form, through licensing, trademarks, and so on. It might therefore be of great interest to reconsider the relation between innovative effort and firm performance.

Recent econometric studies on the relation between innovation and firm performance have estimated empirical structural models to explain the input and output stages of the innovation process as well as its impact on productivity (Crépon et al., 1998; Lööf and Heshmati, 2002, 2006; Griffith et al., 2006; Van Leeuwen and Klomp, 2006), or on turnover and employment growth (Klomp and Van Leeuwen, 2001). Special care was taken to make the models suitable for drawing statistical inferences from a population larger than that of the research-active firms. This was done by adding a probit equation to explain the probability of doing research and so obtain a generalised tobit model. One drawback with these studies, however, is that they tend to neglect the relevance of estimation to the research-active sample, which may, in fact, be appropriate when it comes to gaining information on research-active firms. Gaining information of this type may also be important in interpreting tobit estimates.

Furthermore, irrespective of the firm population that interests us, the least squares or related estimation procedure used in these studies is open to criticism, in the sense that the most it can provide us with is information on the mean behaviour of firms. Management science, however, is just as interested in behaviour at the extremes – that is, in firms that perform exceptionally well, or poorly (in this respect, see the literature based on large case studies, such as Delmar et al., 2003, or O'Regan et al., 2006). This is especially relevant in our case, since open innovators are most likely to be found among the more dynamic firms. The more appropriate estimation method for this purpose is quantile regression analysis. This allows differences to be revealed in the relationship between endogenous and exogenous variables at different points in the conditional distribution of the dependent variable (R&D intensity, turnover, employment), and therefore an identification of systematic behavioural differences between poor and strong innovative performers.

To date only a handful of studies have employed quantile regression analysis to study innovation input or output relations. But each has suffered major shortcomings. Thus, Nahm (2001) only estimates an equation to explain the level of R&D intensity and, in so doing, concentrates on the impact of firm size in the tradition of previous empirical literature. He omits knowledge-related variables of the type considered in the aforementioned structural models, such as incoming knowledge spillovers, appropriability and research cooperation. Ebersberger et al. (2006) present quantile regression estimates of an equation to explain innovative turnover according to level of R&D intensity and other variables, including a measure of appropriability. However, we should expect the impact of appropriability to be mostly indirect through its influence on R&D intensity (Hussinger, 2006). In view of companies' increasing tendency to apply more open innovation strategies, it would be of great interest to discover whether the relative impact of R&D intensity on appropriability, incoming knowledge spillovers and research cooperation differs between strong and weak innovative performers.

In this chapter we will take a more systematic approach than ever before to an analysis of the link between innovation and firm performance. With the help of the Third Community Innovation Survey (CIS3) micro database for Belgium we estimate in several steps a structural model to explain R&D intensity and innovative turnover and their impact on the growth of turnover and employment. In the first step, we perform an estimation on a sample of research-active, product-innovative firms and, from this, draw inference on the corresponding sub-population of firms. In the second step, we widen our scope of interest by extending the sample and adapting the model to draw inference on a larger population that also includes non-innovative firms. In the third step we then turn our attention again to research-active product-innovative firms and perform a quantile regression estimation to identify systematic behavioural differences between strong and weak innovative performers. Finally, we attempt to give a general interpretation by considering the results as a whole.

The chapter is organised as follows: Section 1.2 presents the model and Section 1.3 the data and methods of estimation. The estimates are reproduced in Section 1.4, and Section 1.5 gives a summary and conclusion. Tables containing a description of the variables and descriptive statistics are presented in the Appendix to this chapter.

1.2 THE MODEL

Our model has been designed in the spirit of Klomp and Van Leeuwen (2001). It consists of four equations which relate, respectively, to the input and output stages of innovation and their impact on turnover and employment growth performance.

The dependent variable for the innovation input equation is the log of the R&D intensity level (the proportion of R&D expenditure over turnover), LRDINS. Its first explanatory variable is the growth of turnover, GTURN. The growth of turnover is likely to have a positive impact on R&D intensity, to the extent that it has a positive effect on income expectations in the long run, and therefore R&D investment decisions. Thus it captures a demand-pull effect, in the spirit of Schmookler (1966).

The next four explanatory variables relate to knowledge. They measure the level of incoming knowledge spillovers, INKS, appropriability of in-house knowledge, APPR, research cooperation, COOP, and complexity of knowledge, CKNOW. Incoming spillovers are likely to have a positive effect on the level of R&D intensity, because they increase the firm's knowledge base, from which it is able to draw in its research process, and because the absorption of outside knowledge requires additional in-house R&D efforts, as emphasised by Cohen and Levinthal (1989). A higher capacity in appropriating own knowledge will tend to minimise spillover to competitors and, therefore, increase incentives to undertake R&D efforts (Spence, 1984). In the case of research cooperation the argument has to be drawn out a little, in the sense that it now applies to spillovers to non-partners. Cooperation stimulates R&D efforts to the extent that it raises the efficiency of the research process, and because the internalisation and commercialisation of the knowledge created through cooperation requires further in-house R&D effort from the firms in question (Colombo and Gerrone, 1996; Hagedoorn et al., 2000). Complexity of knowledge may, in turn, be expected to have a positive effect on the level of research effort, to the extent that the greater complexity of basic research requires more effort.

The next explanatory variable is the log of the level of higher education intensity of the personnel, LHEI. This is a human capital variable that reflects the average level of qualification of the workforce. Its impact will be positive, in a straightforward manner, to the extent that it concerns knowledge workers. But this is also likely to be the case for production workers, given that firms will tend to spend more on R&D if they have more skilled workers at their disposal who will be better able to operate in a more complex production environment and can more easily contribute to the innovation process itself, by contributing ideas of their own (Cohen and Levinthal, 1989).

The degree of export orientation is also assumed to have a positive influence on the level of R&D effort, given that firms operating in foreign markets are likely to be subject to stronger competitive pressures. We will, therefore, include the log of export intensity, LEXI, as an additional variable.

The final two variables relate to size. They are the log of size (measured by employment), LSIZE, and a variable to reflect group membership, GROUP. Since the theoretical literature contains arguments implying either positive or negative influences on R&D intensity (Cohen, 1995) we have no a priori expectations as to their sign. We have experimented by including measures of innovation-hampering factors, such as high risks and costs and lack of finance and qualified personnel, which have been used to explain the impact of size. But strong mutual collinearity and collinearity with the LSIZE variable has made their estimates very imprecise. We have, therefore, opted to exclude them and use them merely as additional variables to help construct our instruments when using an instrumental variable estimation procedure.

The innovation output equation concentrates on product innovation. It has as its dependent variable the log of the level of innovative turnover (the proportion of innovative products in the turnover), LITURN. Its main explanatory variable is the log of R&D intensity, LRDINS, which is the key variable in the production function of new knowledge. Its influence on innovation output is not only direct, through the creation of new product innovative ideas, but also indirect, through its contribution to generating the complementary process innovations required to bring product innovation about (Parisi et al., 2006). Since some process innovations are likely to consist of new machinery incorporating new technology purchased elsewhere we also include a measure of process innovation, PCS, as a separate explanatory variable.

Besides this, the innovation output equation includes the same knowledge, human capital and export-related variables as the innovation input equation, given that these may, in each instance, affect innovation output directly, rather than merely indirectly, through their impact on R&D intensity. We also include the size-related variables, to discover whether there is a size-related research efficiency effect, in the sense that firm size affects the level of innovative output for a given level of research intensity.

From our perspective, the main explanatory variable for the growth of turnover equation is the log of the level of innovative turnover, LITURN. We also include the variable on process innovation since this may stimulate market turnover performance by improving a firm's cost competitiveness. The export orientation variable captures the impact of differences between export and domestic market growth conditions.

Finally, the main explanatory variable for the growth of employment equation is growth of turnover, GTURN. Again, we include the process innovation variable but this time expect its impact to be negative, given that the labour-saving bias of most process innovations reduces employment growth.

Additionally, all the equations include an intercept term, C, as well as a set of industry dummies based on the taxonomy used by Pavitt (1984), IDPAV, to depict the general technological environment. Our simultaneous system can thus be summarised as:

LRDINS = f1(C, IDPAV, GTURN, INKS, APPR, COOP, CKNOW, LHEI, LEXI, LSIZE, GROUP) (1.1)
LITURN = f2(C, IDPAV, LRDINS, INKS, APPR, COOP, CKNOW, PCS, LHEI, LEXI, LSIZE, GROUP) (1.2)
GTURN = f3(C, IDPAV, LITURN, PCS, LEXI) (1.3)
GLEMP = f4(C, IDPAV, GTURN, PCS) (1.4)

It should be noted that the model allows for a loop from the level of R&D intensity, LRDINS, over the level of innovative turnover, LITURN, to the growth of turnover, GTURN, and so again to the level of R&D intensity, LRDINS. If confirmed, this would imply a process of cumulative causation.

1.3 DATA AND ESTIMATION METHODS

1.3.1 Data

This research draws on data for Belgium gathered as part of the Third European Community Innovation Survey (CIS3) conducted in 2002 by the various member states of the European Union. The survey relates to activities in the period 1998–2000. For the purpose of international comparability, the innovation survey was conducted in line with the guidelines set out in the second edition of the Oslo Manual (OECD, 1997).

The survey targets the population of enterprises with ten or more employees, located in the country and taking part in economic activities aside from work in the public sector. The survey was stratified according to size and economic activity. Within each stratum a random sample was drawn based on previous CIS results, and care was taken to minimise variance within the strata. The original total sample included 4,482 private enterprises (that is, 32% of the target population). Initial cleaning and careful screening left 2,035 enterprises who had provided useful answers (Spithoven, 2005). From these, it was necessary to exclude a further 48 enterprises due to incomplete data on employment and turnover (25), extreme values for research intensity (16), and positioning in the 'Utilities' sector, which is too small for consideration (7). This left a sample of 1,987 firms. The survey contains information on the product and/or process orientation of the innovations. In total 1,310 enterprises described themselves as innovative. Of these, 951 reported intramural R&D, 945 innovative turnover, and 724 both. Since our main interest is to obtain information on the behaviour of research-active product innovators, we will concentrate on the latter sample. When widening our focus and drawing inference from a larger population, we will consider, in addition, the full sample of 1,987 firms. Table A1.1 in the Appendix gives the precise construction and definition of all the variables used. Here, we confine ourselves to a few necessary comments.

Our innovation input variable is the log of research intensity, LRDINS, measured as the proportion of intramural R&D expenditure over turnover. In addition to intramural R&D, the CIS3 questionnaire includes questions on expenditure related to the acquisition of extramural R&D, advanced machinery and equipment, purchase of the right to use patents, training, marketing and design. We have opted to confine ourselves to intramural R&D rather than total innovation expenditure for two reasons. First, we cannot be certain that all of the respondents understood fully the questions relating to other components of the total innovation expenditure. Second, and more importantly, we believe that intramural R&D is of particular relevance to issues such as knowledge absorption capacity, appropriability and research cooperation.

The measure of incoming spillovers, INKS, rests on questions relating to the importance attached to alternative sources of information, measured on a three-point scale. These sources are: suppliers, clients, competitors, universities, government, professional conferences, and fairs and exhibitions. To generate a firm-specific measure of incoming spillovers, we added together the scores on each of these questions and rescaled the total to a number between 0 and 1. Our measure of incoming spillovers captures not only inter- and intra-sectoral spillovers, but international and national spillovers too. Since it rests on a direct assessment of their importance by the firms themselves, it avoids the difficulties of estimating the technological distance between receiver and transmitter encountered in the indirect method of measuring incoming spillovers by means of external knowledge capital stock constructs, which has regularly been used in the literature on technological diffusion.

The measure of appropriability, APPR, concentrates on methods of strategic protection, rather than on methods of legal protection through patents, because the latter are highly industry-specific (McGaughey, 2002; Hussinger, 2006). Firms were asked whether they made use of strategic protection using secrecy, complexity of design and lead-time advantage over competitors. By giving a score of 1 for yes and 0 for no, an average score between 0 and 1 could be obtained to generate a measure of strategic protection.

The measure of research cooperation, COOP, is a binary variable with the value 1 for cooperation and 0 for otherwise. Cooperation can occur either with suppliers, clients, competitors, research labs, universities or government. We also experimented by replacing cooperation, COOP, with measures of research cooperation with suppliers, clients or competitors on the one hand, and research labs, universities or government on the other. However, even though the statistical significance of the latter variable was somewhat higher than the former, we were unable to rule out the hypothesis of an equal impact on LRDINS, so we confined ourselves to the COOP variable as was.

The measure of complexity of knowledge, CKNOW, is again based on the answers to questions relating to sources of information. It is obtained by taking the ratio of the sum of the scores on information from universities, government and professional conferences over the sum of the scores of all sources of information, as suggested by Cassiman and Veugelers (2002). In other words, it is inferred from the structure of the incoming knowledge flows, on the assumption that this conditions the character of a firm's own research. We should be aware that this might pose the problem of multicollinearity with the variable on incoming knowledge spillovers, INKS.

Innovative performance is the log of innovative turnover, measured by the proportion of new or significantly improved products (goods or services) in turnover. It is a measure of product innovation. It was preferred over a patent-based indicator of innovative output due to the well-known shortcomings of the latter in this respect, as not all patents become innovations, the degree of patenting strongly varies between industries, and few service firms, such as software development, tend to patent (Griliches, 1990).

Finally, to incorporate services, we based industry dummies, capturing the general technological environment, on the taxonomy originally proposed by Pavitt (1984) and later refined and elaborated by Tidd et al. (1997) and Kristensen (1999) . It should be noted that the data from the CIS is limited to the NACE two-digit classification of sectors. This implies that a totally seamless classification is not possible. Despite this, the groupings we present in Table A1.1 in the Appendix translate to Pavitt categories quite well. Table A1.2 presents corresponding descriptive statistics for all variables.

1.3.2 Estimation Methods

The data are cross-sectional and relate to a specific moment in time. Therefore the estimates can only provide a snapshot of the long-term relationships between the variables. They do not enable us to identify lagged responses. Furthermore, as stated above, the model allows for a loop from LRDINS through LITURN to GTURN, and to LRDINS again. If such a feedback does exist it would imply that the variables in question are endogenous when appearing as explanatory variables. In such a case an equation-based estimation by ordinary least squares (OLS) will not provide consistent estimates, and, in view of the cross-sectional nature of the data, the implied simultaneity bias may be severe. A more appropriate estimation method will have to be used. If, however, the feedback loop does break down, the system may be recursive and OLS may still be appropriate, in the sense that it is consistent and the most efficient.

We will, therefore, use the following strategy. We begin estimation by OLS and thereby consider (White) heteroscedasticity-robust standard errors, in order to cope with the possible problem of heteroscedasticity. We then repeat estimation by two instrumental variable (IV) techniques, in order to control for possible endogeneity problems with respect to LRDINS, LITURN and GTURN. The first of these is a two-stage least squares method (2SLS), which constructs the instruments on the basis of their predicted value in the corresponding reduced form equations. In this we again consider (White) heteroscedasticity-robust standard errors. The second IV method we consider is a generalised method of moments (GMM) estimation procedure. It estimates the parameters directly from the moment conditions imposed by the model. This is done by minimising a quadric form in the weighted sample moments, in which the weighting matrix is based on the inverse of a heteroscedasticity-robust estimate of the covariance matrix of the sample moments (Hansen, 1982). From this, we make a first assessment of the endogeneity issue by considering the parameter estimates obtained using the different methods in conjunction. We then seek further confirmation by performing Durbin–Hausman–Wu exogeneity tests on the variables in question (Davidson and MacKinnon, 1993).

After performing estimation on a sample of R&D-active and product-innovative firms, we extend the sample and adapt the model to draw inference from a larger population that also includes non-innovative firms. The model is adapted by including an equation to explain the probability of doing R&D and being product-innovative. By so doing we obtain a generalised tobit II model, consisting of a probit equation with a binary-dependent variable that takes the value 1 in the case of positive R&D and innovative turnover, and the value 0 otherwise, and our considered structural equations. OLS will cease to be appropriate even if no problems of endogeneity arise. This is due to the binary nature of the dependent variable in the probit equation and also to the likely correlation between its error term and the error terms of the structural equations, which will cause a sample selection bias in their parameter estimates. We will, therefore, for each of our equations, estimate the system consisting of the probit equation and the equation in question on the basis of maximum likelihood (ML), while using the Heckman (1979) procedure to choose the initial parameter values in the iterative procedure. The considered standard errors are, again, heteroscedasticity-robust. This method has been shown to provide consistent and efficient estimates (Verbeek, 2000).

In principle, we would like to consider all exogenous variables in our LRDINS and LITURN equations as potential explanatory variables in our probit equation. But, obviously, due to data constraints, we can only enter variables on which data are available over the full sample of both innovative and non-innovative firms. These are, apart from the intercept, the industry dummies (IDPAV), APPR, LHEI, LEXI, LSIZE and GROUP. Though the probit equation we obtained in this way performs quite well, as will be seen below, we will need to exercise caution when interpreting the parameter estimates. Since the results we present in the first three columns of Tables 1.1 to 1.4 below show no evidence of endogeneity of the explanatory variables, we will only present estimates of the equations including the variables as such, rather than in instrumented form.

We then return our attention to R&D-active and product-innovative firms. We perform quantile regression estimation in order to identify systematic differences in behaviour between strong and weak innovative performers. A quantile regression estimates the parameters by minimizing the sum of asymmetrically weighted absolute values for the residuals. Thus, for example, the 0.25 regression quantile estimates lend less weight to the positive residuals than the negative ones. Through this, our attention is drawn, so to speak, to the left-hand tail of the conditional density function for the dependent variable in question. The reverse is true for the 0.75 quantile. Varying the quantile parameter between 0 and 1 generates all the regression quantiles, revealing the conditional distribution of the dependent variable given the explanatory variables. Quantile regression estimates have the additional advantage of being more efficient than least squares estimates when the errors deviate from normal, as they do in our case and as we see below.

Since we cannot rule out the possibility of heteroscedasticity and, therefore, dependence between the quantile regressors and the density of the error terms, we use the bootstrap design method to estimate the standard errors (Buchinsky, 1995, 1998). We performed bootstrap re-sampling using 1,000 repetitions. Again, since the results we present in Tables 1.1 to 1.4 below show no evidence of endogeneity of the explanatory variables in our equations, we concentrate on straight quantile regression, with the variables entered as such, rather than on instrumented quantile regression. Due to constraints of space, the results presented in Tables 1.4 and 1.5 below relate to the 0.25, 0.50 (or median), and 0.75 quantiles.

1.4 ANALYSIS

The first three columns of Tables 1.1 to 1.4 present the estimates of our model on the sample of R&D-active and product-innovative firms. Table 1.1 presents the estimates of the R&D intensity equation.

Table 1.1 Estimates of the equation on R&D intensity

	OLS (i)	2SLS (ii)	GMM (iii)	ML (iv)	ML (v)
Constant	–3.059	–3.030	–2.930	–0.767	–3.356
	(11.3)**	(6.53)**	(6.41)**	(6.06)**	(12.2)**
IDPAV1	–1.397	–1.414	–1.568	–0.120	–1.414
	(6.32)**	(4.35)**	(5.03)**	(1.36)	(6.53)**
IDPAV2	–1.128	–1.141	–1.270	–0.059	–1.133
	(4.91)**	(3.84)**	(4.42)**	(0.63)	(5.03)**
IDPAV3	–0.214	–0.229	–0.364	0.086	–0.205
	(0.85)	(0.69)	(1.13)	(0.78)	(0.83)
IDPAV4	–0.380	–0.394	–0.511	0.250	–0.348
	(1.75)	(1.26)	(1.68)	(1.99)*	(1.61)
GTURN	–0.241	–0.299	–0.466		–0.242
	(1.25)	(0.39)	(0.62)		(1.38)

Table 1.1 (continued)

	OLS (i)	2SLS (ii)	GMM (iii)	ML (iv)	ML (v)
INKS	0.775	0.792	0.745		0.772
	(2.38)**	(2.19)**	(2.08)**		(2.40)**
APPR	0.331	0.334	0.385	1.315	0.512
	(2.10)**	(1.96)**	(2.26)**	(12.6)**	(3.22)**
COOP	0.405	0.407	0.410		0.404
	(3.01)**	(3.04)**	(3.14)**		(3.03)**
CKNOW	0.359	0.352	0.379		0.354
	(1.42)	(1.35)	(1.46)		(1.44)
LHEI	0.178	0.179	0.178	0.150	0.203
	(2.86)**	(2.90)**	(2.88)**	(6.69)**	(3.29)**
LEXI	0.069	0.069	0.062	0.053	0.077
	(2.47)**	(2.50)**	(2.30)**	(4.32)**	(2.75)**
LSIZE	–0.181	–0.184	–0.172	0.171	–0.157
	(3.04)**	(2.84)**	(2.69)**	(6.29)**	(2.78)**
GROUP	–0.240	–0.238	–0.230	–0.034	–0.244
	(1.66)	(1.66)	(1.61)	(0.49)	(1.71)
RHO					0.148
					(2.60)**
Nb. obs.	724	724	724	1,987	724
Adj. R^2	0.214	0.214	0.209		

Notes: The dependent variable is the log of R&D intensity, LRDINS, except in the case of equation (iv) where it is the probability of doing research and of being product innovative, PRDINSAL. RHO stands for the correlation coefficient between the errors of the sample selection and tobit equations. The absolute value of the t-values is presented between brackets. These are based on heteroscedasticity-consistent estimates of the standard errors. The t-tests are one-tailed or two-tailed, according to a priori expectations with respect to the parameter values. Symbols * and ** denote significance at the 5% and 1% levels.

Column (i) gives the estimates by OLS. Statistical performance appears to be decent in terms of statistical fit (adjusted R^2), in view of the disaggregate nature of the data. Closer analysis of the residuals suggests that although there is no evidence of heteroscedasticity, the hypothesis of normality can be rejected. A degree of caution with respect to the precision of the t-values presented is therefore warranted.

Columns (ii) and (iii) give estimates of the equation by 2SLS and by GMM. Their statistical performance is similar and comparison of the parameter estimates also shows remarkable similarity with the OLS estimates. Instrumenting the growth of turnover variable, GTURN, clearly does not affect the results. This variable is always found to be non-significant. In itself this implies that the loop in the model breaks down and that the system is recursive in nature. Applying the Durbin–Hausman–Wu (DHW) test confirms that GTURN is exogenous to the R&D intensity equation.

Let us consider now the other respective parameter estimates in turn. Since the equation includes an intercept, we were forced to drop one of the Pavitt industry dummies to avoid singularity. We dropped IDPAV5, so that its impact is captured by the intercept, C. Estimates of the other industry dummies reflect deviations from this. They are all seen to have a negative sign, and their negative impact appears especially strong in the case of IDPAV1 and IDPAV2. This is perfectly in line with the mean values of the log of R&D intensity, LRDINS, presented in Table A1.2 in the Appendix. It suggests that general industry characteristics imply that the average level of R&D intensity is clearly lowest in the traditional ('resource intensive' and 'scale intensive') industries, and that it is highest in the dynamic hi-tech ('information intensive') sectors.

Of the knowledge-related variables, we can say that incoming spillovers, INKS, appropriability, APPR, and, more particularly, research cooperation, COOP, are clearly significant. The significance of INKS and COOP suggests that incoming spillovers and research cooperation increase the firms' knowledge base and that absorption of the knowledge in question always requires further in-house R&D. The significance of APPR confirms that a higher capacity for appropriating internally-developed knowledge increases the incentives to undertake R&D efforts. In turn, the coefficient of the variable on complexity of knowledge, CKNOW, also has the anticipated positive sign, but is non-significant. Consideration of the direct correlation matrix does suggest that this is due to multicollinearity with the INKS variable, and so there is not much to be concluded from this.

Our human capital and export orientation variables, LHEI and LEXI, are always clearly significant with a positive sign, as expected. This confirms the complementarity between formal R&D and the qualification of workers, and suggests that stronger competitive pressure in foreign markets does, indeed, have a positive effect on research intensity.

Finally, firm size, LSIZE, has a clearly negative impact on R&D intensity, whereas the coefficient of the other size-related variable, group membership, GROUP, is also negative, but non-significant. This negative significance of LSIZE would suggest that, among product innovative R&D performers, R&D expenditure rises less than proportionally with firm size. This contradicts Stylised Facts 2 and 3 of Cohen and Klepper (1996), derived from the earlier empirical literature on this issue, which would imply a proportional increase. Cohen and Klepper (1996) have explained these stylised facts by arguing that the majority of the firms under consideration do not expect to grow rapidly due to innovation. As a consequence, the output over which they can spread the cost of their R&D is closely related to their output at the time when they conduct the R&D. These authors do, however, stress that this argument does not hold for firms that do expect to grow rapidly due to their innovations or that can sell their innovative ideas in disembodied form. Now, this appears to be precisely the case of a majority of the most R&D-active product innovators in Belgium. These are young, dynamic firms, active mainly in hi-tech information-intensive service sectors, which are still relatively small-sized (see Table A1.2 in the Appendix). These firms expect to grow rapidly from their innovations and, in an open innovation logic, they are prepared to sell part of their new ideas, which they do not consider worth exploiting themselves, in disembodied form, by licensing, etc. Therefore their R&D decisions are not greatly affected by their original size constraints or, for that matter, by their short-term turnover performance, and they tend to spend relatively more on R&D in relation to their size than larger, less dynamic firms.

Table 1.2 presents the estimates of the innovative turnover equation.

Table 1.2 Estimates of the innovative turnover equation

	OLS	2SLS	GMM	ML	ML
	(i)	(ii)	(iii)	(iv)	(v)
Constant	–1.004	–0.014	–0.053	–0.763	–1.101
	(6.19)**	(0.02)	(0.07)	(6.06)**	(4.51)**
IDPAV1	–0.490	–0.067	–0.087	–0.120	–0.496
	(4.30)**	(0.19)	(0.25)	(1.36)	(4.38)**
IDPAV2	–0.359	–0.021	–0.039	–0.058	–0.361
	(2.97)**	(0.07)	(0.13)	(0.62)	(3.01)**
IDPAV3	–0.006	0.046	0.047	0.084	–0.002
	(0.05)	(0.31)	(0.32)	(0.76)	(0.02)
IDPAV4	–0.429	–0.325	–0.327	0.253	–0.419
	(3.29)**	(1.96)*	(1.99)*	(2.01)*	(3.22)**
LRDINS	0.102	0.421	0.409		0.102
	(4.95)**	(1.78)*	(1.76)*		(5.00)**
INKS	0.272	0.040	0.022		0.271
	(1.37)	(0.16)	(0.09)		(1.38)
APPR	–0.014	–0.118	–0.111	1.312	0.045
	(0.14)	(0.86)	(0.82)	(12.5)**	(0.33)

Table 1.2 (continued)

	OLS (i)	2SLS (ii)	GMM (iii)	ML (iv)	ML (v)
COOP	0.010	–0.123	–0.109		0.010
	(0.12)	(0.90)	(0.82)		(0.12)
CKNOW	–0.147	–0.274	–0.268		–0.147
	(0.86)	(1.31)	(1.30)		(0.88)
PCS	0.195	0.251	0.240		0.195
	(2.26)**	(2.33)**	(2.28)**		(2.28)**
LHEI	0.004	–0.052	–0.051	0.150	0.012
	(0.12)	(0.93)	(0.92)	(6.70)**	(0.35)
LEXI	–0.021	0.043	–0.043	0.053	–0.019
	(1.49)	(1.56)	(1.56)	(4.37)**	(1.25)
LSIZE	–0.102	–0.050	–0.047	0.170	–0.094
	(3.32)**	(0.93)	(0.91)	(6.28)**	(2.82)**
GROUP	0.075	0.155	0.145	–0.033	0.073
	(0.86)	(1.33)	(1.27)	(0.47)	(0.86)
RHO					0.079
					(0.55)
Nb. obs.	724	724	724	1,987	724
Adj. R^2	0.125	0.072	0.073		

Notes: The dependent variable is the log of innovative turnover, LITURN, except in the case of equation (iv) where it is the probability of doing research and of being product innovative, PRDINSAL. RHO stands for the correlation coefficient between the errors of the sample selection and tobit equations. The absolute value of the t-values is presented between brackets. These are based on heteroscedasticity-consistent estimates of the standard errors. The t-tests are one-tailed or two-tailed, according to a priori expectations with respect to the parameter values. Symbols * and ** denote significance at the 5% and 1% levels.

Here, it is immediately striking that there is a clear difference between the OLS estimates in column (i) and the 2SLS and GMM estimates in columns (ii) and (iii). This can be explained by multicollinearity between the instrumented LRDINS variable and the constant term, the industry dummies (IDPAV) and the LSIZE variable, which were also used to construct the instrument. The breaking-down of the loop in the model, due to the non-significance of GTURN in the R&D intensity equation, would, however, suggest that LRDINS is exogenous in the innovative turnover equation. This is confirmed by a DHW exogeneity test on this hypothesis. We can, therefore, conclude that the OLS estimates are more reliable, because they are more efficient, and concentrate our attention on them.

The evidence of significantly negative IDPAV1, IDPAV2 and IDPAV4 dummies suggests that firms active in 'resource intensive', 'scale intensive' and 'science based' industries are, on average, less product innovative than firms in 'Specialised Suppliers' and 'Information Intensive' industries. This is confirmed by the mean values of LITURN, presented in Table A1.2 in the Appendix.

Of the firm-specific variables, the level of R&D intensity, LRDINS, is, by far, the most significant variable, confirming that it is, indeed, the key variable explaining innovative turnover. There is also evidence of a positive impact of process innovations, PCS, on product innovativeness, even though this is clearly less important.

Interestingly enough, when checking for R&D intensity, there is no evidence that knowledge-related variables, or human capital or export orientation have a direct effect on product innovativeness. This clearly suggests that their influence on the level of product innovation is indirect, through R&D intensity. Or, in other words, incoming spillovers, research cooperation or the ideas of workers do not appear to generate readily usable knowledge: they need first to be absorbed in the knowledge base of the firm through additional R&D efforts. This emphasises the vital importance of in-house intra-mural R&D (Cohen and Levinthal, 1989; Arora and Gambardella, 1994; Chesbrough and Crowther, 2006).

Finally, the negative significance of the LSIZE variable is in line with Stylised Fact 4 of Cohen and Klepper (1996), which states that among R&D-performing firms the number of innovations per dollar of R&D tends to decrease with firm size. In our case this can be directly related to our previous evidence as regards the impact of size. Among R&D-active product innovators, the still relatively small younger firms tend to be more dynamic. Not only, therefore, are they prepared to spend relatively more on R&D in relation to their size, but they are prepared to do so in a more risky manner. Consequently they are able to generate, on average, a higher number of innovations per unit of R&D spend (see Cohen, 1995, for a review of the arguments in this sense).

Tables 1.3 and 1.4 present the estimates of the growth of turnover and the growth of employment equations.

Table 1.3 Estimates of the growth of turnover equation

	OLS (i)	2SLS (ii)	GMM (iii)	ML (iv)	ML (v)
Constant	0.509	1.141	1.048	–0.759	0.527
	(5.30)**	(4.06)**	(4.30)**	(6.52)**	(4.93)**
IDPAV1	–0.294	–0.001	0.006	–0.119	–0.289
	(3.97)**	(0.01)	(0.064)	(1.42)	(3.94)**
IDPAV2	–0.255	–0.013	0.000	–0.057	–0.251
	(3.30)**	(0.14)	(0.000)	(0.65)	(3.31)**
IDPAV3	–0.264	–0.224	–0.204	0.084	–0.266
	(3.06)**	(2.43)**	(2.45)**	(0.82)	(3.08)**
IDPAV4	–0.254	–0.033	–0.011	0.254	–0.258
	(3.21)**	(0.33)	(0.13)	(2.31)**	(3.22)**
LITURN	0.082	0.479	0.442		0.089
	(3.44)**	(3.18)**	(3.85)**		(3.47)**
APPR				1.315	
				(14.7)**	
PCS	0.117	0.057	0.051		0.115
	(2.68)**	(1.10)	(1.03)		(2.63)**
LHEI				0.150	
				(6.95)**	
LEXI	0.002	0.009	0.005	0.053	0.001
	(0.21)	(0.76)	(0.59)	(4.72)**	(0.08)
LSIZE				0.168	
				(6.76)**	
GROUP				–0.031	
				(0.48)	
RHO					–0.041
					(0.69)
Nb. obs.	724	724	724	1,987	724
Adj. R^2	0.070	0.031	0.031		

Notes: The dependent variable is the growth of turnover, GTURN, except in the case of equation (iv) where it is the probability of doing research and of being product innovative, PRDINSAL. RHO stands for the correlation coefficient between the errors of the sample selection and tobit equations. The absolute value of the t-values is presented between brackets. These are based on heteroscedasticity-consistent estimates of the standard errors. The t-tests are one-tailed or two-tailed, according to a priori expectations with respect to the parameter values. Symbols * and ** denote significance at the 5% and 1% levels.

Table 1.4 Growth of employment equation

	OLS (i)	2SLS (ii)	GMM (iii)	ML (iv)	ML (v)
Constant	0.215	0.009	0.003	–0.756	–0.112
	(7.34)**	(0.15)	(0.06)	(7.09)**	(1.68)
IDPAV1	–0.116	0.078	0.085	–0.198	–0.139
	(3.32)**	(1.19)	(1.45)	(2.34)**	(4.00)**
IDPAV2	–0.154	0.011	0.027	–0.158	–0.128
	(5.44)**	(0.21)	(0.56)	(1.65)	(4.18)**
IDPAV3	–0.140	0.009	0.051	0.098	–0.069
	(3.65)**	(0.13)	(0.91)	(0.94)	(1.70)
IDPAV4	–0.169	–0.006	0.007	0.212	–0.040
	(5.47)**	(0.11)	(0.15)	(1.96)*	(0.99)
GTURN	0.195	0.742	0.690		0.160
	(5.94)**	(4.60)**	(4.53)**		(7.81)**
PCS	0.004	–0.067	–0.070		0.038
	(0.16)	(1.87)*	(2.12)**		(1.59)
APPR				0.775	
				(5.76)**	
LHEI				0.101	
				(4.33)**	
LEXI				0.028	
				(2.66)**	
LSIZE				0.164	
				(6.50)**	
GROUP				–0.013	
				(0.25)	
RHO					0.890
					(17.9)**
Nb. obs.	724	724	724	1,987	724
Adj. R^2	0.210	0.166	0.162		

Notes: The dependent variables are the growth of employment, GLEMP, except in the case of equation (iv) where it is the probability of doing research and of being product innovative, PRDINSAL. RHO stands for the correlation coefficient between the errors of the sample selection and tobit equations. The absolute value of the t-values is presented between brackets. These are based on heteroscedasticity-consistent estimates of the standard errors. The t-tests are one-tailed or two-tailed, according to a priori expectations with respect to the parameter values. Symbols * and ** denote significance at the 5% and 1% levels.

As with the innovative output equation, there is always a clear difference between the OLS and the 2SLS and GMM results. Once again, the explanation can be found in the multicollinearity between the instrumented variables, LITURN and GTURN, and the constant term and the industry dummies (IDPAV) that were also used to construct the instrument.

When concentrating on the more reliable OLS estimates we notice that the coefficients for the industry dummies are always negative and statistically significant in the growth of turnover and growth of employment equations. This is plausible and conformant with the evidence in Table A1.2 of the Appendix, which shows that the mean value of GTURN and GLEMP is clearly lower in the industries in question when compared to 'Information Intensive' hi-tech services. As regards the growth of turnover equation, the level of innovative turnover, LITURN, is, of all the firm-specific variables we look at, clearly the key variable, as expected. Moreover, there is evidence that process innovations, PCS, have a significant positive impact, suggesting that cost competitiveness also matters when explaining turnover performance. As regards the growth of employment equation, it is hardly surprising that growth of turnover, GTURN, is the main explanatory variable.

The remaining columns in Table 1.1 to 1.4 present the ML estimates of the generalised tobit model designed to draw statistical inference from a larger population of firms. Column (iv) is the sample selection equation. It is a probit equation to explain the probability of doing R&D and of being product innovative. Column (v) is the corresponding tobit equation, to explain LRDINS, LITURN, GTURN and GLEMP respectively, conditional on the firms doing R&D and being product innovative.

The estimates of the probit equation are always highly similar. In the case of the industry dummies, only IDPAV4 is found to be systematically statistically significant, with a positive sign. This suggests that the probability of being R&D-active and product-innovative is somewhat higher in 'Science based' industries. In addition, only firm-specific characteristics appear to matter. Not surprisingly, appropriability is found to be the main explanatory variable. This confirms that when firms rely on the strategic protection of their ideas through secrecy, complexity of design or lead-time advantage they are more likely to undertake product innovative efforts. The human capital and export orientation variables, LHEI and LEXI, are also always highly statistically significant, with a positive sign. This leads to the plausible suggestion that firms with more qualified personnel and under greater competitive pressure are more likely to be innovative.

Most interestingly, the size variable, LSIZE, is also found to be highly significant with a positive sign. This confirms Stylised Fact 1 put forward by Cohen and Klepper (1996), which states that the likelihood of a firm reporting R&D efforts rises with firm size. Taken in conjunction with our earlier findings, this suggests that a size threshold is necessary for doing formal R&D. But, once this level is exceeded, further size increase does not lead to a proportional increase in R&D for the reasons given above.

The estimated correlation coefficient between the errors of the probit and tobit equations, RHO, is presented at the bottom of column (v). It is found to be positive and statistically significant in the case of R&D intensity and growth of employment, but non-significant otherwise. This is confirmed by the corresponding parameter estimates presented above, which, where these equations are concerned, show evidence of a sample selection effect only. In the case of R&D intensity in Table 1.1, the effect is to somewhat raise the estimates of APPR, LHEI, LEXI and SIZE in the tobit equation, compared to the corresponding OLS estimates in column (i), which do not take account of sample selection. This shows that we are now capturing the combined effect of the impact of the variable in question on the probability of being R&D-active and product-innovative with its impact on the level of R&D intensity once this is the case. Not surprisingly, the increase is strongest in the case of APPR, reflecting its impact in the probit equation. But we should exercise caution in our interpretation given that the relative impact of the other knowledge-related variables, which we could not include in the probit equation due to data constraints, is likely to be underestimated as a result.

As the significance of the RHO value suggests, the strongest sample selection effect can clearly be found in the employment growth equation in Table 1.4. Comparison of the tobit estimates in column (v) against the corresponding OLS estimates in column (i) shows that the main effect is to reduce the size of the intercept significantly. It suggests that the average rate of growth of employment is significantly lower in a larger population of firms than in the R&D-active, product-innovative sub-population. Our final and major question of interest is whether systematic differences in behaviour between strong and weak innovative performers can be identified among R&D-active product innovators. To answer this we have used quantile regression. Tables 1.5 and 1.6 present the results with respect to the 0.25, 0.5 and 0.75 quantiles.

The first part of Table 1.5 presents the estimates of the R&D intensity equation. The first thing we notice is that the intercept, which also captures the influence of the PAV5 dummy, increases substantially from quantile to quantile whereas estimates of the other Pavitt dummies tend to decrease somewhat. This suggests that it is true that the level of R&D intensity is highest, on average, in the 'Information Intensive' sectors, and that this is especially true for the strong R&D performers.

Table 1.5 Quantile regression estimates of the R&D intensity and innovative turnover equations (0.25, 0.50 and 0.75 quantiles)

	R&D intensity equation			Innovative turnover equation		
	Quantiles			Quantiles		
	0.25	0.50	0.75	0.25	0.50	0.75
Constant	–4.227	–3.283	–2.23	–1.525	–0.784	–0.074
	(8.86)**	(10.7)**	(6.98)**	(5.74)**	(3.75)**	(0.35)
IDPAV1	–1.236	–1.460	–1.553	–0.520	–0.482	–0.400
	(3.84)**	(5.30)**	(5.49)**	(3.66)**	(3.67)**	(2.62)**
IDPAV2	–1.013	–1.320	–1.382	–0.514	–0.258	–0.116
	(3.32)**	(4.24)**	(5.13)**	(2.82)**	(1.73)	(0.65)
IDPAV3	–0.015	–0.110	–0.457	0.031	0.017	0.000
	(0.04)	(0.38)	(1.83)	(0.19)	(0.11)	(0.00)
IDPAV4	–0.143	–0.492	–0.874	–0.434	–0.195	–0.345
	(0.43)	(2.18)*	(2.85)**	(1.80)	(1.38)	(1.92)*
GTURN	–0.316	–0.057	–0.054			
	(0.99)	(0.27)	(0.37)			
LRDINS				0.093	0.114	0.147
				(3.21)**	(4.29)**	(4.79)**
INKS	1.004	0.812	0.789	0.253	0.020	0.313
	(1.77)*	(2.00)**	(2.12)**	(0.85)	(0.08)	(1.01)
APPR	0.590	0.352	0.192	0.0470	0.056	–0.050
	(2.40)**	(1.98)**	(1.05)	(0.31)	(0.48)	(0.33)
COOP	0.467	0.394	0.439	0.124	0.005	–0.066
	(2.33)**	(2.28)**	(2.61)**	(1.09)	(0.05)	(0.59)
CKNOW	0.728	0.608	0.310	–0.144	–0.246	–0.292
	(1.69)*	(2.19)**	(0.92)	(0.56)	(1.10)	(1.28)
PCS				0.256	0.127	0.187
				(1.92)*	(1.13)	(1.63)
LHEI	0.231	0.126	0.155	–0.003	0.029	0.027
	(2.31)**	(1.87)*	(1.86)*	(0.10)	(0.62)	(0.77)
LEXI	0.090	0.043	0.023	–0.023	–0.011	–0.044
	(1.93)*	(1.34)	(0.78)	(1.04)	(0.67)	(1.60)
LSIZE	–0.199	–0.166	–0.149	–0.139	–0.086	–0.124
	(1.96)*	(2.57)**	(2.56)**	(3.37)**	(2.74)**	(3.23)**
GROUP	–0.282	–0.063	0.072	0.002	0.065	0.107
	(1.11)	(0.32)	(0.38)	(0.01)	(0.64)	(0.99)
Obs.	724	724	724	724	724	724
Psd. R^2	0.104	0.132	0.167	0.070	0.086	0.091

Notes for Table 1.5: The dependent variables are the log of R&D intensity, LRDINS, and the log of innovative turnover, LITURN. The t-values are based on standard errors obtained by the bootstrap design matrix method. Bootstrap re-sampling was performed using 1,000 repetitions. The t-tests are one-tailed or two-tailed, according to a priori expectations with respect to the parameter values. The pseudo R^2 is obtained as one minus the proportion of the sum of absolute weighted deviations over the raw sum of deviations. Symbols * and ** denote significance at the 5% and 1% levels.

Turning to the firm-specific variables, growth of turnover, GTURN, stays non-significant throughout, confirming average findings in this respect. Of the knowledge-related variables, the point estimates of incoming spillovers, INKS, and research cooperation, COOP, remain fairly constant, but their statistical significance increases somewhat in the higher quantiles. This contrasts sharply with the appropriability variable, APPR, whose parameter estimate decreases considerably, to become completely non-significant in the highest quantile. To a lesser extent this is also the case for our measure of complexity of knowledge, CKNOW. Experimentation has shown that when splitting the APPR variable into its component terms, it is secrecy and to a lesser extent complexity of design that appear to matter most as means of appropriation. And it is the reduction in their impact that helps explain the decrease in influence of APPR across the quantiles. Or, in other words, these factors lose their relevance to incoming spillovers and research cooperation when it comes to explaining R&D intensity among strong R&D investors. This is clearly in line with the predictions of the open innovation model.

Somewhat surprisingly, there is also evidence of a decrease in the relative importance of the impact of our human capital variable, higher education intensity level, LHEI, even though it remains statistically significant. We should be aware, however, that our measure of human capital is still very crude, in the sense that it cannot properly reflect the actual quality of the education concerned, not to mention on-the-job training, which may be particularly important for the most innovative firms. More plausibly, the impact of the export orientation level, LEXI, strongly decreases to become non-significant in the higher quantiles. This suggests that in their R&D investment decisions stronger innovators are clearly less influenced by direct competitive pressures. Finally, the negative impact of size is confirmed in terms of quantiles.

The second part of Table 1.5 presents the estimates of the innovative turnover equation. Here, though the intercept increases once again, the Pavitt dummies we included remain fairly unaltered. This would suggest that, in all industries, we are unable to explain a part of the quantile-to-quantile increase in the level of innovative turnover by means of the explanatory variables under consideration.

Once again, the key firm-specific explanatory variable is level of R&D intensity, as was the case with the average estimates. Most interestingly, its impact can be seen to increase substantially from quantile to quantile. Strong output innovators are, it seems, able to organise their R&D processes more efficiently and so achieve higher returns on their R&D investments. Furthermore, the other knowledge-related variables, as well as worker qualifications and export orientation, are now found to be non-significant too, as was the case in the average estimates. This confirms that their impact on innovative output is indirect, through their influence on the level of R&D intensity.

It seems that evidence for a significant impact from process innovation, PCS, in the average estimates, is present only in the lower quantile. This comes as no surprise, since process innovations that contribute to important product innovation are quite sophisticated and require in-house R&D efforts on the part of the firm. Therefore their impact is likely to be captured by the LRDINS variable itself. Finally, the negative impact of size is confirmed here too across the quantiles.

The first part of Table 1.6 presents the growth of turnover equation. Here, we notice a very sharp increase in the intercept, and thus in the IDPAV5 dummy, and a substantial decrease in the other industry dummies. This suggests that it is true that the average rate of growth of turnover is highest in the 'Information Intensive' hi-tech services, and that this is particularly true for firms with the highest performance in turnover growth. Once again, our firm-specific key explanatory variable is level of innovative turnover, LITURN, as was the case with the average estimates, and its impact is seen to increase substantially from quantile to quantile. This would suggest that product innovativeness becomes especially important in explaining turnover growth performance among the strongest growers.

Finally, the second part of Table 1.6 presents the estimates of the employment growth equation. Here, too, we notice similar tendencies with respect to the quantile-by-quantile evolution of the intercept and the industry dummies we included. This suggests that it is true that the average rate of growth of employment is highest in the 'Information Intensive' high-tech services, and that this is particularly true for the firms with the strongest employment growth performance. Here, our main firm-specific explanatory variable is growth of turnover, GTURN, and now we notice a substantial increase in its impact too.

Table 1.6 Quantile regression estimates of the growth of turnover and growth of employment equations (0.25, 0.50 and 0.75 quantiles)

	Growth of turnover equation			Growth of employment equation		
		Quantiles			Quantiles	
	0.25	0.50	0.75	0.25	0.50	0.75
Constant	0.042	0.248	0.536	0.019	0.192	0.359
	(0.82)	(6.15)*	(7.95)*	(0.71)	(7.56)**	(10.8)**
IDPAV1	–0.044	–0.086	–0.235	–0.044	–0.107	–0.190
	(1.43)	(3.18)*	(3.47)*	(1.60)	(4.13)**	(5.30)**
IDPAV2	0.024	–0.041	–0.168	–0.044	–0.129	–0.205
	(0.81)	(1.60)	(2.28)*	(1.58)	(5.01)**	(6.51)**
IDPAV3	–0.052	–0.052	–0.140	–0.039	–0.126	–0.199
	(1.14)	(1.61)	(–1.79)	(1.36)	(4.86)**	(5.31)**
IDPAV4	0.019	–0.057	–0.174	–0.057	–0.141	–0.184
	(0.52)	(1.66)	(2.52)*	(1.90)	(4.80)**	(5.78)**
LITURN	0.023	0.031	0.052			
	(1.99)**	(3.25)**	(3.63)**			
GTURN				0.124	0.155	0.297
				(4.77)**	(3.42)**	(6.57)**
PCS	0.047	0.022	0.020	0.022	–0.014	–0.052
	(1.55)	(1.04)	(0.58)	(1.04)	(0.90)	(1.82)*
LEXI	–0.005	0.003	0.004			
	(1.60)	(1.19)	(0.74)			
Obs.	724	724	724	724	724	724
Psd. R^2	0.018	0.021	0.056	0.055	0.111	0.190

Notes: The dependent variables are the growth of turnover, GTURN, and the growth of employment, GLEMP. The t-values are based on standard errors obtained by the bootstrap design matrix method. Bootstrap re-sampling was performed using 1,000 repetitions. The t-tests are one-tailed or two-tailed, according to a priori expectations with respect to the parameter values. The pseudo R^2 is obtained as one minus the proportion of the sum of absolute weighted deviations over the raw sum of deviations. Symbols * and ** denote significance at the 5% and 1% levels.

If we remember that product innovation has a particular effect on turnover growth in the case of the strongest growers, the implication is that successful product innovation is of great importance to job creation, as corroborated by other firm-based evidence in this respect (see, for example, Jaumandreu, 2003).

1.5 DISCUSSION AND CONCLUSIONS

Using the CIS3 micro database for Belgium, a structural model was estimated to explain R&D intensity and innovative turnover and their impact on growth of turnover and employment.

Estimation was first performed on a sample of R&D-active product innovators, to gain an insight into the average behaviour of firms in the corresponding sub-population. The estimates show that general technological environment matters in the sense that firms active in 'Information Intensive' hi-tech services show, all things being equal, a higher level of R&D intensity, growth of turnover performance and growth of employment.

Knowledge-related variables, such as incoming knowledge spillovers and research cooperation, have a major positive influence on R&D investment. This confirms that absorption of knowledge calls for substantial, further R&D efforts on the part of enterprises. A higher capacity to appropriate in-house knowledge with respect to non-partners increases incentives to undertake R&D. Higher qualification of workers and greater competitive pressures, measured according to export orientation, also have a positive effect on R&D efforts.

In itself, level of R&D intensity is shown to be the main variable to explain innovative turnover. Knowledge-related and human capital-related variables are all found to be non-significant in the equation on innovative turnover, suggesting that their influence on product innovation is indirect, through R&D. This re-emphasises the need for knowledge absorption first, before it can be utilised in the innovation process.

Firm size is seen to have a negative impact on the level of R&D intensity and, given that level, on innovative turnover. This can be explained by the fact that there are, among the R&D-active product innovators in Belgium, a substantial number of dynamic young firms that pursue more ambitious innovation strategies than the larger firms. They strongly expect to grow as a result of their innovations and are prepared to sell excess knowledge in disembodied form. Therefore, when making R&D decisions, they do not feel highly constrained by their initial size or short-run turnover performance. And, as a result, they are able to generate a greater number of innovations per unit of R&D spend than are larger firms.

We then broadened the focus by considering a larger sample and adapting the model, so as to draw statistical inference from a population that includes non-innovative firms. Estimates of a probit equation showed that the probability of being R&D-active and product-innovative is highly positively affected by level of knowledge appropriability, and somewhat less so by qualification of the workers, degree of competitive pressure measured according to export orientation and firm size. Consequently, the tobit estimates, which control for sample selection effects, showed that the positive impact of appropriability and, to a lesser extent, human capital and export orientation on R&D intensity, is somewhat higher than when considering R&D-active product innovators, whereas the negative impact of size is slightly reduced. The main effect was, however, to reduce the size of the intercept in the employment growth equation, confirming that the average rate of job creation is significantly lower in a larger population of firms than in the R&D-active product innovative sub-population.

Finally, we returned our attention to the R&D-active product innovators and re-estimated the equations using quantile regression, in order to identify systematic behavioural differences between strong and weak innovative performers. Estimates of the industry dummies indicated that firms active in the 'Information Intensive' sectors are particularly over-represented among strong performers in terms of R&D intensity, growth of turnover and growth of employment.

Where firm-specific variables are concerned, the impact of incoming knowledge spillovers and research cooperation on R&D intensity was found to be most significant in the highest quantile. This contrasts sharply with variables such as appropriability or depicting related research characteristics, which tend to become entirely non-significant here.

Furthermore, as was the case with the average estimates, level of R&D intensity, level of innovative turnover, and growth of turnover are the key explanatory variables of the innovative turnover, growth of turnover and growth of employment equations respectively. However, their impact is always seen to rise from quantile to quantile. This clearly suggests that the strongest product innovators have the most efficient R&D processes. The strongest growers are precisely this because they in particular are successful product innovators. And, in turn, the strongest job creators are strong growers.

Taken together, these results seem fairly conformant with the predictions of the open innovation model. The internalisation of external knowledge generated through incoming knowledge spillovers and research cooperation has clearly become one of the major objectives of an innovative enterprise's R&D activities. Also, some of these activities seem to be aimed at absorbing the ideas of workers other than researchers. Though, on average, appropriability does continue to provide an incentive to undertake R&D, quantile regression estimates suggest that this might not be the case among the strongest R&D performers. Presumably it is here that we will find the majority of firms who are prepared to market excess knowledge in disembodied form. Here too, most of the research effort appears to be dedicated to external knowledge absorption. On the whole, this provides implicit evidence of an ongoing and profound reorganisation of the research process.

The heavy impact of R&D intensity on innovative turnover, and the substantial quantile-to-quantile increase of this impact, seems to confirm that this reorganisation is leading to substantial gains in innovative efficiency. This, in turn, appears to have strong positive repercussions on turnover and employment growth performance for the firms in question. It is to be hoped that more accurate data will help us identify the types of jobs created this way.

Finally, our interpretation of the results on the impact of firm size fits in with the logic of open innovation quite well. A substantial portion of the dynamic young firms to which we refer is likely to consist of open innovators. They do not sense that their R&D decisions are highly constrained by their initial size, because they expect to grow rapidly as a result of their innovations. And, as implicitly suggested by our evidence on appropriability, this is also the case because they are prepared to sell excess knowledge in disembodied form, through licensing and so on. It would be of great interest to seek explicit confirmation of this in the form of evidence from complementary case studies.

APPENDIX

Table A1.1 presents a description of the manner in which the variables were constructed from the raw data from the CIS3 questionnaire. Table A1.2 presents descriptive statistics for the variables relating to the total sample, the R&D-active product innovative sample, and the corresponding industry dummies.

Table A1.1 Description of the variables

Variable	Definition
Endogenous variables	
Log of R&D intensity (LRDINS)	Natural log of the ratio of intramural R&D expenditure over turnover in the year 2000
Log of innovative turnover (LITURN)	Natural log of the % of new or significantly improved products (goods and services) in turnover in the year 2000
Growth of turnover (GTURN)	Growth of market turnover (goods and services) from 1998 to 2000
Growth of employment (GLEMP)	Growth of employment from 1998 to 2000
Exogenous variables	
Incoming spillovers (INKS)	Sum of scores of the importance attached to information from suppliers, clients, competitors, universities, government, professional conferences, exhibitions (rescaled between 0 and 1)
Appropriability (APPR)	Average score of the effectiveness of strategic protection through secrecy, complexity of design and lead-time advantage
Research cooperation (COOP)	Cooperation = 1 if the firm cooperates with suppliers, clients, competitors, research labs, universities or government
Complexity of knowledge (CKNOW)	Ratio of the sum of the scores of the importance attached to information from universities, government and professional conferences over the sum of the scores of all sources of information
Log of higher education intensity (LHEI)	Natural log of the % employees with higher education in the year 2000

Table A1.1 (continued)

Variable	Definition
Log of export intensity (LEXI)	Natural log of the proportion of exports in turnover in the year 1998
Log of size (LSIZE)	Natural log of the number of employees in the year 1998
Group membership (GROUP)	Group membership = 1 if the firm is a member of a group
Process innovation (PCS)	Process innovation = 1 if the firm has introduced new or significantly improved production processes during the period 1998–2000
Resource intensive sectors (IDPAV1)	IDPAV1 = 1 if the firm is classified in the 'Resource Intensive' sectors (Mining, Food, Beverages and Tobacco, Wood and Furniture, Recycling, Wholesale)
Scale intensive sectors (IDPAV2)	IDPAV2 = 1 if the firm is classified in the 'Scale Intensive' sectors (Paper and Printing, Rubber and Plastics, Non-metallic Minerals, Ferrous Metals, Non-ferrous Metals, Fabricated Metal Products, Motor Vehicles, Other Transport Equipment)
Specialised supplier sectors (IDPAV3)	IDPAV3 = 1 if the firm is classified in the 'Specialised Supplier' sectors (Non-electrical Machinery, Electronics and Communication, Electrical Machinery)
Science based sectors (IDPAV4)	IDPAV4 = 1 if the firm is classified in the 'Science Based' sectors (Chemicals, Pharmaceuticals, Computers and Office Machinery, Scientific Instruments)
Information intensive sectors (IDPAV5)	IDPAV5 = 1 if the firm is classified in the 'Information Intensive' sectors (Financial Intermediation, Telecommunication, Transport Services, Business Services, Computer and Related Service Activities, Research and Development, Technical Engineering

Table A1.2 Descriptive statistics: mean and standard deviation

Variable	All firms	R&D-active and product innov-ative firms	*Of which*: IDPAV1	IDPAV2	IDPAV3	IDPAV4	IDPAV5
Obs.	1,987	724	163	169	101	97	194
LRDINS		–4.437	–5.298	–5.080	–3.866	–3.998	–3.671
		(1.835)	(1.546)	(1.534)	(1.571)	(1.442)	(2.099)
LITURN		–1.836	–2.178	–2.045	–1.559	–2.015	–1.423
		(1.056)	(1.035)	(1.101)	(0.885)	(1.037)	(0.960)
GTURN	0.168	0.244	0.118	0.181	0.197	0.170	0.467
	(0.548)	(0.605)	(0.304)	(0.238)	(0.399)	(0.288)	(1.025)
GLEMP	0.134	0.162	0.125	0.100	0.117	0.083	0.309
	(0.311)	(0.314)	(0.310)	(0.170)	(0.289)	(0.187)	(0.417)
INKS		0.433	0.416	0.428	0.478	0.455	0.417
		(0.211)	(0.224)	(0.208)	(0.203)	(0.186)	(0.215)
APPR	0.193	0.361	0.290	0.343	0.495	0.422	0.336
	(0.322)	(0.377)	(0.369)	(0.364)	(0.402)	(0.377)	(0.365)
COOP		0.387	0.319	0.367	0.475	0.423	0.397
		(0.487)	(0.467)	(0.483)	(0.502)	(0.497)	(0.490)
CKNOW		0.325	0.304	0.315	0.307	0.400	0.321
		(0.241)	(0.270)	(0.248)	(0.191)	(0.267)	(0.222)
LHEI	–2.313	–1.759	–2.183	–2.388	–1.973	–1.472	–0.884
	(1.838)	(1.300)	(1.226)	(1.094)	(1.291)	(0.934)	(1.220)
LEXI	–3.004	–2.251	–2.316	–1.651	–1.451	–1.388	–3.568
	(2.860)	(2.703)	(2.556)	(2.362)	(2.232)	(2.150)	(3.063)
LSIZE	3.795	4.211	4.301	4.644	4.334	4.412	3.591
	(1.301)	(1.474)	(1.300)	(1.478)	(1.608)	(1.572)	(1.290)
GROUP	0.561	0.656	0.589	0.734	0.604	0.701	0.649
	(0.496)	(0.475)	(0.493)	(0.443)	(0.491)	(0.460)	(0.478)
PCS		0.737	0.736	0.828	0.703	0.701	0.696
		(0.440)	(0.442)	(0.378)	(0.460)	(0.460)	(0.461)

Note: The standard deviation is presented between brackets beneath the mean.

2. Firm-level effects of incoming knowledge spillovers, research collaboration and strategic appropriability

2.1 INTRODUCTION

For a long time the innovation process has been regarded as a closed system: innovation remained largely internal to the firm. Hence the existence of a large body of innovation literature with a 'black box' content when it comes to describing the external knowledge environment in which a company operates (see Tidd et al., 1997; Afuah, 1998; Trott, 1998). It is becoming more widely accepted today that incoming knowledge spillovers and research collaboration make a beneficial contribution to the flexibility of the firm by enhancing its access to knowledge beyond its boundaries (Quinn, 2000; Fritsch and Lukas, 2001; Chesbrough, 2003a; Coombs et al., 2003; Huston and Sakkab, 2006). Moreover, ideas on the incorporation of external knowledge have evolved from cost reduction to actual contribution to innovation efforts.

Additionally, one of the key issues in innovation lies in the appropriation of its results (Teece, 1986; Dosi et al., 2006). The ways in which firms handle knowledge flows, either through incoming spillovers or collaborative results, and the ways in which they are converted into new products or sold in disembodied form, are caught by introducing the role of the business model in capturing value from innovation (Chesbrough and Rosenbloom, 2002; Chesbrough, 2003a).

The aim of this chapter is to incorporate all of these elements into a general framework of analysis pertaining to the relationship between innovative efforts and firm performance and to emphasise that every firm has its own particularity. The majority of empirical studies are about making general statistical statements on the innovative performance of an 'average' firm (see, for example, the recent econometric work estimating structural models to explain both the input and output stages of the innovation process, such as Crépon et al., 1998; Lööf and Heshmati, 2002, 2006; Klomp and Van Leeuwen, 2001; Van Leeuwen and Klomp, 2006; Griffith et al., 2006). Or, as is the case with much of the literature on open innovation, well-documented case studies are put in place to underscore the particularities marking the ways in which a firm converts knowledge to commercial and profitable new products (Chesbrough and Rosenbloom, 2002). This chapter uses econometric quantile regression analysis to bring out the heterogeneity of firm performance in terms of research intensity and turnover realised with new products. In particular, it examines the role played by knowledge-related variables, in order to focus on the differences between weak and strong innovators. Quantile regressions have already been used to estimate equations explaining research intensity, but without incorporating the knowledge-related variables that are deemed most important (Nahm, 2001). Ebersberger et al. (2006) use quantile regression of an equation on innovative turnover which includes a measure of appropriability. However, Hussinger (2006) finds that appropriability has its greatest impact in its influence on research intensity.

This study takes a more systematic approach in analysing the impact of knowledge-related variables on firm performance, allowing for differences between weak and strong innovators. It uses company data from the Third Community Innovation Survey (CIS3) for Belgium.

The chapter is structured as follows: Section 2.2 looks at the conceptual background and gives a review of the literature. Section 2.3 discusses the data, models and estimation strategy. Then, in Section 2.4, results pertaining to the overall effects of the knowledge-related variables are discussed: incoming knowledge spillovers, research collaboration and strategic appropriability. Section 2.5 refines this analysis by using quantile regressions to identify weak and strong innovators. Finally, Section 2.6 summarises and concludes.

2.2 CONCEPTUAL BACKGROUND

Innovation has long been considered a major engine of technical change and economic growth. Therefore, stimulating innovation ranks high on the agenda for business managers and policy makers alike. Thoughts on innovation are constantly evolving as environments change and product innovation becomes more and more complex (Tidd et al., 1997; Afuah, 1998; Trott, 1998). The fact that firms no longer have within the company all the knowledge they need, and recognise that innovations can be commercialised in ways other than through the product market, has recently been captured by Chesbrough's 'imperative' of open innovation (Chesbrough, 2003a). In-house research had implied a strong belief in the do-it-yourself attitude, which is now out of date and has been superseded by active screening of the knowledge and technology that exists beyond the boundaries of the firm. Company strategy has evolved from 'research and develop' to 'connect and develop' (Huston and Sakkab, 2006).

This chapter takes a closer look at the impact of three knowledge-related variables on a company's innovative performance. These variables pertain to incoming knowledge spillovers, research collaboration and appropriability of the benefits of the innovation. Incoming knowledge spillovers have received a great deal of attention in the literature on firm-level innovation (for a review of the literature, see for example von Hippel, 1988; Veugelers and Cassiman, 1999; Monjon and Waelbroeck, 2003; Sena, 2004; Laursen and Salter, 2006). Spillovers occur if research efforts outside the firm are used within the firm to promote in-house innovation. Spillovers are occasionally part of an explicit government strategy, for example through large-scale basic research programmes. In this case, the spillovers are a deliberate phenomenon. But spillovers also arise through the sourcing of ideas that other firms were unable to appropriate for their own benefit (Sena, 2004).

The incoming knowledge spillovers in this chapter can affect both the innovation input side (research intensity) and the output side (turnover realised from the innovations). Veugelers and Cassiman (1999) used a dichotomous variable as innovation output; Laursen and Salter (2006) used the share of innovative turnover as a measure of output. The present chapter extends the latter research by focusing on different types of innovators through quantile regressions. Though interesting, one of the main shortcomings of the literature on incoming spillovers is that these spillovers are often viewed in isolation from other knowledge-related variables such as research collaboration and appropriability issues.

Research collaboration by firms to advance innovative activities has enjoyed a great deal of attention from scholars (for a review of the literature, see Fritsch and Lukas, 2001; Tether, 2002) and public organisations (OECD, 2001). Two separate aspects of research collaboration have been the subject of empirical study. First, in-house research by firms is considered a prerequisite for successful research collaboration due to the creation of so-called absorptive capacity (Cohen and Levinthal, 1990; Veugelers, 1997; Miotti and Sachwald, 2003). Once a certain threshold of qualified knowledge workers has been established, and they are able to cognitively understand the research efforts and then step willingly and knowingly into joint research ventures, the impact of collaboration on research intensity is found to be positive, thereby reflecting the need to internalise the knowledge thus created through additional in-house research efforts (see Veugelers, 1997, for the manufacturing sector). Second, firms consider entering into research collaborations as a means of enhancing their innovative performance. This aspect has received less attention in the literature (Becker and Dietz, 2004). This chapter picks up on both points because, given the paradigm of open innovation, this particular exponent of knowledge-related interaction is being explored further in this new context (Laursen and Salter, 2006).

Moreover, several authors acknowledge the impact of partner type (Fritsch and Lukas, 2001; Tether, 2002; Miotti and Sachwald, 2003; Spithoven and Teirlinck, 2006). Two main categories come under consideration here: inter-firm research collaboration with clients, suppliers and competitors (Fischer and Varga, 2002) and inter-organisational research collaboration with other types of organisations, such as research labs, universities and government research centres (Monjon and Waelbroeck, 2003; Laursen and Salter, 2004).

Once an innovation is conceived, firms adopt a protection strategy to shield the benefits accruing from the new product (Teece, 1986). Recently, the issue of appropriability has received renewed attention (see the special issues on this subject in *Research Policy* in October 2006 and in *Industrial and Corporate Change* in December 2006). There is a huge variety of appropriation channels. The OECD (1997) has divided these channels into formal (patents, trademarks, copyrights) and informal (secrecy, complexity of design and lead-time advantage) appropriation strategies (Hussinger, 2006). Cassiman and Veugelers (2002) refer to the informal methods as 'strategic' protection. Strategic appropriation is preferred for reasons of cost and time efficiency. Also, a formal appropriation strategy implies that at least some elements of the innovation require codification and disclosure, and are therefore rendered publicly available (Liebeskind, 1997). Although most firm-level literature on appropriability is restricted to the use of legal protection strategies (that is, patents), it has long been acknowledged that patenting is not suited to all types of industries or firms (Teece, 1986; Arundel, 2001). Therefore, this chapter follows the tradition of Cassiman and Veugelers (2002) by focusing on strategic appropriability.

An overreliance on appropriation strategies could carry implications for the innovation process's openness to research collaboration and reduce knowledge interactions with clients and suppliers (von Hippel, 1988). As hinted by the open innovation paradigm, this might eventually lead to poorer performance by firms that maintain a tight appropriation regime (Chesbrough, 2003a; Chesbrough and Rosenbloom, 2002).

Innovation performance also differs across industry type. This is due to differences in technological trajectories. Theoretically, the notion of technological trajectories is underpinned by the idea of path-dependency (Pavitt, 1984; Tidd et al., 1997). The key element in this idea is that technical change imposes an internal logic that defines the boundaries within which innovative opportunities might be found. These technological trajectories apply to firms as well as industries and even to countries in which more trajectories are possible (Tidd et al., 1997). Pavitt proposed a taxonomy to regroup several industries according to their main sources of technology and the main uses of this technology. In itself, this classification is quite useful for detecting how innovations differ in different industries (Cesaretto and Mangano, 1992). Five trajectories are used, ranging from hi-tech science-based industries, through information-intensive industries, industries dependent on specialised suppliers and scale-intensive industries to low-tech resource-intensive sectors (see Table A2.1 in the Appendix for a classification and summary of the industries).

The storylines in the literature mentioned in this section were developed by looking at average firm behaviour while controlling for contextual variables, thereby largely neglecting heterogeneity between firms. This chapter aims to fill this gap by using the econometric procedure of quantile regressions.

2.3 DATA, MODEL AND ESTIMATION STRATEGY

2.3.1 Data and Variables

The Community Innovation Survey is a proven source on innovation-related issues. This chapter uses the Third European Community Innovation Survey (CIS3) for Belgium. The survey was conducted in 2002 and the questions it contains relate to the period 1998–2000. All private enterprises in the country with 10 employees or more qualified to receive the questionnaire. The survey sample was stratified with respect to the usual parameters: sector and size. In total, 4,482 questionnaires, covering one third of the target population, were sent, and 2,035 responses were received (Spithoven, 2005). Screening these data with respect to the research problem reduced the sample to 1,987 firms. Reasons for dropping enterprises were that the sector 'gas, water, electricity' had too few observations (7); the variable on research intensity showed a few extreme values as a result of measurement error (16); and, in 25 cases, the data on employment and turnover were incomplete. In total, 1,310 enterprises reported themselves as innovative. Of these, 951 firms reported intramural R&D, 945 reported innovative turnover, and 724 claimed both. Since the main interest is to obtain information on the behaviour of these research-active product innovators, the analysis will concentrate mainly on the latter sample. The Appendix summarises the variables used in the analysis in two tables: Table A2.1 describes all variables and Table A2.2 presents their mean value.

The survey is characterised as subject-oriented because of the involvement of key players in the innovation process within the individual firm. Managing directors, CEOs, CFOs and R&D managers all are (partly) responsible for replying to the questionnaire. Every participating country (or region) has had experience of the innovation survey since the First Community Innovation Survey in 1996, rendering it a reliable and valid survey. One huge advantage is that all major stages of the innovation process are covered: from the input stage on research and development activities, over throughput on research collaboration, to output on innovations realised in terms of turnover.

The major drawback, however, is that very subjectivity on the issue of innovation. The questionnaire is designed to measure the degree of innovativeness at firm-level as well as the human and financial resources engaged in innovative activities. None of the questions relates to the project level of these innovations, obscuring the process from its origin and leaving many aspects largely untouched. No information is collected on the type of interaction when firms are engaging in research collaboration, nor on the way the information sources needed to innovate are handled. So the 'black box' content of innovation remains considerable. Also, the survey was originally designed for the manufacturing sector only and was adapted to incorporate services ex post. However, nowadays the services are innovating to a large extent in the most advanced economies (for example, software development and financial innovations).

This chapter intends to explain both the R&D intensity and the innovative turnover of firms. The R&D intensity, denoted as LRDINS, is a measure of innovation input. It is calculated as the natural log of the share of in-house R&D investments in turnover. In-house R&D is used because it is assumed to be particularly relevant to the firm's capability of using incoming knowledge spillovers, of engaging in research collaboration and of handling the appropriation of innovative results. Innovative turnover, LITURN, is a measure of innovation output. It is measured by the natural log of the share of new or significantly improved products in turnover. The advantage of using product innovation as an output measure over the readily available patent statistics, besides being less industry-specific in general and underestimating the innovation activities in services, is that not all innovations are patented.

Incoming knowledge spillovers, INKS, are captured by the survey question on the importance of the different sources of information used to make the innovation. The sources stem either from inter-firm linkages of the firm – customers, suppliers and competitors (INKS-SCC) – or from inter-organisational linkages of the firm – universities and government research centres (INKS-UG) – or from other publicly available sources – conferences and fairs and exhibitions (INKS-O). By summing the ordinal scores on the use made of each source and rescaling these between zero and unity a firm-level measure of incoming knowledge spillovers is created.

Research collaboration between firms and various partners, COOP, is a dichotomous (binary) variable: unity if collaboration takes place and zero otherwise. Again, various partners for collaboration are considered: inter-firm collaboration linkages with customers, suppliers and competitors (COOP-SCC), and inter-organisational collaboration linkages with research labs, universities and governmental research facilities (COOP-RUG).

Appropriability, APPR, measures the degree of strategic protection. Secrecy (APPR-SE), complexity of design (APPR-CD) and lead-time advantage (APPR-LT) are considered as strategic protection methods. These variables are all dichotomous and take the value of unity when the method is used and zero otherwise. Averaging these scores provides an average measure of strategic protection, APPR, ranging between zero and unity.

The reliance on complex knowledge, CKNOW, is measured using the firm's sources of information. If a firm sources from universities, government and professional conferences, this is viewed as an indication that it performs basic research. As suggested by Cassiman and Veugelers (2002), the ratio of these scores in the total scores of all information sources may be interpreted as a measure of basic research. Because the measure of incoming knowledge spillovers is itself derived from the firm's sources of information, the possibility of multicollinearity cannot be ruled out.

The other firm-related variables control for the influence of demand, qualified labour, the competitive pressure, firm size, group affiliation and process innovation. Demand is brought in as the growth of total turnover, GTURN, between 1998 and 2000; qualified labour, LHEI, is measured by the natural log of the share of employees with higher education; competitive pressure, LEXI, is captured by the natural log of the share of exports in turnover; size, LSIZE, is the natural log of the number of employees in the base year; group affiliation, GROUP, and process innovation, PCS, are dichotomous variables assuming the value of one if the characteristic is present and zero otherwise.

Industry-specific dummy variables are often used to detect systematic differences in innovative behaviour by firms between industries. Use is made of the classification developed by Pavitt (1984), revised and elaborated by Tidd et al. (1997) to incorporate the services economy. This classification is based on the characteristics of the technologies used in the production system; the skills needed to operate these technologies; the sources about these technologies and the strategy to monitor and exploit the technologies. Five industrial sectors are identified: resource intensive (IDPAV1), scale-intensive (IDPAV2), specialised suppliers (IDPAV3), science-based (IDPAV4), and information-intensive (IDPAV5). Each classification entails a good deal of subjectivity, yet the alternative classification in high-medium-low technology (OECD, 1997) is far less concerned with the underlying technology because it exclusively rests on the ratio of R&D investment over turnover. Converting the industrial sectors, which are gathered at two digits NACE, into the Pavitt trajectories, also introduces some subjectivity. Despite this inevitable shortcoming the conversion, as indicated in the Appendix, proved satisfactory.

2.3.2 Model and Estimation Strategy

The CIS3 data refer to the period 1998–2000 and are cross-sectional in nature. They do not, therefore, allow the introduction of lagged responses between the variables. The model in question consists of two structural equations, which explain, respectively, the level of R&D intensity, LRDINS, and the level of innovative turnover, LITURN. The model equations include common variables: a constant term, C, an array of industry-specific dummy variables describing the overall technical environment, and several specific control variables: basic research, qualified labour force, competitive pressure, firm size and group affiliation. The main variables of interest are incoming spillovers, research collaboration and appropriability. They are first entered as such and then split, to detect the precise origin of their contribution to research intensity and innovative turnover. The equations can be presented as follows.

LRDINS = f1(C, IDPAV1, IDPAV2, IDPAV3, IDPAV4, IDPAV5, GTURN, INKS, COOP, APPR, CKNOW, LHEI, LEXI, LSIZE, GROUP) (2.1)

LITURN = f2(C, IDPAV1, IDPAV2, IDPAV3, IDPAV4, IDPAV5, LRDINS, INKS, COOP, APPR, CKNOW, PCS, LHEI, LEXI, LSIZE, GROUP) (2.2)

Since innovative turnover is likely to influence market turnover performance, and therefore growth of turnover, there may be feedback from LRDINS over LITURN to GTURN, and so again to LRDINS. Provided such feedback exists, this would imply that the variables concerned are endogenous when appearing as explanatory variables. In this case, equation-based estimation by ordinary least squares (OLS) will not provide consistent estimates and the implied simultaneity bias may be severe in view of the cross-sectional nature of the data. A more appropriate estimation method will have to be used. If the feedback loop does, however, break down, the system may be recursive in nature and OLS may still be appropriate, in the sense of being consistent and the most efficient.

Our strategy will therefore be as follows. We begin estimation by OLS, while considering thereby (White) heteroscedasticity-robust standard errors to cope with the possible problem of heteroscedasticity. We then repeat estimation by an instrumental variable technique, to control for possible problems of endogeneity with respect to GTURN and LRDINS. This is the generalised method of moments estimation procedure (GMM). It estimates the parameters directly from the moment conditions that are imposed by the model. This is done by minimising a quadric form in the weighted sample moments, in which the weighting matrix is based on the inverse of a heteroscedasticity-robust estimate of the covariance matrix of the sample moments (Hansen, 1982). From this, we make a first assessment of the endogeneity issue by considering the parameter estimates obtained by using the different methods in conjunction. We then seek further confirmation in this respect by performing Durbin–Hausman–Wu exogeneity tests on the variables concerned (Davidson and MacKinnon, 1993).

Following Crépon et al. (1998), recent econometric work on the relationship between innovation and firm performance has devoted a great deal of attention to the issue of sample selection. After estimating the baseline equations on a sample of research-active and product-innovative firms, we will, as a check of robustness, extend the sample and adapt the model in order to draw inference about a larger population that also includes non-innovative firms. The model is adapted by including an equation explaining the probability of doing research and of being product-innovative. In so doing, we obtain a generalised tobit II model, consisting of a probit equation, with a binary dependent variable that takes the value 1 in the case of positive research and innovative turnover and the value 0 otherwise, and our considered structural equations. OLS will cease to be appropriate even if no problems of endogeneity arise, as seen below. This is due to the binary nature of the dependent variable in the probit equation and to the possible correlation between its error term and the error terms of the structural equations, which will cause a sample selection bias in their parameter estimates. We will, therefore, for each of our equations, estimate the system consisting of the probit equation and the equation in question on the basis of maximum likelihood (ML), while using the Heckman (1979) procedure to choose the initial parameter values in the iterative procedure. The considered standard errors are, again, heteroscedasticity-robust. This method has been shown to provide consistent and efficient estimates (Verbeek, 2000).

We then return our attention to the sample of research-active product-innovative firms and investigate the consequences of splitting incoming knowledge spillovers, research collaboration and appropriability by comparison with the baseline estimates.

Finally, we perform quantile regression estimation to identify systematic behavioural differences between strong and weak innovative performers. A quantile regression estimates the parameters by minimising the sum of asymmetrically-weighted absolute values for the residuals. Thus, for example, the 0.25 regression quantile estimates lend less weight to positive residuals than negative ones. Attention is therefore drawn, so to speak, to the left-hand tail of the conditional density function of the dependent variable in question. The reverse is true for the 0.75 quantile. Varying the quantile parameter between 0 and 1 generates all the regression quantiles, revealing the conditional distribution of the dependent variable given the explanatory variables. Quantile regression estimates have the additional advantage of being more efficient than least squares estimates when the errors deviate from normality, which is the case here.

Since we cannot exclude the possibility of heteroscedasticity, and, therefore, dependence between the quantile regressors and the density of the error terms, we use the bootstrap design method to estimate the standard errors (Buchinsky, 1995, 1998). We performed bootstrap re-sampling using 1,000 repetitions. Due to space constraints, the results presented in Tables 2.3 and 2.4 below relate to the 0.25, 0.50 (or median), and 0.75 quantiles.

2.4 AVERAGE EFFECTS OF INCOMING KNOWLEDGE SPILLOVERS, RESEARCH COLLABORATION AND STRATEGIC APPROPRIABILITY

Table 2.1 presents the baseline estimates of equation (2.1) explaining research intensity, and Table 2.2 gives the baseline estimates of equation (2.2) explaining innovative turnover.

Consider the results with respect to research intensity in Table 2.1. Estimation on the sample of the 724 research-active product innovators shows that the estimates by OLS in column (i) and by GMM in column (ii) are not substantially different. Instrumenting the growth of turnover variable, GTURN, has, in other words, hardly any effect at all. In fact, this variable is always found to be non-significant. Demand conditions do not seem to play a vital role in research intensity, indicating that technical change within firms may be partly science or technology-driven, which ties in with finding that the more technologically advanced industries are most heavily involved in research activities.

Table 2.1 Research intensity: baseline estimates

	Research intensity			
	OLS	GMM	ML	ML
	(i)	(ii)	(iii)	(iv)
Constant	–3.059	–2.930	–0.767	–3.356
	(11.3)**	(6.41)**	(6.06)**	(12.2)**
IDPAV1	–1.397	–1.568	–0.120	–1.414
	(6.32)**	(5.03)*	(1.36)	(6.53)**
IDPAV2	–1.128	–1.270	–0.059	–1.133
	(4.91)**	(4.42)**	(0.63)	(5.03)**
IDPAV3	–0.214	–0.364	0.086	–0.205
	(0.85)	(1.13)	(0.78)	(0.83)
IDPAV4	–0.380	–0.511	0.250	–0.348
	(1.75)	(1.68)	(1.99)*	(1.61)
GTURN	–0.241	–0.466		–0.242
	(1.25)	(0.62)		(1.38)
INKS	0.775	0.745		0.772
	(2.38)**	(2.08)**		(2.40)**
COOP	0.405	0.410		0.404
	(3.01)**	(3.14)**		(3.03)**
APPR	0.331	0.385	1.315	0.512
	(2.10)**	(2.26)**	(12.6)**	(3.22)**
CKNOW	0.359	0.379		0.354
	(1.42)	(1.46)		(1.44)
LHEI	0.178	0.178	0.150	0.203
	(2.86)**	(2.88)**	(6.69)**	(3.29)**
LEXI	0.069	0.062	0.053	0.077
	(2.47)**	(2.30)**	(4.32)**	(2.75)**
LSIZE	–0.181	–0.172	0.171	–0.157
	(3.04)**	(2.69)**	(6.29)**	(2.78)**
GROUP	–0.240	–0.230	–0.034	–0.244
	(1.66)	(1.61)	(0.49)	(1.71)
RHO				0.148
				(2.60)**
Obs.	724	724	1,987	724
Adj. R^2	0.214	0.209		

Notes for Table 2.1: The dependent variable is the log of research intensity, LRDINS, in columns (i), (ii) and (iv), the log of innovative turnover, LITURN, in columns (v), (vi) and (viii), and the probability of doing research and of being product innovative, PRDTURNIN, in columns (iii) and (vii). RHO stands for the correlation coefficient between the errors of the sample selection and tobit equations. The absolute value of the t-values is presented between brackets. These are based on heteroscedasticity-consistent estimates of the standard errors. The t-tests are one-tailed or two-tailed, according to a priori expectations with respect to the parameter values. Symbols * and ** denote significance at the 5% and 1% levels.

It would also imply that the potential feedback from research intensity over innovative turnover to demand growth, and so again to research intensity, breaks down and that the model is recursive in nature.

Consider now the other respective parameter estimates in turn. Because the equation includes an intercept, we had to drop one of the industry dummies to avoid singularity. We dropped IDPAV5, so that its impact is captured by the intercept, C. Therefore, the estimates of the remaining industry dummies should be interpreted as reflecting deviations from the information-intensive sectors. Since their estimates are all negative, this means that these industries are less research-intensive than the information-intensive sectors. This is particularly the case for the industries that are resource-intensive, IDPAV1, and scale-intensive, IDPAV2.

Besides demand conditions, two other elements have no significant influence on research intensity: basic research and group affiliation. In the case of basic research this may reflect the difficulty that many firms have in distinguishing between useful sources of information from the universities and public research labs and information stemming from other sources, which implies more heterogeneous networking by innovative firms (see also Spithoven and Teirlinck, 2006). Part of the explanation may, however, also be found in the existence of multicollinearity with the measure of incoming knowledge spillovers, INKS, which is based on more or less the same underlying data. Group membership also has a non-significant influence on the research intensity of innovators, indicating that the resources of the group to which the subsidiary belongs are of minor importance to product development within the firm (see also Spithoven, 2005).

The important findings pertain to the knowledge-related explanatory variables. Incoming knowledge spillovers, INKS, research collaboration, COOP, and strategic appropriability, APPR, are all highly significant. Research intensity is higher for firms actively sourcing information to develop product innovations, and is even more so when it comes to engaging in research collaboration. Appropriating the benefits accruing from innovation further stimulates in-house research intensity. In the remainder of the text these three knowledge-related variables are broken down to analyse this issue in greater depth.

The quality of the labour force, LHEI, and competitive pressure, LEXI, have a positive sign as expected. People are extremely important in the case of innovative activities. The labour force needs to be highly skilled in order to be able to contribute to the innovation process and to absorb external knowledge (see also Frantzen, 2000). Furthermore, the pressure resulting from exposure to foreign competition has a positive effect on the need for research investment to remain competitive in a globalised world.

The size of the innovative firm has a negative impact on research intensity, thus favouring small entities in terms of higher research intensities. This finding contradicts stylised facts 2 and 3 of Cohen and Klepper (1996), which were derived from the earlier empirical literature on this issue, thereby implying a proportional increase of R&D expenditure with firm size. This might be due to the fact that a major portion of technical change nowadays stems from small-scale firms active in either information-intensive or science-based industries, and so far less in the resource-intensive and scale-intensive sectors that were much more dominant before the ICT revolution. These firms were, indeed, capable of enjoying economies of scale in the field of technical progress. The new technology-based firms (NTBFs), which are heavily stimulated by the regional governments in Belgium, are explicitly meant to be fast-growing in the early and therefore small-scale phases of their existence. The CIS3 survey contains no information on the age of the firms, meaning that this topic requires further investigation in the future.

Table 2.2 reports on the equation on innovative turnover, LITURN. In this case, there is a clear difference between the OLS estimates (in column (v)) and the GMM estimates (in column (vi)). The explanation is the existence of strong multicollinearity between the instrumented variable on research intensity, LRDINS, and the intercept, the industry dummies, and firm size, LSIZE, which were also used in the construction of the instrument, in the case of the estimation by GMM. But, since the model was found to be recursive in nature, the OLS estimation is the most efficient and thus more reliable. Therefore, the focus is on the OLS estimates.

First consider the industry-dummies: these are negative, indicating that the innovative turnover from the information-intensive industries, IDPAV5, is much higher than that of the specialised-supplier, IDPAV3, and the scale-intensive industries, IDPAV2, as in the case of research intensity. The negative sign for the science-based industries, IDPAV4, is somewhat surprising since these industries are also characterised as dynamic and hi-tech. The explanation is probably that the development of new products in industries such as biotech and pharmaceuticals is characterised by long gestation periods (Bruneel et al., 2007).

Table 2.2 Innovative turnover: baseline estimates

	Innovative turnover			
	OLS	GMM	ML	ML
	(v)	(vi)	(vii)	(viii)
Constant	–1.004	–0.053	–0.763	–1.101
	(6.19)**	(0.07)	(6.06)**	(4.51)**
IDPAV1	–0.490	–0.087	–0.120	–0.496
	(4.30)**	(0.25)	(1.36)	(4.38)**
IDPAV2	–0.359	–0.039	–0.058	–0.361
	(2.97)**	(0.13)	(0.62)	(3.01)**
IDPAV3	–0.006	0.047	0.084	–0.002
	(0.05)	(0.32)	(0.76)	(0.02)
IDPAV4	–0.429	–0.327	0.253	–0.419
	(3.29)**	(1.99)*	(2.01)*	(3.22)**
LRDINS	0.102	0.409		0.102
	(4.95)**	(1.72)*		(5.00)**
INKS	0.272	0.022		0.271
	(1.37)	(0.09)		(1.38)
COOP	0.010	–0.109		0.010
	(0.12)	(0.82)		(0.12)
APPR	–0.014	–0.111	1.312	0.045
	(0.14)	(0.82)	(12.5)**	(0.33)
CKNOW	–0.147	–0.268		–0.147
	(0.86)	(1.30)		(0.88)
PCS	0.195	0.240		0.195
	(2.26)**	(2.28)**		(2.28)**
LHEI	0.004	–0.051	0.150	0.012
	(0.12)	(0.92)	(6.70)**	(0.35)
LEXI	–0.021	–0.043	0.053	–0.019
	(1.49)	(1.56)	(4.37)**	(1.25)
LSIZE	–0.102	–0.047	0.170	–0.094
	(3.32)**	(0.91)	(6.28)**	(2.82)**
GROUP	0.075	0.145	–0.033	0.073
	(0.86)	(1.27)	(0.47)	(0.86)
RHO				0.079
				(0.55)
Obs.	724	724	1,987	724
Adj. R^2	0.125	0.073		

Notes for Table 2.2: The dependent variable is the log of innovative turnover, LITURN, in columns (v), (vi) and (viii), and the probability of doing research and of being product innovative, PRDTURNIN, in column (vii). RHO stands for the correlation coefficient between the errors of the sample selection and tobit equations. The absolute value of the t-values is presented between brackets. These are based on heteroscedasticity-consistent estimates of the standard errors. The t-tests are one-tailed or two-tailed, according to a priori expectations with respect to the parameter values. Symbols * and ** denote significance at the 5% and 1% levels.

The equation on innovative turnover confirms the importance of the link with research intensity on the one hand, and, to a lesser extent, the impact of process innovation on the other. The knowledge-related variables seem to affect innovative turnover only in an indirect way, through their impact on research intensity. This suggests that the knowledge in question has first to be internalised within the firm. It is awkward, in this context, to find the quality of the labour force, LHEI, to be non-significant as well, even when we consider that the effect is largely indirect over research intensity. This finding might be explained by recalling that the sample under consideration is confined to those cases in which the firms surveyed reported in-house research activities.

The remaining columns in Table 2.2 present the ML estimates of the generalised tobit model aimed at drawing statistical inference from a larger population of firms. Columns (iii) and (vii) present the sample selection equation. It is a probit equation explaining the probability of doing research and of being product-innovative. The choice of its variables was conditioned by data availability on the total sample. Columns (iv) and (viii) present the corresponding tobit equations, explaining respectively the level of R&D intensity and innovative turnover, LRDINS and LITURN, conditional upon the firms doing research and being product-innovative.

The estimates of the probit equation are very similar. In the case of the industry dummies, only the science-based firms are found to be systematically statistically significant, and with a positive sign. This suggests that the probability of being research-active and product-innovative is somewhat higher in the science based industries. Beyond that, only firm-specific characteristics appear to matter. Not surprisingly, appropriability is found to be the main explanatory variable. This confirms that firms who rely on strategic protection for their ideas are more likely to undertake product-innovative efforts. The quality of the labour force and competitive pressure variables, LHEI and LEXI, are also always highly statistically significant with a positive sign. This leads to the plausible suggestion that firms with more qualified personnel and under greater competitive pressure are more likely to be innovative.

Most interestingly, the size variable, LSIZE, turns out to be highly significant with a positive sign. This confirms stylised fact 1 of Cohen and Klepper (1996), which states that the likelihood of a firm reporting research efforts rises with firm size. When considered in conjunction with our earlier findings, this suggests that a threshold level of size is necessary for doing formal research. But, once this level is passed, further size increase does not lead to a proportional increase in research for the reasons given above.

The estimated correlation coefficient between the errors of the probit and tobit equations, RHO, is presented at the bottom of column (viii). It is found to be non-significant in the case of innovative turnover. This is confirmed by the corresponding parameter estimates presented above, which show evidence of a sample selection effect only in the case of the research intensity equation. Its effect is to raise somewhat the estimates of APPR, LHEI, LEXI and LSIZE, in the tobit equation, compared to the corresponding OLS estimates in column (i), which do not take account of sample selection. This shows that we are now capturing the combined effect of the impact of the variable in question on the probability of being research-active and product-innovative on the one hand, and of its impact on level of research intensity, once this is the case, on the other. Not surprisingly, the increase is sharpest in the case of APPR, reflecting its impact in the probit equation. However, we should exercise caution in our interpretation since the relative impact of the other knowledge-related variables, which could not be included in the probit equation due to data constraints, is likely to be underestimated as a result. On the whole it would appear that accounting for the sample selection effect does not affect any of the conclusions reached previously.

Tables 2.3a and 2.3b concentrate again on the research-active product-innovative sample and considers the effect of splitting the knowledge-related variables in the equation on research intensity.

To facilitate comparison, the baseline estimates of the equation by OLS, presented in Table 2.1, are reproduced in column (i). The analysis is then refined by subdividing the explanatory knowledge-related variables on incoming knowledge spillovers (INKS, equations ii-v), research collaboration (COOP, equations vi-viii) and the appropriability (APPR, equations ix-xii) into distinctive elements. Finally, all of these elements are brought in separately (equation xiii).

Table 2.3a Research intensity: splitting incoming knowledge spillovers, research collaboration and appropriability (models i-vii)

	OLS (i)	OLS (ii)	OLS (iii)	OLS (iv)	OLS (v)	OLS (vi)	OLS (vii)
Constant	–3.059	–3.105	–2.867	–2.892	–3.050	–3.060	–2.958
	(11.3)**	(11.2)**	(10.9)**	(10.8)**	(10.8)**	(11.4)**	(10.9)**
IDPAV1	–1.397	–1.393	–1.377	–1.387	–1.382	–1.371	–1.411
	(6.32)**	(6.31)**	(6.26)**	(6.27)**	(6.25)**	(6.18)**	(6.46)**
IDPAV2	–1.128	–1.123	–1.102	–1.121	–1.101	–1.094	–1.149
	(4.91)**	(4.89)**	(4.81)**	(4.85)**	(4.78)**	(4.76)**	(5.09)**
IDPAV3	–0.214	–0.203	–0.196	–0.175	–0.213	–0.175	–0.238
	(0.85)	(0.81)	(0.78)	(0.69)	(0.85)	(0.69)	(0.96)
IDPAV4	–0.380	–0.379	–0.368	–0.374	–0.371	–0.353	–0.418
	(1.75)	(1.75)	(1.71)	(1.73)	(1.71)	(1.62)	(1.96)*
GTURN	–0.241	–0.234	–0.203	–0.218	–0.223	–0.241	–0.241
	(1.25)	(1.20)	(1.03)	(1.11)	(1.13)	(1.26)	(1.26)
INKS	0.775					0.708	0.710
	(2.38)**					(2.19)**	(2.17)**
INKS_SCC		0.486			0.501		
		(2.08)**			(1.72)*		
INKS_UG			0.553		0.401		
			(2.03)**		(1.40)		
INKS_O				0.089	–0.226		
				(0.40)	(0.86)		
COOP	0.405	0.416	0.403	0.419	0.401		
	(3.01)**	(3.09)**	(2.99)**	(3.10)**	(2.98)**		
COOP_SCC						0.473	
						(3.50)**	
COOP_RUG							0.536
							(3.70)**
APPR	0.331	0.336	0.336	0.386	0.297	0.340	0.312
	(2.10)**	(2.13)**	(2.15)**	(2.48)**	(1.85)*	(2.18)**	(1.99)**
APPR_SE							
APPR_CD							
APPR_LT							
CKNOW	0.359	0.681	0.285	0.525	0.604	0.410	0.238
	(1.42)	(2.71)**	(1.02)	(1.97)**	(1.74)*	(1.63)	(0.92)

Table 2.3a (continued)

	OLS (i)	OLS (ii)	OLS (iii)	OLS (iv)	OLS (v)	OLS (vi)	OLS (vii)
LHEI	0.178	0.177	0.179	0.182	0.176	0.182	0.179
	(2.86)**	(2.84)**	(2.88)**	(2.92)**	(2.85)**	(2.93)**	(2.94)**
LEXI	0.069	0.070	0.066	0.070	0.067	0.068	0.067
	(2.47)**	(2.48)**	(2.36)**	(2.50)**	(2.36)**	(2.43)**	(2.41)**
LSIZE	−0.181	−0.179	−0.177	−0.172	−0.180	−0.184	−0.182
	(3.04)**	(3.01)**	(2.99)**	(2.94)**	(3.03)**	(3.10)**	(3.09)**
GROUP	−0.240	−0.247	−0.241	−0.237	−0.254	−0.235	−0.238
	(1.66)	(1.70)	(1.67)	(1.65)	(1.76)	(1.64)	(1.66)
Obs.	724	724	724	724	724	724	724
Adj. R^2	0.214	0.213	0.212	0.208	0.213	0.218	0.218

Notes: The dependent variable is the log of research intensity, LRDINS. Symbols * and ** denote significance at the 5% and 1% levels.

Table 2.3b Research intensity: splitting incoming knowledge spillovers, research collaboration and appropriability (models viii-xiii)

	OLS (i)	OLS (viii)	OLS (ix)	OLS (x)	OLS (xi)	OLS (xii)	OLS (xiii)
Constant	−3.059	−2.995	−3.056	−3.034	−3.039	−3.060	−3.014
	(11.3)**	(11.1)**	(11.3)*	(11.1)**	(11.2)**	(11.3)**	(10.7)**
IDPAV1	−1.397	−1.384	−1.389	−1.398	−1.402	−1.388	−1.360
	(6.32)**	(6.31)**	(6.29)**	(6.31)**	(6.31)**	(6.29)**	(6.20)**
IDPAV2	−1.128	−1.115	−1.134	−1.12	−1.122	−1.133	−1.092
	(4.91)**	(4.94)**	(4.94)**	(4.88)**	(4.86)**	(4.92)**	(4.80)**
IDPAV3	−0.214	−0.209	−0.196	−0.197	−0.185	−0.204	−0.196
	(0.85)	(0.84)	(0.78)	(0.77)	(0.73)	(0.81)	(0.79)
IDPAV4	−0.380	−0.383	−0.391	−0.359	0.366	−0.388	−0.377
	(1.75)	(1.80)	(1.81)	(1.66)	(1.69)	(1.80)	(1.75)
GTURN	−0.241	−0.244	−0.232	−0.224	−0.237	−0.229	−0.216
	(1.25)	(1.28)	(1.22)	(1.14)	(1.21)	(1.19)	(1.12)
INKS	0.775	0.685	0.799	0.800	0.830	0.785	
	(2.38)**	(2.11)**	(2.46)**	(2.45)**	(2.58)**	(2.40)**	
INKS_SCC							0.501
							(1.74)*
INKS_UG							0.302
							(1.01)

Table 2.3b (continued)

	OLS (i)	OLS (viii)	OLS (ix)	OLS (x)	OLS (xi)	OLS (xii)	OLS (xiii)
INKS_O							–0.246
							(0.95)
COOP	0.405		0.397	0.435	0.426	0.401	
	(3.01)**		(2.93)**	(3.26)**	(3.18)**	(2.99)**	
COOP_SCC		0.287					0.323
		(1.57)					(1.71)*
COOP_RUG		0.334					0.277
		(1.70)*					(1.32)
APPR	0.331	0.312					
	(2.10)**	(1.99)**					
APPR_SE			0.260			0.236	0.213
			(2.02)**			(1.51)	(1.37)
APPR_CD				0.191		0.072	0.094
				(1.24)		(0.398)	(0.51)
APPR_LT					0.127	–0.008	–0.035
					(1.02)	(0.05)	(0.21)
CKNOW	0.359	0.286	0.359	0.367	0.368	0.361	0.591
	(1.42)	(1.09)	(1.43)	(1.45)	(1.45)	(1.43)	(1.72)*
LHEI	0.178	0.179	0.178	0.182	0.182	0.177	0.176
	(2.86)**	(2.93)**	(2.86)**	(2.89)**	(2.88)**	(2.84)**	(2.90)**
LEXI	0.069	0.067	0.070	0.070	0.071	0.070	0.065
	(2.47)**	(2.41)**	(2.52)**	(2.47)**	(2.54)**	(2.46)**	(2.33)**
LSIZE	–0.181	–0.188	–0.184	–0.177	0.180	–0.183	–0.189
	(3.04)**	(3.17)**	(3.09)**	(2.96)**	(2.99)**	(3.03)**	(3.14)**
GROUP	0.240	–0.241	–0.237	–0.221	–0.222	–0.237	–0.253
	(1.66)	(1.68)	(1.65)	(1.54)	(1.54)	(1.64)	(1.76)
Obs.	724	724	724	724	724	724	724
Adj. R^2	0.214	0.220	0.215	0.212	0.211	0.213	0.217

Notes: The dependent variable is the log of research intensity, LRDINS. Symbols * and ** denote significance at the 5% and 1% levels.

The industry-specific dummies show similar negative and significant estimates as in the baseline equation in column (i). The only exception, at a 5% level of significance, relates to the science-based industries in equation (vii). This finding is discussed below. Also, the estimates of the parameters of the variables on the quality of the labour force, LHEI, competitive pressure, LEXI, firm size, LSIZE, and group affiliation, GROUP, are consistent with the baseline estimates. However, the parameter estimate of complexity of knowledge, CKNOW, is influenced by the subdivision of the knowledge-related variable pertaining to the incoming knowledge spillovers and deserves an explanation in this context. It has already been pointed out that multicollinearity may be a factor, due to the measurement of basic research using particular sources of information. From equation (iii), where INKS-UG is used, this does indeed appear to be the case since CKNOW is no longer significant. But if INKS-UG is not used, its effect is captured by the basic research variable.

Research collaboration, in equation (i), affects the research intensity of product-innovating firms positively, as was expected. Equation (vi) looks at the separate effect of inter-firm collaboration (COOP-SCC) and equation (vii) brings in inter-organisational collaboration with universities and government research labs (COOP-RUG). Note that, in this particular case, the science-based industry dummy is negatively significant, which means that the information-intensive industries are positively influenced by research collaboration with universities and government labs. Bringing both types in separately to the equation on research intensity (viii) renders inter-firm collaboration non-significant and diminishes the positive impact of inter-organisational collaboration. These results point to multicollinearity, and thus substitutability rather than complementarity between inter-firm and inter-organisational types of partners. The correlation coefficient between these two types of research collaboration is about 0.65, which is high. This issue requires further investigation.

In the baseline estimate, equation (i), strategic appropriation is found to have a highly significant impact on the research intensity of the firm. Splitting these into various channels yields the equations (ix), (x), and (xi). Only secrecy, APPR-SE, is seen to have a positive significant effect on research intensity. Complexity of design, APPR-CD, and lead time advantage, APPR-LT, have no influence. Bringing all three channels into the equation (xii) even renders the mechanism of secrecy non-significant. Again, this might be due to substitution effects between the appropriation strategies.

Can the effects identified in the preceding paragraphs be generalised to cover all types of firms? Or are there marked differences between different types of innovators? We will tackle these questions in the next section.

2.5 SCRUTINIZING FOR DIFFERENCES: STRONG AND WEAK INNOVATORS

As stated above, the key idea behind quantile regression is to screen for differences in behaviour between weak and strong innovators. Table 2.4 presents the estimates of the quantile regressions for research intensity and innovative turnover.

Table 2.4 Research intensity and innovative turnover: quantile regressions

	Research intensity			Innovative turnover		
	Quantiles			Quantiles		
	0.25	0.50	0.75	0.25	0.50	0.75
Constant	–4.227	–3.283	–2.23	–1.525	–0.784	–0.074
	(8.86)**	(10.7)**	(6.98)**	(5.74)**	(3.75)**	(0.35)
IDPAV1	–1.236	–1.460	–1.553	–0.520	–0.482	–0.400
	(3.84)**	(5.30)**	(5.49)**	(3.66)**	(3.67)**	(2.62)**
IDPAV2	–1.013	–1.320	–1.382	–0.514	–0.258	–0.116
	(3.32)**	(4.24)**	(5.13)**	(2.82)**	(1.73)	(0.65)
IDPAV3	–0.015	–0.110	–0.457	0.031	0.017	0.000
	(0.04)	(0.38)	(1.83)	(0.19)	(0.11)	(0.00)
IDPAV4	–0.143	–0.492	–0.874	–0.434	–0.195	–0.345
	(0.43)	(2.18)*	(2.85)**	(1.80)	(1.38)	(1.92)
GTURN	–0.316	–0.057	–0.054			
	(0.99)	(0.27)	(0.37)			
LRDINS				0.093	0.114	0.147
				(3.21)**	(4.29)**	(4.79)**
INKS	1.004	0.812	0.789	0.253	0.020	0.313
	(1.77)*	(2.00)**	(2.12)**	(0.85)	(0.08)	(1.01)
COOP	0.467	0.394	0.439	0.124	0.005	–0.066
	(2.33)**	(2.28)**	(2.61)**	(1.09)	(0.05)	(0.59)
APPR	0.590	0.352	0.192	0.0470	0.056	–0.050
	(2.40)**	(1.98)*	(1.05)	(0.31)	(0.48)	(0.33)
CKNOW	0.728	0.608	0.310	–0.144	–0.246	–0.292
	(1.69)*	(2.19)**	(0.92)	(0.56)	(1.10)	(1.28)
PCS				0.256	0.127	0.187
				(1.92)*	(1.13)	(1.63)
LHEI	0.231	0.126	0.155	–0.003	0.029	0.027
	(2.31)**	(1.87)*	(1.86)*	(0.10)	(0.62)	(0.77)

Table 2.4 (continued)

	Research intensity			Innovative turnover		
	Quantiles			Quantiles		
	0.25	0.50	0.75	0.25	0.50	0.75
LEXI	0.090	0.043	0.023	–0.023	–0.011	–0.044
	(1.93)*	(1.34)	(0.78)	(1.04)	(0.67)	(1.60)
LSIZE	–0.199	–0.166	–0.149	–0.139	–0.086	–0.124
	(1.96)*	(2.57)**	(2.56)**	(3.37)**	(2.74)**	(3.23)**
GROUP	–0.282	–0.063	0.072	0.002	0.065	0.107
	(1.11)	(0.32)	(0.38)	(0.01)	(0.64)	(0.99)
Obs.	724	724	724	724	724	724
Psd. R^2	0.104	0.132	0.167	0.070	0.086	0.091

Notes: The dependent variables are the log of research intensity, LRDINS, and the log of innovative turnover, LITURN. The t-values are based on standard errors obtained by the bootstrap design matrix method. Bootstrap re-sampling was performed using 1,000 repetitions. The t-tests are one-tailed or two-tailed, according to a priori expectations with respect to the parameter values. The pseudo R^2 is obtained as one minus the proportion of the sum of absolute weighted deviations over the raw sum of deviations. Symbols * and ** denote significance at the 5% and 1% levels.

Consider research intensity (the left-hand side of Table 2.4). It seems that industry dummies play an increasing role for firms with higher research intensities. Growth in turnover, as in the baseline estimates in Table 2.1, has no impact on research intensity. The same applies to the group affiliation of the innovative firm.

Of greatest interest, however, is the behaviour of the knowledge-related variables. What is striking is that incoming knowledge spillovers for strong innovators gain in importance when statistical significance is taken into account, despite a slight decrease in the parameter estimate. Also, the significance of research collaboration is highest in the highest quantile, though the parameter estimate remains more or less unaltered.

The results for strategic appropriability are in sharp contrast to the results for incoming knowledge spillovers and research collaboration. Strategic appropriability differs markedly across the quantiles, with a pronounced impact on research intensity in the lower quantiles focusing on weaker innovators, and successively becoming non-significant. This is also the case, to a lesser extent, for the measure of complex knowledge. These findings are fully in line with the process of opening up innovative and research-active firms. Here, knowledge-sharing becomes part and parcel of innovation management and leads to higher research intensities. These dynamic firms embrace the new adage "use it or lose it".

The presence of a qualified labour force in the firm has a positive effect on research intensity. Somewhat surprisingly, however, we see its impact decrease in the higher quantiles. This may be due to the fact that the measure of labour qualification is still very crude, in the sense that it cannot properly reflect the actual quality of the education in question, not to mention on-the-job training, which may be particularly important in the case of the most innovative firms. More plausibly, the impact of level of competitive pressure sharply diminishes to become non-significant in the higher quantiles. This suggests that the stronger innovators are clearly less influenced in their research investment decisions by direct competitive pressure. Size, exerting a negative influence, again contradicts the traditional discourses that large firms are the most research-intensive. Beyond a threshold level of size, which is necessary for doing research, small firms often prove more dynamic in this respect and this is most pronounced in the higher quantiles of stronger innovators.

Consider, next, innovative turnover (the right-hand side of Table 2.4). The influence exerted by the less technologically-involved industries active in resource-intensive and scale-intensive sectors is negative compared to the information-intensive ones captured by the constant term. In the case of weaker innovators in particular, firms with a less technology-intensive trajectory perform fairly poorly. As we see in Table A2.2 in the Appendix, many firms belong to these sectors (46% of the research-active product innovators), the implication being that innovative activities in these so-called traditional sectors do not pay off in terms of innovative turnover. More successful technological paths do clearly exist. Yet, most advanced economies are locked-in to their specific trajectory and it may not be easy to convert the economy rapidly to a knowledge-based hi-tech environment.

The impact of a firm's research intensity on its innovative turnover is highly positive and this relation gains strength if firms are strong innovators in the higher quantile. None of the knowledge-related variables seem to play a direct role in innovative performance in terms of turnover. Rather, as was the case with the baseline estimates in Table 2.1, the influence works indirectly through their effect on research intensity.

Evidence that process innovation, PCS, has a significant impact on innovative turnover in the baseline estimates in Table 2.2, appears confined to the lower quantile. This is not surprising, since the process innovations that contribute to important product innovations are quite sophisticated and require in-house research efforts on the part of the firm. Therefore their impact is likely to be captured by the LRDINS variable itself. Finally, as in the case of research intensity, the negative impact of size is confirmed quantile by quantile.

The finding on strategic appropriability in Table 2.4 calls for further investigation. This is done in Tables 2.5a and 2.5b, which shows the quantile regression results on research intensity that focus on the separate means by which strategic appropriability occurs.

Table 2.5a Research intensity equation with appropriability split: quantile regressions

	Appropriability through secrecy			Appropriability through complexity of design		
	Quantiles			Quantiles		
	0.25	0.50	0.75	0.25	0.50	0.75
Constant	–4.088	–3.197	–2.261	–4.196	–3.190	–2.311
	(9.40)**	(9.68)**	(7.59)**	(9.78)**	(10.1)**	(7.63)
IDPAV1	–0.985	–1.507	–1.604	–1.189	–1.49	–1.611
	(3.05)**	(5.65)**	(5.99)**	(3.82)**	(5.43)**	(5.89)**
IDPAV2	–0.926	–1.385	–1.369	–0.839	–1.307	–1.352
	(2.93)**	(4.57)**	(5.14)**	(2.82)**	(4.23)**	(5.16)**
IDPAV3	–0.095	–0.206	–0.485	0.011	–0.100	–0.527
	(0.26)	(0.74)	(2.01)*	(0.03)	(0.34)	(2.09)*
IDPAV4	–0.186	–0.594	–0.780	–0.078	–0.524	–0.804
	(0.60)	(2.41)**	(2.57)**	(0.25)	(2.10)*	(2.71)**
GLTURN	–0.397	–0.045	–0.022	–0.397	–0.128	–0.004
	(1.32)	(0.21)	(0.15)	(1.25)	(0.62)	(0.03)
INKS	0.789	0.643	0.753	0.875	0.824	0.777
	(1.49)	(1.57)	(2.12)**	(1.63)	(2.04)**	(2.09)**
COOP	0.285	0.370	0.432	0.413	0.397	0.500
	(1.37)	(2.35)**	(2.58)**	(2.16)**	(2.46)**	(3.01)**
APPR_SE	0.533	0.269	0.207			
	(2.68)**	(1.84)*	(1.36)			
APPR_CD				0.457	0.286	0.144
				(1.99)**	(1.70)*	(0.82)
APPR_LT						
CKNOW	0.583	0.585	0.313	0.733	0.526	0.252
	(1.36)	(2.08)*	(0.98)	(1.66)*	(1.89)*	(0.77)
LHEI	0.214	0.100	0.145	0.235	0.130	0.153
	(2.08)**	(1.46)	(1.72)*	(2.22)**	(1.92)*	(1.89)*
LEXI	0.086	0.042	0.029	0.082	0.038	0.016
	(1.84)*	(1.32)	(1.01)	(1.82)*	(1.15)	(0.55)

Table 2.5a (continued)

	Appropriability through secrecy			Appropriability through complexity of design		
	Quantiles			Quantiles		
	0.25	0.50	0.75	0.25	0.50	0.75
LSIZE	–0.207	–0.165	–0.125	–0.184	–0.165	–0.122
	(2.15)*	(2.56)**	(2.10)*	(1.88)*	(2.47)**	(2.02)*
GROUP	–0.298	–0.058	–0.149	–0.273	–0.035	–0.062
	(1.18)	(0.29)	(0.84)	(1.13)	(0.18)	(0.35)
Obs.	724	724	724	724	724	724
Psd. R^2	0.106	0.131	0.168	0.102	0.130	0.167

Notes: The dependent variable is the log of research intensity, LRDINS. The t-values are based on standard errors obtained by the bootstrap design matrix method. Bootstrap re-sampling was performed using 1,000 repetitions. The t-tests are one-tailed or two-tailed, according to a priori expectations with respect to the parameter values. The pseudo R^2 is obtained as one minus the proportion of the sum of absolute weighted deviations over the raw sum of deviations. Symbols * and ** denote significance at the 5% and 1% levels.

Table 2.5b Research intensity equation with appropriability split: quantile regressions (continued)

	Appropriability through lead-time advantage			Appropriability through secrecy, complexity of design and lead-time advantage		
	Quantiles			Quantiles		
	0.25	0.50	0.75	0.25	0.50	0.75
Constant	–4.097	–3.072	–2.205	–4.199	–3.218	–2.266
	(8.78)**	(9.16)**	(7.05)**	(9.23)**	(9.75)**	(7.15)**
IDPAV1	–1.149	–1.524	–1.609	–1.010	–1.499	–1.586
	(3.43)**	(5.57)**	(6.24)**	(3.09)**	(5.35)**	(5.75)**
IDPAV2	–0.769	–1.371	–1.343	–0.918	–1.421	–1.410
	(2.56)**	(4.45)**	(5.07)**	(2.98)**	(4.56)**	(5.25)**
IDPAV3	0.071	–0.043	–0.475	–0.083	–0.231	–0.504
	(0.20)	(0.15)	(1.97)*	(0.24)	(0.82)	(2.00)*
IDPAV4	0.064	–0.594	–0.854	–0.076	–0.581	–0.823
	(0.20)	(2.40)**	(2.75)**	(0.24)	(2.37)**	(2.71)**

Table 2.5b (continued)

	Appropriability through lead-time advantage			Appropriability through secrecy, complexity of design and lead-time advantage		
		Quantiles			Quantiles	
	0.25	0.50	0.75	0.25	0.50	0.75
GLTURN	–0.567	–0.084	–0.022	–0.446	–0.090	–0.028
	(1.63)	(0.38)	(0.15)	(1.54)	(0.42)	(0.15)
INKS	0.851	0.707	0.808	0.796	0.646	0.754
	(1.47)	(1.71)*	(2.25)**	(1.52)	(1.53)	(1.96)**
COOP	0.454	0.422	0.437	0.301	0.388	0.427
	(2.19)**	(2.54)**	(2.65)**	(1.46)	(2.40)**	(2.68)**
APPR_SE				0.442	0.234	0.160
				(1.96)**	(1.45)	(0.83)
APPR_CD				0.272	0.147	0.028
				(0.99)	(0.72)	(0.14)
APPR_LT	0.056	0.129	0.159	–0.133	–0.036	0.045
	(0.28)	(0.84)	(1.06)	(0.58)	(0.23)	(0.24)
CKNOW	0.393	0.588	0.225	0.639	0.653	0.339
	(0.91)	(2.03)**	(0.73)	(1.50)	(2.22)**	(1.01)
LHEI	0.242	0.124	0.147	0.222	0.109	0.142
	(2.18)**	(1.89)*	(1.77)*	(2.21)**	(1.54)	(1.67)*
LEXI	0.070	0.045	0.031	0.082	0.034	0.029
	(1.51)	(1.42)	(1.03)	(1.82)*	(1.09)	(0.94)
LSIZE	–0.162	–0.175	–0.136	–0.179	–0.160	–0.133
	(1.59)	(2.55)**	(2.38)**	(1.97)*	(2.50)**	(2.32)**
GROUP	–0.378	–0.111	–0.115	–0.334	–0.067	–0.124
	(1.50)	(0.54)	(0.60)	(1.42)	(0.33)	(0.68)
Obs.	724	724	724	724	724	724
Psd. R^2	0.097	0.128	0.167	0.108	0.132	0.168

Notes: The dependent variable is the log of research intensity, LRDINS. The t-values are based on standard errors obtained by the bootstrap design matrix method. Bootstrap re-sampling was performed using 1,000 repetitions. The t-tests are one-tailed or two-tailed, according to a priori expectations with respect to the parameter values. The pseudo R^2 is obtained as one minus the proportion of the sum of absolute weighted deviations over the raw sum of deviations. Symbols * and ** denote significance at the 5% and 1% levels.

The results are particularly interesting because they provide insights into the ways in which weak and strong innovators try to benefit from the product innovations they have made. Firstly, we look at the protection of innovations through secrecy. Almost half of the firms in the sample of research-active product innovators (44.9%) reported that they rely on this and, as seen in Tables 2.3a and 2.3b above, this mode of protection has, by far, the greatest impact on research intensity. The quantile regression estimates confirm that weak innovators seem to adhere to secrecy much more to enhance research intensity than the stronger innovators present in the higher quantiles. Other knowledge-related variables are affected by a consideration of this channel of strategic appropriability. Both incoming knowledge spillovers and research collaboration become non-significant for weak innovators, but, as in Table 2.4 where a measure of average appropriability, APPR, was used, they are again highly significant for strong innovators. This confirms that strong innovators are far less concerned with the issue of secrecy to stimulate research intensity. This finding underlines the 'connect and develop' strategy of open innovation (Chesbrough, 2003a; Huston and Sakkab, 2006). At the same time, the fact that appropriability through secrecy shows a stronger statistical performance in the lower quantile than measure of average appropriability suggests that the latter is somewhat less appropriate for weaker innovators. The actual quantile-by-quantile increase of the impact of incoming knowledge spillovers and research collaboration may, therefore, be stronger than suggested by Table 2.4.

Complexity of design used much less as a means of knowledge appropriation by research-active product innovators (only 22.4% of the firms in the sample used it). As in the case with secrecy, the estimates suggest that its impact on research intensity is significant only for weak innovators. Lead-time advantage is a channel of appropriability in 41.1% of the cases. However, this channel has no significant impact on research intensity. Finally, the separate channels of strategic appropriation are brought into the equation simultaneously. The sample contains 119 firms that use all three informal appropriability channels simultaneously. Due to multicollinearity, appropriability through secrecy is all that remains significant here, and only in the case of weak innovators.

2.6 DISCUSSION AND CONCLUSIONS

An econometric model to explain research intensity and turnover from product innovation was estimated on CIS3 firm-level data from Belgium. Key explanatory variables on knowledge were used: incoming knowledge spillovers, research collaboration and strategic appropriability. The estimates showed first and foremost that the impact of these variables on innovative turnover is indirect, and works through research intensity. This emphasises the need for knowledge absorption first, prior to use in the innovation process. This conclusion remained unaltered when we accounted for a sample selection effect.

Each of the knowledge-related variables was then broken down into elements. In the case of incoming knowledge spillovers the effect of inter-firm spillovers on research intensity proved to be the most robust. A consideration of the types of partners in research collaboration revealed that inter-organisational collaboration has the greatest impact on research intensity. It appears that strategic appropriability influences research intensity mainly through the use of secrecy. However, due to multicollinearity we were unable to clearly identify its separate impact when combined with the other means of strategic appropriation.

Quantile regression takes the analysis one step further by breaking down various types of innovating firms into weak and strong performers in terms of research intensity and innovative turnover. Again, the effects of all knowledge-related variables on innovative turnover remain indirect for all quantiles. In terms of statistical significance, incoming knowledge spillovers are more important in explaining research intensity in the case of strong innovators. Also, research collaboration is most significant in the highest quantile, even though the parameter estimate remains as good as unaltered. In sharp contrast to this, strategic appropriation seems to be highly relevant for weak innovators but loses significance for strong innovators.

A breakdown of the means of appropriation showed that secrecy and, to a lesser extent, complexity of design were responsible for this result. A measure of appropriability based exclusively on secrecy even outperforms the average measure of appropriability in the lower quantile, and its inclusion tends to sharpen the heightened significance of incoming knowledge spillovers and research collaboration that we observed. These results would suggest that, in line with the predictions of open innovation theory, it is among the strong innovators that we can find the majority of firms who are prepared to market excess knowledge in disembodied form and devote a major share of their research effort to external knowledge absorption.

On the whole, effects on research intensity seem to give rise to a typology of firms. This has important consequences for innovation policy, which focuses too much on the overall population of firms, although recent policy measures to enhance new technology-based firms (academic and corporate spin-offs, independent hi-tech start-ups) have been created. These relate to the provision of venture capital in particular, whereas the results of the present analysis show that knowledge-related variables are also vital in stimulating a firm's research intensity.

APPENDIX

Table A2.1 presents a description of the manner in which the variables were constructed from the raw data of the CIS3 questionnaire. Table A2.2 presents the mean value of the variables with respect to the total sample, the sample of research-active product innovators, and the corresponding Pavitt groups.

Table A2.1 Description of the variables

Variable	Definition
Log of research intensity (LRDINS)	Natural log of the ratio of intramural R&D expenditure over turnover in the year 2000
Log of innovative turnover (LITURN)	Natural log of the % of new or significantly improved products (goods and services) in turnover in the year 2000
Growth of turnover (GTURN)	Growth of market turnover (goods and services) from 1998 to 2000
Incoming knowledge spillovers (INKS)	Sum of scores of the importance attached to information from suppliers, clients, competitors, universities, government, professional conferences, exhibitions (rescaled between 0 and 1)
Research collaboration (COOP)	Collaboration = 1 if the firm collaborates either with suppliers, clients, competitors, research labs, universities or government
Appropriability (APPR)	Average score of the effectiveness of strategic protection through secrecy, complexity of design and lead-time advantage
Complexity of knowledge (CKNOW)	Ratio of the sum of the scores of the importance attached to information from universities, government and professional conferences over the sum of the scores of all sources of information
Log of qualified labour force (LHEI)	Natural log of the % employees with higher education in the year 2000
Log of competitive pressure (LEXI)	Natural log of the proportion of exports in turnover in the year 1998
Log of size (LSIZE)	Natural log of the number of employees in the year 1998
Group membership (GROUP)	Group membership = 1 if the firm is member of a group

Table A2.1 (continued)

Variable	Definition
Process innovation (PCS)	Process innovation = 1 if the firm has introduced new or significantly improved production processes during the period 1998–2000
IDPAV1	IDPAV1 = 1 if the firm is classified in the 'Resource-intensive' sectors (mining, food, beverages and tobacco, wood and furniture, recycling, wholesale)
IDPAV2	IDPAV2 = 1 if the firm is classified in the 'Scale-intensive' sectors (paper and printing, rubber and plastics, non-metallic minerals, ferrous metals, non-ferrous metals, fabricated metal products, motor vehicles, other transport equipment)
IDPAV3	IDPAV3 = 1 if the firm is classified in the 'Specialised supplier' sectors (non-electrical machinery, electronics and communication, electrical machinery)
IDPAV4	IDPAV4 = 1 if the firm is classified in the 'Science-based' sectors (chemicals, pharmaceuticals, computers and office machinery, scientific instruments)
IDPAV5	IDPAV5 = 1 if the firm is classified in the 'Information-intensive' sectors (financial intermediation, telecommunication, transport services, business services, computer and related service activities, research and development, technical engineering)

Table A2.2 Means of the variables

Variable	All firms	R&D-active and product innovative firms	*Of which:* IDPAV1	IDPAV2	IDPAV3	IDPAV4	IDPAV5
Obs.	1,987	724	163	169	101	97	194
LRDINS		–4.437	–5.298	–5.080	–3.866	–3.998	–3.671
LITURN		–1.836	–2.178	–2.045	–1.559	–2.015	–1.423
GTURN	0.168	0.244	0.118	0.181	0.197	0.170	0.467
INKS		0.433	0.416	0.428	0.478	0.455	0.417
INKS_SCC		0.534	0.521	0.529	0.596	0.523	0.521
INKS_UG		0.220	0.192	0.196	0.280	0.273	0.206
INKS_O		0.494	0.481	0.509	0.496	0.532	0.472
COOP		0.387	0.319	0.367	0.475	0.423	0.397
COOP_SCC		0.334	0.276	0.301	0.386	0.350	0.376
COOP_RUG		0.266	0.202	0.260	0.356	0.360	0.231
APPR	0.193	0.361	0.290	0.343	0.495	0.422	0.336
APPR_SE	0.244	0.449	0.337	0.455	0.554	0.577	0.417
APPR_CD	0.117	0.224	0.171	0.207	0.386	0.226	0.196
APPR_LT	0.217	0.411	0.362	0.366	0.544	0.463	0.397
CKNOW		0.325	0.304	0.315	0.307	0.400	0.321
LHEI	–2.313	–1.759	–2.183	–2.388	–1.973	–1.472	–0.884
LEXI	–3.004	–2.251	–2.316	–1.651	–1.451	–1.388	–3.568
LSIZE	3.795	4.211	4.301	4.644	4.334	4.412	3.591
GROUP	0.561	0.656	0.589	0.734	0.604	0.701	0.649
PCS		0.737	0.736	0.828	0.703	0.701	0.696

3. A comparative analysis of open innovation and the relation between research intensity and product innovativeness

3.1 INTRODUCTION

In recent years firms have brought about significant change in their innovation management. They used to invest heavily in specialised in-house R&D departments to bring about innovation. This model has recently ceded ground to a more 'open' innovation model, in which the research strategies of innovative firms increasingly rely on external knowledge. This encourages active external information sourcing and, in particular, research cooperation, and implies a profound reorganisation of the research process with a view to internalising external ideas (Chesbrough and Rosenbloom, 2002; Chesbrough, 2003a; Chesbrough and Crowther, 2006; Huston and Sakkab, 2006). Open innovation also implies a changing attitude to the appropriability of new ideas, since this may occur not only through product innovation, but through their sale in disembodied form, through licensing, trademarks, and so on. It might, therefore, be of great interest to us to reconsider the relation between research intensity and product innovativeness.

Since 2000 a new, econometric literature has improved our understanding of the relationship between innovation and firm performance by estimating structural models to explain both the input and output stages of the innovation process, as well as its impact on productivity or turnover and employment growth (see, especially, the influential studies by Crépon et al., 1998; Lööf and Heshmati, 2002, 2006; Klomp and Van Leeuwen, 2001; Van Leeuwen and Klomp, 2006; and Griffith et al., 2006). It also pays due attention to the problems of endogeneity and sample selection, which have hampered earlier work.

One of the drawbacks with these studies is that they tend, by concentrating on manufacturing, to neglect some of the factors that are most relevant to the emergent, ICT-based 'new economy', such as knowledge intensive services. Worse still, the estimation methods used allow only general statements about the innovative performance of an 'average firm'. The fiction of a 'representative firm' may, however, be highly misleading in view of the heterogeneity in innovative behaviour. This is highly relevant since open innovators are likely to be found among the most dynamic firms in particular.

A few recent attempts have been made to identify systematic behavioural differences between strong and weak innovators, by means of quantile regression analysis. Thus, Spithoven et al. (2010) have found evidence of a quantile-by-quantile increase in the relative impact of incoming knowledge spillovers and research cooperation on research intensity in relation to that appropriability, and in the impact of research intensity on innovative turnover. This would suggest that strong innovators have a more open and, thus, more efficient innovation process than weaker ones. But, with the focus on a single country (Belgium), it is unclear whether these results can be generalised. Can a similar pattern be found for other countries, and, if so, to what extent? Also, are the results robust when it comes to using different measures of research intensity? This could turn out to be very relevant if significant differences are observed between the levels of technological development in the countries under consideration.

This study aims to answer these and related questions, by means of a comparative analysis of five European countries – Belgium, Germany, Norway, Spain and Portugal – on the basis of harmonised firm-level data from the Third Community Innovation Survey (CIS3). To do so, we concentrate on a model explaining research intensity and innovative turnover. The countries were chosen for reasons of data availability and because of their distinct general features. Belgium is an industrialised, small open economy in the heart of Europe. Germany is Europe's main economy. Norway is an advanced Nordic economy. Spain is a major Latin European economy that has succeeded in catching up since joining the EU. Finally, Portugal is a smaller, somewhat less industrialised Latin European economy, still going through a spectacular catch-up process. Attention will always focus on the average behaviour of innovative firms as well as on systematic behavioural differences between strong and weak innovators.

The remainder of the chapter is organised as follows. Section 3.2 presents the model and Section 3.3 discusses the data and the estimation strategy. Section 3.4 presents the estimates on the average behaviour of firms and Section 3.5 refines this analysis by using quantile regression to identify behavioural differences between strong and weak innovators.

Section 3.6 summarises and concludes. Tables with a description of the variables and mean values are presented in the Appendix.

3.2 THE MODEL

Our model is inspired by earlier empirical models of the relation between innovation and firm performance. It concentrates on the core: the innovation input and the innovation output equations.

The innovation input equation explains the research intensity of firms. The level of research intensity is measured alternately by the level of in-house research intensity and the level of total research intensity. The level of in-house research intensity, LRDINS, is calculated as the log of the share of in-house R&D expenditure in turnover, and the level of total research intensity, LTRDINS, is calculated as the log of the share of total innovation expenditure in turnover. The explanatory variables are a set of knowledge-related variables, measures of human capital, competitive pressure and firm size, and industry dummies depicting the technological environment.

The knowledge-related variables measure the level of incoming knowledge spillovers, INKS, research cooperation, COOP, appropriability of in-house knowledge, APPR, and complexity of knowledge, CKNOW. Incoming spillovers are expected to have a positive effect on research intensity because they increase the firm's knowledge base, from which it can draw in its research process, and because the absorption of external knowledge requires additional in-house research efforts (Cohen and Levinthal, 1989, 1990). Cooperation on research is thought to stimulate R&D efforts to the extent that it raises the efficiency of the research process, and because the internalisation and commercialisation of the knowledge created through cooperation requires further research efforts by the firms concerned (Colombo and Gerrone, 1996; Hagedoorn et al., 2000; Monjon and Waelbroeck, 2003; Becker and Dietz, 2004; Veugelers and Cassiman, 2005). A higher capacity in appropriating in-house knowledge will tend to reduce spillovers to competitors and, thereby, increase incentives to undertake research efforts (Spence, 1984; Teece, 1986). In turn, the basic character of the research can be expected to have a positive effect on the level of research effort, to the extent that the greater complexity of basic research requires more effort than applied research.

It should be noted that where research cooperation is involved, the appropriability argument has to be attenuated a little, in the sense that it now applies to spillovers to non-partners only (Cassiman and Veugelers, 2002). In fact, in the context of the increasing openness of the innovation process we see the development of a changing attitude to the creation of value by new ideas, as this may occur not only through product innovation, but through sale in disembodied form, through licensing, trademarks, and so on (Chesbrough and Rosenbloom, 2002). In as far as open innovators are likely to be found among strong innovative performers in particular, the a priori expectations are that the joint relative impact of incoming knowledge spillovers and research cooperation on research intensity will tend to increase quantile-by-quantile in relation to that of appropriability.

The next explanatory variable in the research intensity equation is a measure of human capital. It is the log of the level of higher education intensity of the personnel, LHEI, and it reflects the average level of qualification of the workforce. Firms will tend to spend more on research if they have highly qualified workers at their disposal who are able to operate in a more complex production environment and may more easily contribute to the innovation process itself, by contributing ideas of their own (Cohen and Levinthal, 1989, 1990; Redding, 1996; Frantzen, 2000).

As a measure of competitive pressure, the log of export intensity, LEXI, is included, since firms operating on foreign markets are likely to be subject to stronger competition. The need to remain competitive is assumed to affect research efforts positively.

In order to investigate whether firm size impacts on research intensity, two size-related variables are included. These are the log of size (measured by employment in the base year), LSIZE, and a variable reflecting group membership, GROUP. Since the theoretical literature contains arguments implying either positive or negative influences on R&D intensity (Cohen, 1995) we have no a priori expectations as to their sign.

Finally, we added industry dummy variables based on Pavitt's taxonomy, IDPAV, depicting the general technological environment. These reflect inter-industry differences in technological trajectories and are further described below.

The innovation output equation concentrates on product innovation. It can be conceived as a production function of new knowledge and it has as a dependent variable the log of the level of innovative turnover (the proportion of innovative products in turnover), LITURN. Its explanatory variables are the level of research intensity, a measure of process innovation, and the same knowledge, human capital, competitive pressure and size-related variables, as well as industry dummies, included in the innovation input equation.

The influence of research intensity on innovation output is not only direct, through the creation of new product-innovative ideas, but also indirect, through the contribution to generating the complementary process innovations required for the implementation of product innovation (Parisi et al., 2006). Since some of the process innovations involve production reorganisations that do not require formal research efforts, a measure of process innovation, PCS, is included as a separate variable.

The inclusion of variables already present in the innovation input equation allows us to discover whether there is also a direct influence from these variables on innovation output, rather than merely an indirect one, through their impact on research intensity. The size-related variables will, in this case, indicate whether there is a size-related research efficiency effect.

Both equations include, in addition, an intercept, C. The model can thus be summarised as follows.

L(T)RDINS = f1(C, IDPAV, INKS, COOP, APPR, CKNOW, LHEI, LEXI, LSIZE, GROUP) (3.1)

LITURN = f2(C, IDPAV, L(T)RDINS, INKS, COOP, APPR, CKNOW, PCS, LHEI, LEXI, LSIZE, GROUP) (3.2)

3.3 DATA AND METHODS

3.3.1 Data

This study draws on data for Belgium, Germany, Norway, Spain and Portugal, gathered as part of the Third European Community Innovation Survey, CIS3, released by Eurostat. The survey was conducted on a country-by-country basis in 2002 and relates to the period 1998–2000. The target population was every enterprise with ten or more employees and the survey was stratified according to size and economic activity. After eliminating enterprises with incomplete data, this left us with 1,223 firms for Belgium, 2,813 for Germany, 3,403 for Norway, 7,826 for Spain and 1,743 for Portugal. Of these, respectively, 492, 1,117, 962, 1,773 and 368 reported in-house R&D, and 390, 931, 821, 1,367 and 282 both in-house R&D and innovative turnover. In the case of total innovation expenditure the respective figures are 697, 1,749, 1,390, 3,079 and 746. And of these respectively 531, 1,321, 1,124, 1,983 and 479 reported both total innovation expenditure and innovative turnover.

For reasons of confidentiality, these data were released by Eurostat in micro-aggregated form. In the case of Belgium we also had the original data at our disposal and this related to a larger set of firms (Spithoven, 2005). As a check of robustness, we re-performed estimation for Belgium on the latter data. Although not presented here, the results showed very similar parameter estimates. However, the statistical fit was slightly worse than when use is made of micro-aggregated data. This confirms that micro-aggregation does not affect the structure of the data and suggests that, if anything, it helps to reduce somewhat the noise due to errors of measurement.

Table A3.1 in the Appendix presents the precise construction and definition of all variables used and Table A3.2 their mean values. We confine ourselves here to a few necessary comments.

The measure of innovative input is the level of research intensity. This is captured, alternately, by the level of in-house research intensity, LRDINS, and the level of total research intensity, LTRDINS. The level of in-house research intensity is calculated as the (natural) log of the share of in-house R&D expenditure in turnover, whereas the level of total research intensity is obtained as the (natural) log of the share of total innovation expenditure in turnover. This covers, in addition to in-house R&D, expenditure on the acquisition of advanced machinery, the purchase of the right to use patents, licenses and trademarks, training, and marketing of new products. The advantage of focusing on in-house research intensity is that the composition of in-house R&D expenditure is more homogenous across the countries and, therefore, more comparable. In-house research is assumed to be particularly relevant to the capability of firms to absorb incoming knowledge and engage in research cooperation. The advantage with total research intensity is that total innovation expenditure is more complete. Firms from countries with a technological lag are likely to have a lower proportion of in-house R&D in total innovation expenditure and a higher proportion of outlay on the acquisition of (imported) advanced machinery, than firms from countries on or closer to the technological frontier. In their case, total research intensity, LTRDINS, may be a more relevant measure of innovation input in the innovation output equation than is in-house research intensity, LRDINS. It is therefore useful to consider either measure of research intensity alternately.

The measure of incoming knowledge spillovers, INKS, is based on questions on the importance attached to alternative sources of information on a three-point scale. These sources are: suppliers, clients, competitors, universities, government, professional conferences, and fairs and exhibitions. To generate a firm-specific measure of incoming spillovers, we summed the scores on each of these questions and rescaled the total score to a number between 0 and 1.

The measure of research cooperation, COOP, is a binary variable that takes the value 1 in the case of cooperation and 0 otherwise. Cooperation may take place either with suppliers, clients, competitors, research labs, universities or government.

The measure of appropriability, APPR, concentrates on methods of strategic protection, rather than on methods of legal protection through patents, because the latter are highly industry-specific (Griliches, 1990). The firms were asked whether they made use of strategic protection through secrecy, complexity of design or lead-time advantage over competitors. By giving a score of 1 if they did and 0 otherwise, an average score between 0 and 1 could be obtained to generate a measure of strategic protection.

The measure of complex knowledge, CKNOW, is again based on the answers to questions on sources of information. It is obtained by taking the ratio of the sum of the scores with regard to information from universities, government and professional conferences over the sum of the scores of all sources of information, as suggested by Cassiman and Veugelers (2002). In other words, it is inferred from the structure of the incoming knowledge flows, on the assumption that this conditions the character of the firm's in-house research.

Innovation output is measured by the level of innovative turnover, LITURN. It is calculated as the log of the proportion of new or significantly improved products (goods or services) in turnover. It was preferred over a patent-based indicator of innovative output because of the well-known shortcomings of the latter in this respect, since not all patents become innovations, the degree of patenting varies sharply between industries, and few service firms, such as software development, tend to patent (Griliches, 1990).

The research intensity and innovative turnover equations always include industry dummies capturing the general technological environment. These are based on the taxonomy originally proposed by Pavitt (1984) and revised and elaborated by Tidd et al. (1997) to incorporate the service economy. This classification reflects differences in technological trajectories between groups of industries. It is based on the general characteristics of the technologies used in the production system, the skills needed to operate them, the sources and the monitoring strategy. Five groups of sectors are identified: resource-intensive (IDPAV1), scale-intensive (IDPAV2), specialised suppliers (IDPAV3), science-based (IDPAV4) and information-intensive (IDPAV5). The two-digit NACE industries included in these groupings are given in Table A3.1 in the Appendix. As Table A3.2 shows, the relative number of research-active firms in the respective IDPAV dummies in the data clearly differs from country to country. This reflects the differences in production specialisation.

3.3.2 Estimation Methods

The CIS3 data refer to the period 1998–2000 and are cross-sectional. Therefore the parameter estimates can provide only a snapshot of the long-term relations between the variables. The data do not allow for the introduction of lagged responses.

Country-based estimation will first be performed to get an idea of the average behaviour of research-active firms in the case of the research intensity equation, and the research-active and product-innovative firms in the case of the innovative turnover equation. An equation-based estimation by ordinary least squares (OLS) will produce consistent parameter estimates provided the explanatory variables are exogenous. Although this may seem plausible for most of the explanatory variables under consideration, doubts may arise over some. Therefore, it can be argued that since firms undertake research efforts to be able to absorb external knowledge, higher research intensity may, in itself, make firms more likely to source outside information and cooperate in research with others. This would imply that there is also a reverse causation from research intensity, L(T)RDINS, to incoming knowledge spillovers, INKS, and research cooperation, COOP, and that these variables are, therefore, endogenous to the research intensity equation. Furthermore, there may be feedback from level of innovative turnover, LITURN, over growth of turnover to level of research intensity, L(T)RDINS, so that this variable may be endogenous to the innovative turnover equation. To check this, we performed Durbin–Hausman–Wu (DHW) exogeneity tests on the variables in question, on the assumption that the other explanatory variables are exogenous (Davidson and MacKinnon, 1993). As mentioned below, the results systematically confirmed exogeneity, so it would appear that there is no need for instrumenting. Therefore, estimation starts by OLS and (White) robust standard errors are deemed, as a result, to cope with potential heteroscedasticity problems.

Following Crépon et al. (1998), recent econometric work on the relation between innovation and firm performance has devoted a great deal of attention to the issue of sample selection. After estimating the research intensity and innovative turnover equations respectively on samples of research-active and both research-active and product-innovative firms, we decided, as a check of robustness, to extend these samples and always complement the equations with a sample selection equation, to draw inference from a larger population that also includes non-innovative firms. This sample selection equation is a probit equation explaining the probability of doing research, or of doing research and being product-innovative. When considered together with the corresponding research intensity or innovative turnover equation, it always forms a generalised tobit model, consisting of a probit and a tobit equation. Though we do not present them due to constraints of space, we will occasionally refer to their estimates by maximum likelihood (ML). Suffice it to say here that the sample selection effects we identified were no more than modest.

After considering the average behaviour of firms in the research-active or research-active and product-innovative sub-populations, we will then repeat estimation by quantile regression in order to identify systematic differences in behaviour between strong and weak innovators. A quantile regression estimates the parameters by minimising the sum of the asymmetrically-weighted absolute values of the residuals. Thus, for example, the 0.25 regression quantile estimates lend less weight to the positive residuals than to the negative ones. Therefore, our attention is drawn, so to speak, to the left-hand tail of the conditional density function of the dependent variable in question. For the 0.75 quantile, the reverse is true. Varying the quantile parameter between 0 and 1 generates all the regression quantiles, revealing the conditional distribution of the dependent variable given the explanatory variables. The additional advantage with quantile regression estimates is that they are more efficient than least squares estimates when the errors deviate from normality, which is the case.

Since the possibility of heteroscedasticity and therefore dependency between the quantile regressors and the density of the error terms cannot be ruled out, we use the bootstrap design method to estimate the standard errors (Buchinsky, 1995, 1998). We performed bootstrap resampling using 1,000 repetitions. Due to space constraints, the results below relate to the 0.25, 0.50 (or median), and 0.75 quantiles.

3.4 COMPARATIVE ANALYSIS

Tables 3.1 to 3.5 present the country-based estimates of the research intensity equation.

Table 3.1 Research intensity: Belgium – average and quantile behaviour

	With in–house research intensity				With total research intensity			
	OLS	Quantile regression			OLS	Quantile regression		
		0.25	0.50	0.75		0.25	0.50	0.75
	(i)	(ii)	(iii)	(iv)	(i)	(ii)	(iii)	(iv)
Constant	0.606	1.662	2.028	0.618	0.299	0.545	1.500	0.766
	(0.56)	(1.56)	(1.68)	(0.61)	(1.30)	(0.47)	(1.50)	(0.79)
IDPAV1	–1.219	–0.978	–1.376	–1.623	–0.427	–0.611	–0.771	–0.746
	(4.83)**	(2.64)**	(4.18)**	(5.26)**	(1.64)	(2.06)*	(2.67)**	(2.76)**
IDPAV2	–1.300	–1.033	–1.526	–1.636	–0.326	–0.367	–0.787	–0.832
	(5.10)**	(3.37)**	(4.19)**	(6.47)**	(1.29)	(1.21)	(3.47)**	(3.22)**
IDPAV3	–0.185	0.005	0.011	–0.303	–0.091	0.170	–0.154	–0.310
	(0.70)	(0.01)	(0.04)	(1.46)	(0.33)	(0.48)	(0.67)	(1.51)

Table 3.1 (continued)

	With in–house research intensity				With total research intensity			
	OLS	Quantile regression			OLS	Quantile regression		
		0.25	0.50	0.75		0.25	0.50	0.75
	(i)	(ii)	(iii)	(iv)	(i)	(ii)	(iii)	(iv)
IDPAV4	–0.262	–0.131	–0.523	–0.938	0.160	0.138	–0.404	–0.282
	(0.97)	(0.32)	(1.70)	(2.64)**	(0.51)	(0.38)	(1.61)	(0.68)
INKS	1.075	1.639	0.860	0.806	1.272	1.293	0.952	0.837
	(2.91)**	(3.26)**	(1.97)**	(2.07)**	(3.00)**	(2.67)**	(2.66)**	(2.43)**
COOP	0.437	0.626	0.635	0.563	0.478	0.558	0.609	0.515
	(2.98)**	(2.91)**	(3.65)**	(3.37)**	(3.24)**	(2.65)**	(3.79)**	(3.52)**
APPR	0.506	0.600	0.463	0.231	0.813	0.904	0.691	0.525
	(3.01)**	(2.47)**	(1.96)**	(1.22)	(4.10)**	(3.89)**	(3.50)**	(2.83)**
CKNOW	0.429	–0.208	0.394	0.526	0.462	–0.184	–0.083	0.364
	(1.12)	(0.39)	(1.17)	(1.43)	(1.42)	(0.40)	(0.28)	(1.21)
LHEI	0.187	0.291	0.201	0.126	0.147	0.204	0.180	0.085
	(3.28)**	(3.02)**	(3.07)**	(2.70)**	(3.37)**	(2.96)**	(3.56)**	(1.78)*
LEXI	0.022	0.022	0.030	0.012	–0.003	0.027	0.038	0.027
	(0.55)	(0.46	(0.75)	(0.38)	(0.01)	(0.68)	(1.08)	(0.81)
LSIZE	–0.166	–0.212	–0.226	–0.145	–0.183	–0.183	–0.178	–0.165
	(2.16)*	(2.82)**	(3.20)**	(2.13)**	(2.05)*	(2.70)**	(3.08)**	(2.79)**
GROUP	–0.189	–0.278	–0.165	0.009	–0.324	–0.349	–0.334	–0.163
	(0.85)	(1.21)	(0.67)	(0.05)	(1.54)	(1.44)	(2.17)*	(1.05)
Obs.	492	492	492	492	697	697	697	697
(Psd.) R^2	0.302	0.201	0.225	0.242	0.178	0.127	0.141	0.150

Notes: The dependent variable is the log of in-house research intensity, LRDINS, in the left-hand side of the table and the log of total research intensity, LTRDINS, in the right-hand side of the table. In the case of the OLS regression estimates in column (i) the t-values are based on heteroscedasticity-consistent standard errors. In the case of quantile regression estimates in columns (ii), (iii) and (iv) the t-values are based on standard errors obtained by the bootstrap design matrix method. Bootstrap re-sampling was performed using 1,000 repetitions. The t-tests are one-tailed or two-tailed, according to a priori expectations with respect to the parameter values. In the case of quantile regression the R^2 is a pseudo R^2 obtained as one minus the proportion of the sum of absolute weighted deviations over the raw sum of deviations. Symbols * and ** denote significance at the 5% and 1% levels.

Table 3.2 Research intensity: Germany – average and quantile behaviour

	With in-house research intensity				With total research intensity			
	OLS	Quantile regression			OLS	Quantile regression		
		0.25	0.50	0.75		0.25	0.50	0.75
	(i)	(ii)	(iii)	(iv)	(i)	(ii)	(iii)	(iv)
Constant	1.900	2.145	1.856	1.569	1.533	1.296	1.680	1.867
	(3.43)**	(2.20)*	(2.13)*	(3.13)**	(3.72)**	(2.19)*	(2.92)**	(4.45)**
IDPAV1	–0.420	–0.310	–0.404	–0.606	–0.400	–0.502	–0.527	–0.235
	(2.29)*	(1.12)	(1.85)	(2.79)**	(3.37)**	(2.87)**	(3.86)**	(1.68)
IDPAV2	–0.290	–0.383	–0.186	–0.383	0.132	0.046	0.040	0.216
	(1.99)*	(1.35)	(0.99)	(2.72)**	(1.27)	(0.29)	(0.29)	(1.79)
IDPAV3	0.140	0.252	0.324	0.067	0.182	0.228	0.100	0.114
	(0.94)	(1.04)	(1.92)	(0.41)	(1.68)	(1.47)	(0.71)	(0.93)
IDPAV4	0.266	0.204	0.021	–0.032	0.502	0.323	0.385	0.360
	(1.37)	(0.72)	0.08	(0.14)	(2.91)**	(1.36)	(1.72)	(1.78)
INKS	0.408	0.342	0.574	0.297	0.727	1.004	0.571	0.706
	(1.64)*	(0.87)	(1.91)*	(1.00)	(3.72)**	(3.56)**	(2.24)**	(3.18)**
COOP	0.377	0.250	0.331	0.434	0.275	0.301	0.265	0.356
	(3.75)**	(1.52)	(2.51)**	(3.95)**	(3.36)**	(2.49)**	(3.01)**	(3.65)**
APPR	0.326	0.419	0.344	0.406	0.581	0.725	0.684	0.520
	(2.63)**	(2.24)**	(1.96)**	(3.18)**	(5.90)**	(5.67)**	(5.60)**	(3.95)**
CKNOW	0.080	0.168	–0.102	0.297	–0.106	–0.301	–0.103	–0.180
	(0.33)	(0.41)	(0.31)	(1.06)	(0.59)	(1.08)	(0.42)	(0.83)
LHEI	0.241	0.318	0.286	0.200	0.139	0.133	0.150	0.170
	(7.11)**	(3.52)**	(4.01)**	(6.18)**	(7.24)**	(5.49)**	(5.45)**	(7.95)**
LEXI	0.128	0.097	0.101	0.126	0.090	0.120	0.112	0.058
	(5.54)**	(2.97)**	(3.98)**	(5.03)**	(5.56)**	(5.19)**	(5.90)**	(3.29)**
LSIZE	–0.174	–0.188	–0.137	0.127	–0.216	–0.262	–0.197	–0.161
	(6.93)**	(3.82)**	(4.41)**	(4.56)**	(10.8)**	(8.31)**	(8.24)**	(8.71)**
GROUP	–0.317	–0.233	–0.284	–0.428	–0.229	–0.111	–0.266	–0.264
	(3.22)**	(1.61)	(2.36)**	(3.60)**	(3.00)**	(0.92)	(2.93)**	(2.86)**
Obs.	1,117	1,117	1,117	1,117	1,749	1,749	1,749	1,749
(Psd.) R^2	0.262	0.150	0.156	0.166	0.246	0.162	0.152	0.118

Notes: See Table 3.1. Symbols * and ** denote significance at the 5% and 1% levels.

Table 3.3 Research intensity: Norway – average and quantile behaviour

	With in–house research intensity				With total research intensity			
	OLS	Quantile regression			OLS	Quantile regression		
		0.25	0.50	0.75		0.25	0.50	0.75
	(i)	(ii)	(iii)	(iv)	(i)	(ii)	(iii)	(iv)
Constant	4.455	3.568	4.629	6.779	3.715	2.844	4.019	5.757
	(8.35)**	(3.96)**	(7.03)**	(9.01)**	(8.27)**	(4.43)**	(6.80)**	(11.3)**
IDPAV1	–1.060	–0.953	–1.165	–1.082	–0.744	–0.683	–0.706	–0.814
	(7.36)**	(4.71)**	(6.26)**	(5.39)**	(6.43)**	(4.52)**	(4.93)**	(5.92)**
IDPAV2	–1.115	–0.972	–1.127	–1.153	–0.818	–0.754	–0.755	–1.017
	(8.09)**	(4.84)**	(6.07)**	(6.46)**	(7.37)**	(5.89)**	(5.15)**	(6.94)**
IDPAV3	–0.217	–0.028	–0.192	–0.258	–0.181	0.001	–0.101	–0.177
	(1.45)	(0.13)	(0.96)	(1.56)	(1.38)	(0.00)	(0.60)	(1.07)
IDPAV4	–0.247	–0.186	–0.360	–0.328	–0.083	–0.057	0.013	–0.178
	(1.13)	(0.54)	(1.32)	(1.28)	(0.46)	(0.27)	(0.05)	(0.71)
INKS	1.164	1.078	1.142	1.011	1.138	1.026	1.244	1.332
	(4.47)**	(2.83)**	(3.75)**	(3.29)**	(5.12)**	(3.29)**	(4.11)**	(4.92)**
COOP	0.291	0.357	0.242	0.197	0.410	0.490	0.446	0.336
	(3.20)**	(2.93)**	(2.09)**	(1.99)**	(5.03)**	(5.02)**	(4.29)**	(3.23)**
APPR	0.781	0.822	0.916	0.641	0.860	1.131	0.925	0.625
	(6.65)**	(4.77)**	(6.59)**	(4.51)**	(8.30)**	(7.55)**	(6.67)**	(4.46)**
CKNOW	0.113	0.395	0.107	–0.048	0.253	0.451	0.246	0.133
	(0.52)	(1.19)	(0.41)	(0.18)	(1.40)	(2.16)**	(0.90)	(0.63)
LHEI	0.227	0.183	0.184	0.353	0.201	0.168	0.206	0.270
	(6.44)**	(4.24)**	(4.35)**	(4.72)**	(7.00)**	(5.11)**	(5.95)**	(6.80)**
LEXI	0.067	0.075	0.064	0.072	0.057	0.082	0.049	0.042
	(4.01)**	(2.89)**	(3.54)**	(3.71)**	(4.19)**	(4.47)**	(2.88)**	(2.58)**
LSIZE	–0.349	–0.387	–0.399	–0.322	–0.323	–0.347	–0.349	–0.338
	(8.90)**	(6.52)**	(8.46)**	(7.29)**	(10.1)**	(9.25)**	(8.62)**	(9.37)**
GROUP	0.006	–0.090	0.170	0.208	0.012	–0.173	0.116	0.223
	(0.049)	(0.53)	(1.30)	(1.42)	(0.12)	(1.53)	(0.88)	(1.95)
Obs.	962	962	962	962	1,390	1,390	1,390	1,390
(Psd.) R^2	0.445	0.262	0.282	0.297	0.373	0.220	0.235	0.258

Notes: See Table 3.1. Symbols * and ** denote significance at the 5% and 1% levels.

Table 3.4 Research intensity: Spain – average and quantile behaviour

	With in–house research intensity				With total research intensity			
	OLS	Quantile regression			OLS	Quantile regression		
		0.25	0.50	0.75		0.25	0.50	0.75
	(i)	(ii)	(iii)	(iv)	(i)	(ii)	(iii)	(iv)
Constant	3.581	2.089	3.576	3.851	2.243	1.915	2.144	2.539
	(7.08)**	(4.49)**	(7.27)**	(7.93)**	(7.00)**	(4.66)**	(5.03)**	(6.35)**
IDPAV1	–0.995	–0.809	–1.040	–1.259	–0.379	–0.423	–0.362	–0.444
	(8.00)**	(5.35)**	(7.16)**	(6.80)**	(4.30)**	(3.38)**	(3.70)**	(3.04)**
IDPAV2	–0.735	–0.453	–0.822	–1.121	–0.056	–0.077	–0.068	–0.084
	(6.08)**	(3.04)**	(6.08)**	(6.12)**	(0.63)	(0.64)	(0.64)	(0.55)
IDPAV3	–0.194	0.187	–0.261	–0.476	–0.008	0.105	0.066	–0.093
	(1.54)	(1.15)	(1.58)	(2.68)**	(0.084)	(0.75)	(0.55)	(0.61)
IDPAV4	–0.206	0.095	–0.221	–0.594	0.049	0.251	0.056	–0.161
	(1.53)	(0.55)	(1.48)	(3.01)**	(0.44)	(1.53)	(0.43)	(1.04)
INKS	0.438	0.253	0.429	0.369	0.767	0.808	0.868	0.659
	(2.56)**	(1.21)	(2.11)**	(1.97)**	(5.33)**	(3.98)**	(5.49)**	(3.06)**
COOP	0.596	0.553	0.635	0.554	0.513	0.459	0.525	0.500
	(7.84)**	(6.71)**	(5.53)**	(5.64)**	(7.17)**	(4.40)**	(5.57)**	(6.24)**
APPR	0.225	0.252	0.195	0.205	0.289	0.372	0.272	0.129
	(2.95)**	(3.10)**	(1.97)**	(2.06)**	(4.59)**	(3.80)**	(3.67)**	(1.64)*
CKNOW	0.167	0.257	0.071	0.133	–0.155	–0.055	–0.296	–0.152
	(1.14)	(1.38)	(0.37)	(0.87)	(1.27)	(0.32)	(1.62)	(0.85)
LHEI	0.053	0.026	0.023	0.076	0.016	0.026	0.020	0.007
	(3.03)**	(1.59)	(1.25)	(3.55)**	(1.35)	(1.66)	(1.57)	(0.42)
LEXI	0.087	0.084	0.077	0.055	0.033	0.048	0.036	0.007
	(5.29)**	(4.66)**	(4 77)**	(3.34)**	(2.96)**	(2.90)**	(2.37)**	(0.43)
LSIZE	–0.442	–0.434	–0.459	0.372	–0.384	–0.411	–0.369	–0.348
	(14.8)**	(16.0)**	(16.1)**	(13.4)**	(17.2)**	(16.6)**	(13.4)**	(13.0)**
GROUP	0.137	0.161	0.100	0.088	0.028	0.011	–0.004	0.101
	(1.73)	(1.57)	(1.00)	(1.03)	(0.45)	(0.12)	(0.06)	(1.14)
Obs.	1,773	1,773	1,773	1,773	3,079	3,079	3,079	3,079
(Psd.) R^2	0.344	0.184	0.197	0.225	0.205	0.120	0.114	0.099

Notes: See Table 3.1. Symbols * and ** denote significance at the 5% and 1% levels.

Table 3.5 Research intensity: Portugal – average and quantile behaviour

	With in–house research intensity				With total research intensity			
	OLS	Quantile regression			OLS	Quantile regression		
		0.25	0.50	0.75		0.25	0.50	0.75
	(i)	(ii)	(iii)	(iv)	(i)	(ii)	(iii)	(iv)
Constant	2.426	1.328	1.744	4.169	1.103	–0.747	0.092	2.276
	(2.42)**	(0.92)	(1.55)	(3.52)**	(1.53)	(0.71)	(0.11)	(2.82)**
IDPAV1	–0.643	–0.921	–0.636	–0.903	0.062	0.181	0.286	–0.176
	(2.53)**	(2.48)**	(1.98)*	(2.55)**	(0.36)	(0.69)	(1.04)	(0.82)
IDPAV2	–0.741	–1.037	–0.547	–1.218	0.050	0.143	0.389	–0.177
	(2.90)**	(2.86)**	(1.74)	(3.23)**	(0.27)	(0.56)	(1.38)	(0.79)
IDPAV3	–0.165	–1.025	–0.143	0.056	0.048	0.249	0.449	–0.033
	(0.45)	(2.30)*	(0.29)	(0.12)	(0.21)	(0.79)	(1.32)	(0.13)
IDPAV4	0.090	–0.036	0.169	–0.156	0.027	0.413	0.202	–0.314
	(0.31)	(0.10)	(0.50)	(0.44)	(0.13)	(1.21)	(0.66)	(1.31)
INKS	0.958	1.246	0.638	0.549	1.536	2.124	1.022	0.916
	(2.31)**	(1.86)*	(1.41)	(0.96)	(5.08)**	(4.37)**	(2.34)**	(2.81)**
COOP	0.334	0.354	0.536	0.465	0.214	–0.010	0.208	0.162
	(1.73)*	(1.54)	(2.29)**	(1.97)**	(1.63)	(0.04)	(1.32)	(1.01)
APPR	0.594	0.837	0.777	0.480	0.420	0.782	0.422	0.031
	(2.77)**	(2.98)**	(3.21)**	(1.91)*	(2.96)**	(3.70)**	(2.46)**	(0.18)
CKNOW	0.327	0.406	0.276	0.471	0.306	0.405	0.592	0.132
	(0.96)	(0.78)	(0.71)	(1.03)	(1.33)	(0.99)	(1.32)	(0.45)
LHEI	0.065	0.074	0.086	0.100	0.020	–0.022	0.035	0.031
	(1.90)*	(1.74)*	(1.77)*	(2.21)**	(0.88)	(0.68)	(1.22)	(1.31)
LEXI	0.057	0.080	0.054	0.075	0.001	–0.017	–0.006	–0.003
	(1.68)*	(1.82)*	(1.36)	(1.69)*	(0.03)	(0.46)	(0.18)	(0.12)
LSIZE	–0.470	–0.430	–0.410	–0.472	–0.363	–0.370	–0.296	–0.336
	(6.65)**	(4.09)**	(5.15)**	(6.49)**	(7.00)**	(4.93)**	(5.60)**	(6.28)**
GROUP	0.315	0.245	0.305	0.479	0.135	0.180	0.181	0.216
	(1.78)	(1.06)	(1.49)	(2.32)*	(0.97)	(0.87)	(0.88)	(1.55)
Obs.	368	368	368	368	746	746	746	746
(Psd.) R^2	0.285	0.150	0.168	0.188	0.153	0.098	0.080	0.092

Notes: See Table 3.1. Symbols * and ** denote significance at the 5% and 1% levels.

The left-hand side of the tables presents the estimates of the equation explaining in-house research intensity on the samples of firms reporting in-house research activity, LRDINS, whereas the right-hand side of the tables presents the estimates of the equation explaining total research intensity, LTRDINS, on the samples of firms reporting innovation expenditure.

Columns (i) always present the estimates by OLS and provide an idea of the average behaviour of research-active firms. Columns (ii), (iii) and (iv) present the corresponding quantile regression estimates on the 0.25, 0.50 and 0.75 quantiles and help to identify systematic differences in behaviour between strong and weak research performers.

Consider first the estimates by OLS of the equation explaining in-house research intensity on the left-hand side of the tables in columns (i). The statistical performance appears to give a decent statistical fit (adjusted R^2), in view of the errors of measurement in the data. Durbin–Hausman–Wu (DHW) tests confirmed the exogeneity of incoming knowledge spillovers, INKS, and research cooperation, COOP, justifying the use of these variables as they are rather than in instrumented form.

In turn, consider the respective parameter estimates. Since the equation includes an intercept, one of the industry dummies had to be dropped to avoid singularity. We dropped the information-intensive services dummy, IDPAV5, so that its impact is captured, among other elements, by the intercept, C. The estimates of the other industry dummies reflect deviations from this. The results in columns (i) indicate that there is always a statistically significant negative impact in the case of IDPAV1 and IDPAV2. This suggests that general technological characteristics ensure that resource-intensive and scale-intensive industries are, on average, not as in-house research intensive as the specialised supplier, science-based and information-intensive sectors.

Of the knowledge-related variables, the two variables relating to external knowledge, incoming knowledge spillovers, INKS, and research cooperation, COOP, are statistically significant with a positive sign for all countries. In-house research intensity is higher for firms actively sourcing external information and engaging in research cooperation. This suggests that incoming knowledge spillovers and research collaboration increase the firms' knowledge base and that the absorption of the knowledge concerned always requires further in-house R&D. The confirmation by the DHW test that INKS and COOP are, indeed, exogenous in the research intensity equation suggests, at the same time, that the estimates are not suffering an upward endogeneity bias, due to a reverse causation. This seems plausible, since information sourcing and research cooperation are deliberate activities, implying a re-organisation of the research process, and should not be assumed to be some automatic by-product of higher in-house research efforts.

Of the two remaining knowledge-related variables, only appropriability is found always to be statistically significant, with a positive sign, conforming to expectations. The capacity for appropriating the benefits of internally-developed knowledge increases incentives to undertake research efforts. The coefficient of the variable on complex knowledge, CKNOW, also has the expected positive sign, but is non-significant. A consideration of the correlation matrix suggests that this may be due to multicollinearity with the INKS variable, which was constructed from the same underlying data, so that there is not much to conclude from this.

The human capital variable, LHEI, is also clearly significant, with the positive sign as expected for each country. With the exception of Belgium, this is also the case for export orientation, LEXI. This confirms the complementarity between formal R&D and the qualification of workers, and suggests that the stronger competitive pressure in foreign markets does, indeed, positively affect in-house research intensity. The lack of significance for export orientation in the case of Belgium can be explained by the fact that most research-active firms are highly export-oriented here. This implies not only a substantially higher mean value for this variable here (as shown in Table A3.2 in the Appendix), but a lower variability between firms (lower variance) and, therefore, explanatory power in the cross-section.

Finally, where the size-related variables are concerned, firm size, LSIZE, always has a clearly negative impact on in-house research intensity, whereas, if significant, this is also the case for group membership, GROUP (in the case of Germany). This negative significance of size would suggest that in-house R&D expenditure among research performers rises less than proportionally with firm size. This contradicts stylised facts 2 and 3 of Cohen and Klepper (1996), derived from the earlier empirical literature on this issue, which would imply a proportional increase. This might be due to the fact that nowadays an important part of technical change stems from small-scale firms mainly active in either information-intensive, science-based or specialised supplier industries, and therefore much less in the resource-intensive and scale-intensive sectors that were far more dominant before the ICT revolution. These new technology-based firms expect to grow rapidly from their innovations and, in an open innovation logic, they are prepared to sell part of their new ideas, which they do not consider worth exploiting themselves, in disembodied form, by licensing, etc. Therefore their R&D decisions are not greatly affected by their original size constraints or, for that matter, by their short-term turnover performance, and they tend to spend relatively more on in-house R&D in relation to their size than larger, less dynamic firms.

Though we do not present them here due to space constraints, the ML estimates of the corresponding generalised tobit model aimed at drawing statistical inference from a larger population that also includes non-research-active firms, do, by and large, confirm the above conclusions.

Consider now the corresponding quantile regression estimates of the equation to explain in-house research intensity, presented in columns (ii),

(iii) and (iv) on the left-hand side of Tables 3.1 to 3.5. As far as the industry dummies are concerned, it is striking that the negative impact of the IDPAV1 and IDPAV2 dummies, observed in the average estimates by OLS in columns (i), tends to become more explicit, quantile-by-quantile, in four of the five countries under consideration (Belgium, Germany, Norway and Spain). This suggests that it is true that the level of in-house research intensity is, on average, higher in the hi-tech information-intensive, specialised supplier and science-based industries than in the more traditional resource-intensive and scale-intensive sectors, and especially true for the strong research performers. Or, to put it differently, the most dynamic R&D investors are found particularly in sectors with a technology intensive trajectory.

As suggested by the average estimates in columns (i), it is the case that of the knowledge-related variables, only incoming knowledge spillovers, INKS, research cooperation, COOP, and appropriability, APPR, are always found to be statistically significant in at least one of the quantiles, whereas complex knowledge, CKNOW, is systematically found to be non-significant. Most interestingly, in four of the five countries under consideration (Belgium, Germany, Spain and Portugal) the joint relative impact of incoming knowledge spillovers, INKS, and research cooperation, COOP, tends to increase quantile-by-quantile in relation to that of appropriability, APPR, whereas in the remaining country (Norway) it remains fairly constant. In the case of Belgium, this largely occurs through a decrease from quantile to quantile in the impact of appropriability, APPR. In the case of Germany, it does so through an increase from quantile to quantile in the impact of research cooperation, COOP. In the case of Spain, it takes place both through a significant increase in the impact of incoming knowledge spillovers, INKS, and a slight decrease in that of appropriability, APPR, between the lower and the median quantiles. Whereas in the case of Portugal, it is due to a strong increase in the influence of research cooperation, COOP, between the lower and the median quantiles and a decrease in the impact of appropriability, APPR, between the median and the higher ones.

Therefore it would appear that, though there are differences between countries, the quantile regression estimates for the knowledge-related variables suggest the innovation process tends to become more open among the stronger in-house research performers than among the weaker ones. The parameter estimates for the remaining variables by and large confirm the average findings in columns (i), while showing no quantile-by-quantile pattern.

Consider now the right-hand side of Tables 3.1 to 3.5 with the corresponding estimates of the equation explaining the level of total research intensity, LTRDINS. Consider first the estimates by OLS in columns (i). The statistical fit (R^2) for each country is slightly worse than in the equation explaining in-house research intensity, in columns (i) of the left-hand side. This is mainly due to a reduction in the negative significance of the IDPAV1 and IDPAV2 industry dummies, which suggests that the differences in total research intensity between sectors are less pronounced than those in in-house research intensity. This reflects the fact that firms operating in the less technological sectors spend a larger part of their total innovation outlay on components other than in-house R&D, such as the purchase of advanced machinery. The parameter estimates and statistical significance of the other explanatory variables are, however, quite similar to those in the equation explaining in-house research intensity, in columns (i) of the left-hand side.

Similar conclusions can also be reached with regard to the quantile regression estimates, in columns (ii), (iii) and (iv) of the right-hand side of Tables 3.1 to 3.5. There is always a worse statistical fit, due to a reduction in the negative significance of the IDPAV1 and IDPAV2 industry dummies. And now too, the pattern of the parameter estimates for the other variables is quite similar to that in the equation explaining in-house research intensity, in columns (ii), (iii) and (iv) of the left-hand side. Once again, there is clear evidence of a quantile-by-quantile increase in the joint impact of incoming knowledge spillovers, INKS, and research cooperation, COOP, in relation to that of appropriability, APPR, and this is true for all of the countries under consideration.

Table 3.6 Innovative turnover: Belgium – average and quantile behaviour

	With in–house research intensity				With total research intensity			
	OLS	Quantile regression			OLS	Quantile regression		
		0.25	0.50	0.75		0.25	0.50	0.75
	(i)	(ii)	(iii)	(iv)	(i)	(ii)	(iii)	(iv)
Constant	–0.849	–1.490	–0.526	1.009	–1.037	–1.494	–0.487	0.464
	(1.80)	(1.98)*	(0.98)	(1.89)	(2.40)**	(2.47)**	(0.99)	(0.83)
IDPAV1	–0.688	–0.668	–0.651	–0.585	–0.503	–0.511	–0.448	–0.546
	(4.32)**	(3.02)**	(3.59)**	(2.46)**	(3.92)**	(3.01)**	(2.93)**	(3.29)**
IDPAV2	–0.436	–0.640	–0.413	–0.145	–0.312	–0.282	–0.347	–0.237
	(2.48)**	(2.03)*	(1.96)*	(0.59)	(2.23)*	(1.40)	(1.81)	(1.28)
IDPAV3	–0.100	–0.026	–0.058	–0.107	–0.111	0.033	–0.001	–0.261
	(0.68)	(0.09)	(0.35)	(0.47)	(0.83)	(0.18)	(0.01)	(1.45)

Table 3.6 (continued)

	With in–house research intensity				With total research intensity			
	OLS	Quantile regression			OLS	Quantile regression		
		0.25	0.50	0.75		0.25	0.50	0.75
	(i)	(ii)	(iii)	(iv)	(i)	(ii)	(iii)	(iv)
IDPAV4	–0.641	–0.803	–0.383	–0.423	–0.602	–0.655	–0.313	–0.510
	(3.66)**	(2.31)*	(1.94)	(1.84)	(3.68)**	(2.53)**	(1.56)	(2.35)**
LTRDINS	0.077	0.064	0.086	0.114	0.079	0.049	0.079	0.118
	(2.80)**	(1.64)*	(2.36)**	(2.54)**	(3.15)**	(1.36)	(2.64)**	(3.63)**
INKS	0.113	–0.173	–0.015	0.069	0.304	0.209	0.051	0.219
	(0.43)	(0.44)	(0.05)	(0.20)	(1.39)	(0.66)	(0.18)	(0.77)
COOP	0.076	0.118	0.088	0.054	0.019	0.073	0.030	0.077
	(0.68)	(0.68)	(0.74)	(0.38)	(0.19)	(0.48)	(0.27)	(0.52)
APPR	–0.033	–0.039	0.002	–0.221	–0.060	–0.193	0.018	0.016
	(0.24)	(0.17)	(0.01)	(1.05)	(0.51)	(1.12)	(0.12)	(0.09)
CKNOW	–0.133	–0.010	–0.039	–0.153	–0.057	0.027	0.037	–0.118
	(0.55)	(0.03)	(0.12)	(0.55)	(0.28)	(0.09)	(0.15)	(0.48)
PCS	0.312	0.328	0.234	0.374	0.173	0.180	0.113	0.092
	(2.99)**	(1.96)**	(1.68)*	(2.63)**	(1.98)**	(1.37)	(1.07)	(0.68)
LHEI	–0.023	–0.030	–0.035	–0.007	–0.020	–0.022	–0.017	–0.016
	(0.97)	(0.94)	(0.93)	(0.17)	(0.90)	(0.87)	(0.61)	(0.63)
LEXI	–0.002	–0.012	0.019	0.020	–0.013	–0.008	–0.001	–0.012
	(0.11)	(0.43)	(0.83)	(0.66)	(0.84)	(0.38)	(0.04)	(0.49)
LSIZE	–0.060	–0.057	–0.083	–0.112	–0.053	–0.067	–0.081	–0.091
	(2.00)*	(1.27)	(2.46)**	(2.78)**	(2.03)*	(2.02)*	(2.61)**	(2.43)**
GROUP	0.047	–0.062	0.067	0.107	0.080	0.059	0.152	0.135
	(0.41)	(0.39)	(0.53)	(0.63)	(0.84)	(0.50)	(1.30)	(1.11)
Obs.	390	390	390	390	531	531	531	531
(Psd.) R^2	0.127	0.100	0.113	0.106	0.088	0.052	0.069	0.088

Notes: The dependent variable is the log of innovative turnover, LITURN. In the case of the OLS regression estimates in column (i) the t-values are based on heteroscedasticity-consistent standard errors. In the case of quantile regression estimates in columns (ii), (iii) and (iv) the t-values are based on standard errors obtained by the bootstrap design matrix method. Bootstrap re-sampling was performed using 1,000 repetitions. The t-tests are one-tailed or two-tailed, according to a priori expectations with respect to the parameter values. In the case of quantile regression the R^2 is a pseudo R^2 obtained as one minus the proportion of the sum of absolute weighted deviations over the raw sum of deviations. Symbols * and ** denote significance at the 5% and 1% levels.

Table 3.7 Innovative turnover: Germany – average and quantile behaviour

	With in–house research intensity				With total research intensity			
	OLS	Quantile regression			OLS	Quantile regression		
		0.25	0.50	0.75		0.25	0.50	0.75
	(i)	(ii)	(iii)	(iv)	(i)	(ii)	(iii)	(iv)
Constant	−0.214	−0.725	−0.076	0.290	−0.353	−0.748	−0.132	0.147
	(0.67)	(1.44)	(0.21)	(0.51)	(1.37)	(2.07)*	(0.47)	(0.32)
IDPAV1	−0.040	−0.059	−0.003	−0.033	−0.077	−0.113	−0.121	−0.161
	(0.37)	(0.38)	(0.02)	(0.18)	(0.96)	(1.23)	(1.32)	(1.15)
IDPAV2	0.094	0.113	0.143	0.061	0.055	0.060	0.070	0.037
	(1.02)	(0.74)	(1.31)	(0.40)	(0.73)	(0.53)	(0.85)	(0.30)
IDPAV3	0.135	0.138	0.137	0.103	0.080	0.058	0.049	0.054
	(1.57)	(0.91)	(1.34)	(0.78)	(1.05)	(0.42)	(0.52)	(0.50)
IDPAV4	−0.118	−0.128	−0.197	−0.246	−0.160	−0.146	−0.297	−0.243
	(0.93)	(0.67)	(1.14)	(1.25)	(1.37)	(0.95)	(1.91)	(1.39)
LTRDINS	0.119	0.119	0.092	0.129	0.138	0.133	0.141	0.158
	(6.30)**	(4.02)**	(3.48)**	(4.08)**	(7.67)**	(5.99)**	(6.84)**	(5.58)**
INKS	0.019	0.110	−0.028	0.050	0.070	0.151	0.068	0.120
	(0.13)	(0.42)	(0.15)	(0.22)	(0.55)	(0.81)	(0.45)	(0.57)
COOP	0.085	0.099	0.087	0.184	0.089	0.097	0.088	0.134
	(1.42)	(1.07)	(1.14)	(1.92)*	(1.63)	(1.19)	(1.27)	(1.75)*
APPR	0.063	0.006	0.089	0.065	0.012	−0.063	0.026	−0.018
	(0.81)	(0.06)	(0.88)	(0.56)	(0.18)	(0.75)	(0.31)	(0.16)
CKNOW	0.088	0.016	0.080	−0.010	0.070	0.186	0.024	−0.050
	(0.53)	(0.06)	(0.40)	(0.04)	(0.55)	(1.05)	(0.15)	(0.24)
PCS	0.241	0.281	0.232	0.245	0.191	0.272	0.192	0.157
	(3.95)**	(2.88)**	(3.24)**	(2.66)**	(3.74)**	(3.70)**	(3.29)**	(1.91)*
LHEI	0.031	0.023	0.327	0.053	0.035	0.036	0.028	0.048
	(1.52)	(0.77)	(1.42)	(1.54)	(2.60)**	(1.97)**	(2.00)**	(2.17)**
LEXI	0.003	0.006	0.017	0.002	0.011	0.023	0.028	0.007
	(0.24)	(0.27)	(1.01)	(0.14)	(1.04)	(1.31)	(2.45)**	(0.38)
LSIZE	−0.032	−0.040	−0.043	−0.005	−0.019	−0.030	−0.031	0.004
	(2.53)**	(2.21)*	(2.66)**	(0.22)	(1.62)	(2.22)*	(2.05)*	(0.19)
GROUP	−0.208	−0.175	−0.190	−0.269	−0.179	−0.119	−0.157	−0.170
	(3.43)**	(2.04)*	(2.55)**	(2.82)**	(3.43)**	(1.75)	(2.84)**	(2.03)*
Obs.	931	931	931	931	1,321	1,321	1,321	1,321
(Psd.) R^2	0.119	0.087	0.073	0.071	0.127	0.078	0.072	0.067

Notes: See Table 3.6. Symbols * and ** denote significance at the 5% and 1% levels.

Table 3.8 Innovative turnover: Norway – average and quantile behaviour

	With in–house research intensity				With total research intensity			
	OLS	Quantile regression			OLS	Quantile regression		
		0.25	0.50	0.75		0.25	0.50	0.75
	(i)	(ii)	(iii)	(iv)	(i)	(ii)	(iii)	(iv)
Constant	0.299	0.271	–0.108	0.463	0.355	0.221	0.639	0.844
	(0.81)	(0.50)	(0.27)	(0.79)	(1.15)	(0.41)	(1.94)	(2.09)*
IDPAV1	–0.152	–0.268	–0.262	0.047	–0.126	–0.165	–0.200	–0.032
	(1.42)	(1.66)	(1.93)	(0.32)	(1.50)	(1.20)	(1.81)	(0.29)
IDPAV2	–0.250	–0.305	–0.367	–0.246	–0.254	–0.217	–0.416	–0.214
	(2.36)**	(2.00)*	(3.07)**	(1.84)	(2.93)**	(1.66)	(4.11)**	(1.81)
IDPAV3	0.032	–0.036	0.038	0.203	0.118	0.115	0.170	0.230
	(0.27)	(0.16)	(0.29)	(1.63)	(1.07)	(0.62)	(1.35)	(2.35)**
IDPAV4	–0.663	–0.339	–0.626	–0.543	–0.496	–0.129	–0.391	–0.466
	(3.34)**	(0.77)	(3.23)**	(2.72)**	(2.83)**	(0.34)	(2.39)**	(2.79)**
LTRDINS	0.113	0.067	0.153	0.167	0.106	0.069	0.113	0.143
	(3.81)**	(1.48)	(4.43)**	(4.81)**	(4.74)**	(2.01)**	(3.77)**	(5.46)**
INKS	0.344	0.288	–0.011	0.277	0.415	0.356	0.161	0.404
	(1.61)	(0.80)	(0.04)	(1.17)	(2.29)**	(1.14)	(0.66)	(2.04)**
COOP	–0.189	–0.175	–0.127	–0.105	–0.135	–0.097	–0.050	–0.035
	(1.56)	(1.50)	(1.42)	(1.26)	(1.40)	(0.97)	(0.59)	(0.44)
APPR	0.156	0.152	0.208	0.066	0.226	0.227	0.242	0.054
	(1.63)	(0.97)	(1.87)*	(0.63)	(2.62)**	(1.78)*	(2.21)**	(0.55)
CKNOW	–0.044	–0.031	–0.120	–0.295	–0.086	–0.010	–0.100	–0.264
	(0.22)	(0.10)	(0.52)	(1.61)	(0.51)	(0.04)	(0.53)	(1.51)
PCS	0.032	0.083	–0.075	–0.123	0.001	0.103	–0.121	–0.125
	(0.39)	(0.61)	(0.79)	(1.48)	(0.01)	(0.87)	(1.33)	(1.61)
LHEI	–0.020	0.007	–0.017	0.032	–0.004	0.008	–0.012	–0.011
	(0.83)	(0.21)	(1.40)	(0.64)	(0.23)	(0.31)	(0.50)	(0.34)
LEXI	0.037	0.059	0.031	0.019	0.030	0.049	0.025	0.024
	(2.78)**	(2.60)**	(2.20)**	(1.31)	(2.60)**	(2.27)**	(2.21)**	(2.06)**
LSIZE	–0.114	–0.132	–0.095	–0.073	–0.118	–0.150	–0.119	–0.099
	(4.87)**	(3.70)**	(3.25)**	(2.81)**	(5.77)**	(5.04)**	(5.11)**	(4.06)**
GROUP	0.079	0.070	0.121	0.070	0.084	0.109	0.053	0.152
	(0.92)	(0.55)	(0.97)	(0.77)	(1.15)	(1.12)	(0.54)	(1.90)
Obs.	821	821	821	821	1,124	1,124	1,124	1,124
(Psd.) R^2	0.138	0.048	0.102	0.131	0.138	0.058	0.099	0.124

Notes: See Table 3.6. Symbols * and ** denote significance at the 5% and 1% levels.

Table 3.9 Innovative turnover: Spain – average and quantile behaviour

	With in–house research intensity				With total research intensity			
	OLS	Quantile regression			OLS	Quantile regression		
		0.25	0.50	0.75		0.25	0.50	0.75
	(i)	(ii)	(iii)	(iv)	(i)	(ii)	(iii)	(iv)
Constant	–0.233	–0.577	–0.009	0.297	–0.112	–0.386	0.292	0.721
	(0.68)	(1.18)	(0.02)	(0.73)	(0.38)	(0.83)	(0.71)	(2.07)*
IDPAV1	0.021	0.038	–0.049	–0.108	–0.017	–0.046	–0.096	–0.165
	(0.19)	(0.32)	(0.33)	(0.78)	(0.21)	(0.41)	(0.83)	(1.45)
IDPAV2	–0.048	–0.031	–0.072	–0.139	–0.037	–0.041	–0.077	–0.166
	(0.46)	(0.22)	(0.53)	(1.09)	(0.44)	(0.35)	(0.70)	(1.78)
IDPAV3	0.188	0.321	0.172	0.001	0.215	0.346	0.134	0.022
	(1.85)	(2.46)**	(1.24)	(0.01)	(2.47)**	(2.86)**	(1.11)	(0.20)
IDPAV4	–0.338	–0.246	–0.369	–0.486	–0.344	–0.233	–0.421	–0.531
	(2.91)**	(1.53)	(2.42)**	(3.50)**	(3.42)**	(1.65)	(3.20)**	(4.42)**
LTRDINS	0.163	0.211	0.192	0.140	0.158	0.180	0.166	0.141
	(5.95)**	(5.20)**	(6.11)**	(3.98)**	(8.78)**	(6.22)**	(7.00)**	(5.76)**
INKS	0.071	0.230	0.240	–0.275	0.128	0.212	0.262	–0.070
	(0.44)	(1.24)	(1.12)	(1.32)	(0.96)	(1.19)	(1.35)	(0.39)
COOP	0.024	0.013	0.041	–0.036	0.045	–0.014	0.027	–0.022
	(0.33)	(0.14)	(0.51)	(0.41)	(0.74)	(0.16)	(0.36)	(0.26)
APPR	0.060	0.172	0.067	–0.060	0.055	0.223	0.095	–0.053
	(0.92)	(1.97)**	(0.86)	(0.71)	(0.98)	(2.60)**	(1.28)	(0.71)
CKNOW	–0.128	–0.224	–0.208	–0.122	–0.060	–0.143	–0.097	–0.003
	(0.87)	(1.06)	(0.97)	(0.63)	(0.52)	(0.73)	(0.66)	(0.02)
PCS	0.168	0.220	0.202	0.067	0.023	0.054	0.063	–0.039
	(2.68)**	(2.73)**	(2.43)**	(0.90)	(0.46)	(0.77)	(0.94)	(0.60)
LHEI	0.013	0.019	0.011	0.009	0.013	0.014	0.027	0.023
	(0.90)	(0.96)	(0.67)	(0.36)	(1.18)	(0.83)	(2.19)**	(1.45)
LEXI	0.012	0.030	0.016	0.008	0.009	0.018	0.014	0.006
	(0.93)	(1.72)*	(0.93)	(0.49)	(0.81)	(1.22)	(0.96)	(0.43)
LSIZE	–0.030	–0.032	–0.039	0.001	–0.044	–0.064	–0.053	–0.025
	(1.33)	(0.94)	(1.14)	(0.03)	(2.31)*	(2.19)*	(1.95)	(1.04)
GROUP	–0.093	–0.167	–0.028	–0.151	–0.098	–0.100	–0.043	–0.116
	(1.38)	(1.93)	(0.30)	(1.69)	(1.70)	(1.19)	(0.58)	(1.53)
Obs.	1,367	1,367	1,367	1,367	1,983	1,983	1,983	1,983
(Psd.) R^2	0.100	0.075	0.075	0.058	0.093	0.057	0.062	0.054

Notes: See Table 3.6. Symbols * and ** denote significance at the 5% and 1% levels.

Table 3.10 Innovative turnover: Portugal – average and quantile behaviour

	With in–house research intensity				With total research intensity			
	OLS	Quantile regression			OLS	Quantile regression		
		0.25	0.50	0.75		0.25	0.50	0.75
	(i)	(ii)	(iii)	(iv)	(i)	(ii)	(iii)	(iv)
Constant	1.207	1.309	1.631	2.531	1.449	1.750	1.073	2.412
	(1.58)	(0.96)	(1.64)	(2.43)**	(2.53)**	(1.73)	(1.48)	(3.18)**
IDPAV1	–0.260	–0.363	–0.248	–0.603	–0.117	–0.041	–0.099	–0.291
	(1.20)	(0.88)	(0.95)	(2.37)**	(0.71)	(0.13)	(0.51)	(1.58)
IDPAV2	–0.244	–0.289	–0.241	–0.323	–0.157	–0.184	–0.071	–0.236
	(1.05)	(0.63)	(0.88)	(1.30)	(0.88)	(0.48)	(0.32)	(1.24)
IDPAV3	–0.247	–0.132	–0.299	–0.396	–0.342	–0.071	–0.398	–0.453
	(0.89)	(0.28)	(0.99)	(1.16)	(1.54)	(0.18)	(1.53)	(1.89)
IDPAV4	–0.841	–0.894	–0.776	–0.950	–0.647	–0.377	–0.623	–1.015
	(3.24)**	(1.96)*	(2.67)**	(3.01)**	(3.14)**	(0.96)	(2.83)**	(4.09)**
LTRDINS	0.023	0.054	0.063	0.043	0.086	0.098	0.105	0.117
	(0.43)	(0.73)	(1.10)	(0.53)	(2.23)**	(1.17)	(2.32)**	(2.94)**
INKS	0.255	–0.118	0.231	0.492	0.322	0.071	0.370	0.467
	(0.76)	(0.18)	(0.53)	(1.28)	(1.19)	(0.13)	(1.15)	(1.62)
COOP	0.065	–0.057	0.270	0.230	0.165	0.171	0.147	0.207
	(0.41)	(0.18)	(1.50)	(1.22)	(1.31)	(0.65)	(0.95)	(1.33)
APPR	0.250	0.097	0.184	0.083	0.286	0.297	0.270	0.147
	(1.51)	(0.31)	(0.98)	(0.35)	(2.15)**	(1.14)	(1.68)*	(1.00)
CKNOW	–0.393	–0.657	–0.712	–0.576	–0.242	–0.032	–0.354	–0.262
	(1.51)	(1.46)	(1.55)	(1.41)	(1.16)	(0.09)	(1.52)	(0.93)
PCS	0.133	0.466	–0.069	–0.059	0.109	0.311	0.011	–0.055
	(0.83)	(1.51)	(0.35)	(0.31)	(0.96)	(1.41)	(0.08)	(0.48)
LHEI	–0.025	–0.017	–0.077	–0.063	0.027	–0.039	–0.034	–0.049
	(0.96)	(0.47)	(1.51)	(1.56)	(1.50)	(1.20)	(1.47)	(1.56)
LEXI	0.016	0.021	–0.033	0.030	0.032	0.038	–0.008	0.035
	(0.54)	(0.44)	(0.85)	(0.83)	(1.44)	(0.96)	(0.28)	(1.32)
LSIZE	–0.195	–0.211	–0.240	–0.237	–0.212	–0.290	–0.184	–0.221
	(3.45)**	(2.25)*	(3.49)**	(3.06)**	(5.18)**	(4.02)**	(3.50)**	(4.31)**
GROUP	0.111	0.125	0.069	0.000	0.173	0.148	0.130	0.142
	(0.76)	(0.52)	(0.39)	(0.00)	(1.51)	(0.64)	(0.35)	(1.05)
Obs.	282	282	282	282	479	479	479	479
(Psd.) R^2	0.089	0.070	0.108	0.136	0.106	0.072	0.082	0.107

Notes: See Table 3.6. Symbols * and ** denote significance at the 5% and 1% levels.

This confirms the robustness of our findings in this respect, and lends further support to our interpretation, in that the innovation process is becoming increasingly open among strong research performers.

Tables 3.6 to 3.10 present the country-based estimates of the innovative turnover equation. The left-hand side of the tables presents the estimates of the equation including in-house research intensity, LRDINS, as an explanatory variable on the samples of firms who report both in-house research activity and being product-innovative, whereas the right-hand side presents the estimates of the equations including total research intensity, LTRDINS, on the somewhat larger samples of firms who report innovation expenditure and being product-innovative.

Consider the left-hand side of the tables. Columns (i) present the estimates by OLS. DHW tests confirmed, for all countries, the exogeneity of the level of in-house research intensity, LRDINS, in the innovative turnover equation, implying that the variable can be used as such rather than in instrumented form.

First consider the industry dummies: to the extent that they are found statistically significant, the parameter estimates are negative. This is the case of IDPAV1 for Belgium, IDPAV2 for Belgium and Norway, and IDPAV4 for Belgium, Norway, Spain and Portugal. This suggests that, when controlling for the impact of in-house research intensity, the innovative output is lower in the industries concerned than in the information-intensive industries. The evidence that this is most predominantly the case in science-based industries can be explained by the fact that the development of new products in industries such as biotech and pharmaceuticals is characterised by long and variable gestation periods (Bruneel et al., 2007).

The level of in-house research intensity, LRDINS, is found to be highly statistically significant in four of the five countries under consideration (Belgium, Germany, Norway and Spain), confirming that this is a key variable in explaining innovative turnover. The confirmation by the DHW test that LRDINS is exogenous in the innovative turnover equation suggests, at the same time, that the estimates are not suffering an upward endogeneity bias, due to feedback from the level of innovative turnover, LITURN, over the growth of turnover to the level of in-house research intensity, LRDINS. Further confirmation was sought by experimenting with an extended model that also includes an equation explaining the growth of turnover (between 1998 and 2000) and by including the latter variable as an additional explanatory variable in the research intensity equation. As was the case in earlier studies with respect to Belgium (Spithoven et al., 2010), the results always indicate that whereas the level of innovative turnover has a significant impact on the growth of turnover, the latter has, however, no impact on the level of research intensity. This would suggest that the feedback loop breaks down here and that the model is recursive in nature.

In the case of the remaining country, Portugal, in-house research intensity is however found to be non-significant in the innovative turnover equation. As seen in Table A3.2 in the Appendix, not only is the number of firms reporting in-house R&D relatively modest in Portugal, but, among these, the mean value of the level of in-house research intensity is much lower than in the other countries under consideration. It would therefore appear that in-house research intensity is less relevant in explaining product innovations in a catch-up economy such as Portugal, which is still operating well below the technological frontier in most of its productive activities.

Apart from this, there is also evidence of a positive net impact of process innovations, PCS, on product innovativeness in three of the five countries under consideration (Belgium, Germany and Spain). This would suggest that the process innovations concerned are production re-organisations that do not require formal in-house research efforts. It should, however, be noted that, due to the binary nature of this variable, its impact is likely to be underestimated. This is especially likely to be true for a country such as Portugal, where, as we see in Table A3.2 in the Appendix, no fewer than 75% of the research-active firms reported process innovations. In this case there is insufficient variation in this variable between firms and, therefore, explanatory power in the cross-section.

There is, however, no evidence that the knowledge-related variables or human capital directly affect product innovativeness, when controlling for in-house research intensity. This clearly suggests that their influence on the level of product innovation is indirect, working through research intensity. Incoming spillovers, research cooperation or the ideas of workers do not appear to generate readily usable knowledge: they need first to be absorbed in the knowledge base of the firm through additional R&D efforts (Cohen and Levinthal 1989, 1990; Chesbrough and Crowther, 2006). In the case of export orientation only one out of five countries (Norway) shows evidence of a direct positive impact on product innovativeness. Also for this variable, the impact is mainly indirect, through research intensity, further stressing the importance of in-house R&D efforts.

As far as the size-related variables are concerned, the impact of firm size, LSIZE, is again systematically negative and significantly so in four of the five countries under consideration (Belgium, Germany, Norway and Portugal). In the case of Germany, there is again, in addition, a negative impact of group membership, GROUP. These results would imply a decrease in efficiency of innovation process with firm size. This can be explained by the fact that among the in-house research-active product innovators the hitherto relatively small, new technology-based firms tend to be more dynamic. They are, therefore, prepared not only to spend relatively more on in-house R&D in relation to their size, but to do so in a more risky manner. They are able, as a result, to generate a higher number of innovations per unit of R&D spending on average (see Cohen 1995, for a review of arguments in this sense). It should be noted that the CIS3 innovation survey contains no information on the age of firms, so this topic needs further investigation.

These conclusions were again confirmed, by and large, by the ML estimates for the corresponding generalised tobit model that were not presented.

Columns (ii), (iii) and (iv) of the left-hand side of Tables 3.6 to 3.10 present the corresponding quantile regression estimates. As in the case of the average estimates by OLS in columns (i), there is evidence, as far as the industry dummies are concerned, of a negative impact of IDPAV1 for Belgium and of IDPAV2 for Belgium and Norway. This negative influence always seems most pronounced in the lower quantiles, suggesting that among weak innovators in particular, those firms active in sectors with a less technology-intensive trajectory perform poorly in the countries concerned. As with the average estimates there is also evidence that IDPAV4 has a negative impact in the case of Belgium, Norway, Spain and Portugal, but there is no clear quantile-by-quantile pattern here.

As with the average estimates in columns (i), the level of in-house research intensity, LRDINS, is the key explanatory variable for most of the countries under consideration (Belgium, Germany, Norway and Spain). In two of these countries, Belgium and Norway, its impact is seen to increase quantile by quantile, whereas it remains fairly constant in the others. This suggests that, if anything, strong output innovators are able to organise their R&D processes more efficiently and, in so doing, achieve higher returns on their R&D investments. In the technologically less advanced economy, Portugal, the lack of significance of in-house research intensity, found in the average estimates in column (i), is confirmed in all quantiles. This would suggest that in-house research intensity is still less relevant, even among the strongest product innovators, in explaining product innovations in a catch-up economy such as Portugal.

The evidence of a significant influence of process innovation, PCS, in the average estimates in three of the five countries (Belgium, Germany and Spain) is confirmed in these countries in nearly all quantiles.

Apart from the odd exception, none of the knowledge-related variables, INKS, COOP, APPR and CKNOW, or labour qualification, LHEI, seem to have a direct impact on innovative output. Rather, as with the average estimates in columns (i), their influence on innovative output works indirectly, through in-house research intensity. Again, there is only one country, Norway, in which we see evidence of a net direct impact of export orientation. This appears to be due to the lower quantiles, suggesting that, to the extent that competitive pressure directly affects innovative performance, it only does so in the case of the weaker innovators.

Finally, as for the average estimates, there is evidence that firm size, LSIZE, has a negative impact in four of the five countries under consideration (Belgium, Germany, Norway and Portugal) and this is the case in most quantiles. Now, too, the negative impact of firm size is re-enforced by a negative influence of group membership, GROUP, in the case of Germany in all quantiles.

Next, consider the estimates of the innovative turnover equation including total research intensity, LTRDINS, as an explanatory variable, presented on the right-hand side of Tables 3.6 to 3.10. Consider first the estimates by OLS in columns (i) and compare these with the estimates of the innovative turnover equation including in-house research intensity, LRDINS, in columns (i) in the left-hand side. The Pavitt industry dummies are left largely unaltered. The same holds, it seems, for the parameter estimates and statistical significance of total research intensity, LTRDINS, in the four countries in which in-house research intensity, LRDINS, was found to be significant (Belgium, Germany, Norway and Spain). In the case of the remaining country, Portugal, there is, however, a substantial difference in the sense that total research intensity, LTRDINS, is found to be highly significant, whereas in-house research intensity, LRDINS, was non-significant. Table A3.2 in the Appendix shows that in Portugal the number of firms reporting total R&D (innovation expenditure) is more than twice as large as those reporting in-house R&D. The mean value of total research intensity here is also nearly as high as in the other countries, despite a much lower level of in-house research intensity. It would therefore appear that in a catch-up economy, such as Portugal, firms devote a relatively larger part of their total innovation expenditure to components other than in-house R&D – particularly the purchase of new machinery imported from technologically more advanced countries. And it is by doing this that they succeed in bringing about product innovations.

When compared to columns (i) on the left-hand side, the parameter estimates and the significance of the process innovation variable, PCS, clearly decreases in the countries where it was found significant. This suggests that part of the process innovations in question involve the introduction of new, more advanced machinery, captured by the total research intensity variable, LTRDINS. This appears to be particularly the case in Spain, where PCS is now completely non-significant. The much stronger performance of total research intensity, LTRDINS, than in-house research intensity, LRDINS, in the case of Portugal can also be interpreted as an indication of the relevance of this type of process innovation in that country. But, for the reasons given above, the PCS variable could not provide useful information in this case.

Apart from this, the parameter estimates and significance of the remaining labour qualification, export orientation and size-related variables, LHEI, LEXI, LSIZE and GROUP, are much alike in columns (i) on the right-hand side and left-hand side of the tables. In the case of Germany, higher education intensity, LHEI, which nearly had a statistically significant positive net impact, now has it, suggesting that in this case workers' qualifications may have a direct impact on product innovativeness. And in the case of Spain, the negative impact of firm size, LSIZE, is now somewhat more pronounced and becomes statistically significant, as it is elsewhere.

Consider, finally, the quantile regression estimates on the right-hand side of Tables 3.6 to 3.10. Comparing columns (ii), (iii) and (iv) to those in the left-hand side of the tables, shows that the impact of the industry dummies is very similar. In the case of total research intensity, LTRDINS, there is, again, evidence of a quantile-by-quantile increase in significance for Belgium and Norway, as was the case with in-house research intensity, LRDINS. But now this is also the case for Portugal, where in-house research intensity was found to be non-significant. This would tend to reinforce somewhat the previous conclusion that there is, if anything, evidence of a gain in efficiency of the research process among the strong product innovators.

Finally, in the case of the remaining variables, the quantile regression parameter estimates are, again, quite similar to what they were on the left-hand side of the tables, when in-house research intensity was included. The slight differences, noted in the case of the average estimates in columns (i), are also confirmed.

3.5 DISCUSSION AND CONCLUSIONS

A model explaining the levels of research intensity and innovative turnover was estimated on a country-basis on harmonised firm-level data from the Third Community Innovation Survey for five European countries: Belgium, Germany, Norway, Spain and Portugal.

Estimation by OLS provides an insight into the average behaviour of firms. The results show that in the case of in-house research in particular, general technological environment matters: firms active in hi-tech industries have, on average, a higher level of research intensity. Further, irrespective of the measure of research intensity under consideration, knowledge-related variables, such as incoming knowledge spillovers, research cooperation and appropriability, always exercise a major influence on R&D investment. This confirms that the absorption of external knowledge requires substantial further research efforts by firms and that the capacity to appropriate knowledge with respect to non-partners increases incentives to undertake R&D. Higher qualification of workers and stronger competitive pressures, measured by export orientation, also positively affect research efforts. Finally, the impact of firm size on the level of research intensity is always negative, showing that there is among research-active firms in the countries under consideration a substantial number of young dynamic technology-based firms, pursuing ambitious innovation strategies.

In four of the five countries under consideration (Belgium, Germany, Norway and Spain) the level of research intensity, regardless of how it is measured, is itself the main variable explaining innovative turnover. Process innovations also matter. The other knowledge and human capital-related variables are always found to be non-significant in the equation on innovative turnover, suggesting that their influence on product innovation is indirect, working through research intensity. This emphasises the necessity of knowledge absorption firstly, before it can become usable in the innovation process.

These results are confirmed in the case of the remaining country, Portugal, except that here only total research intensity appears to have a significant impact on innovative turnover, contrary to in-house research intensity. This shows that, in this technologically somewhat less-developed catch-up economy, the majority of firms still devote a relatively greater part of their total innovation expenditure to components other than in-house R&D, such as the purchase of imported new machinery, than is the case in the technologically more-advanced economies. And it is mainly through such process innovations that they bring about product innovations. The relatively greater impact of in-house research intensity in the other countries under consideration emphasises the increasing importance of in-house R&D as we approach or reach the technological frontier.

Quantile regression estimates show that the average behaviour of firms conceals substantial differences between strong and weak innovators. Firms active in information-intensive, specialised supplier and science-based sectors are over-represented among strong in-house R&D investors. Although there are differences in the precise manner in which this occurs, in the case of both in-house and total research intensity, the joint relative impact of incoming knowledge spillovers and research cooperation tends to increase quantile-by-quantile in relation to that of appropriability. Furthermore, if anything, the impact of research intensity on innovative turnover also tends to increase.

When considered as a whole, these results appear to confirm the assumption that there is an increasing openness of the innovation process, especially among strong innovators. The internalisation of external knowledge generated through incoming knowledge spillovers and research cooperation has, clearly, become one of the major purposes of the research activities of innovative firms. Although appropriability does, on average, continue to provide an incentive to undertake research, this is apparently relatively less important in the case of the strongest R&D performers. This would suggest that it is here that we can expect to find most of the firms that are prepared to market excess knowledge in disembodied form. The greater impact of research intensity on innovative turnover among strong product innovators would tend to confirm that these have the most efficient R&D processes.

APPENDIX

Table A3.1 presents a description of the manner in which the variables were constructed from the raw data of the CIS3 questionnaire and Table A3.2 presents the mean values of the variables with respect to the total and in-house research-active samples, and the corresponding Pavitt industry groups.

Table A3.1 Description of the variables

Variable	Definition
Log of in-house research intensity (LRDINS)	Natural log of the ratio of intramural R&D expenditure over turnover in the year 2000
Log of total research intensity (LTRDINS)	Natural log of the ratio of total innovation expenditure over turnover in the year 2000
Log of innovative turnover (LITURN)	Natural log of the % of new or significantly improved products (goods and services) in turnover in the year 2000
Incoming knowledge spillovers (INKS)	Sum of scores of the importance attached to information from suppliers, clients, competitors, universities, government, professional conferences, exhibitions (rescaled between 0 and 1)
Research cooperation (COOP)	Cooperation = 1 if the firm collaborates either with suppliers, clients, competitors, research labs, universities or government
Appropriability (APPR)	Average score of the effectiveness of strategic protection through secrecy, complexity of design and lead-time advantage
Complexity of knowledge (CKNOW)	Ratio of the sum of the scores of the importance attached to information from universities, government and professional conferences over the sum of the scores of all sources of information
Log of higher education intensity (LHEI)	Natural log of the number of employees with higher education over the level of turnover in the year 2000
Log of export intensity (LEXI)	Natural log of the proportion of exports in turnover in the year 1998
Log of size (LSIZE)	Natural log of the level of turnover in the year 1998
Group affiliation (GROUP)	Group affiliation = 1 if the firm is a member of a group
Process innovation (PCS)	Process innovation = 1 if the firm has introduced new or significantly improved production processes during the period 1998–2000
IDPAV1	IDPAV1 = 1 if the firm is classified in the 'Resource-intensive' sectors (mining, food, beverages and tobacco, textiles, wearing apparel, leather products, wood and furniture, recycling, wholesale)

Table A3.1 (continued)

Variable	Definition
IDPAV2	IDPAV2 = 1 if the firm is classified in the 'Scale-intensive' sectors (paper and printing, rubber and plastics, non-metallic minerals, ferrous metals, non-ferrous metals, fabricated metal products, motor vehicles, other transport equipment)
IDPAV3	IDPAV3 = 1 if the firm is classified in the 'Specialised supplier' sectors (non-electrical machinery, electronics and communication, electrical machinery)
IDPAV4	IDPAV4 = 1 if the firm is classified in the 'Science-based' sectors (chemicals, pharmaceuticals, computers and office machinery, scientific instruments)
IDPAV5	IDPAV5 = 1 if the firm is classified in the 'Information-intensive' sectors (financial intermediation, telecommunication, transport services, business services, computer and related service activities, research and development, technical engineering)

Table A3.2 Mean values of the variables

Variable	Total research-active#	In-house research-active*	*Of which:* Res.-intensive	Scale-intensive	Special. supplier	Science-based	Infor.-intensive
Belgium							
Obs.	697	492	96	114	102	68	112
LTRDINS	-4.068						
LRDINS		-4.546	-5.497	-5.400	-3.927	-4.337	-3.552
LITURN	-1.921	-1.912	-2.354	-2.067	-1.637	-2.265	-1.492
INKS	0.389	0.419	0.379	0.424	0.476	0.410	0.402
COOP	0.312	0.380	0.312	0.403	0.402	0.456	0.348
APPR	0.346	0.415	0.344	0.427	0.474	0.426	0.405
CKNOW	0.302	0.328	0.276	0.304	0.360	0.373	0.340
PCS	0.641	0.669	0.698	0.754	0.618	0.676	0.598
LHEI	-14.358	-14.190	-15.102	-14.721	-14.079	-14.277	-12.841
LEXI	-2.326	-2.024	-1.896	-1.571	-1.421	-1.834	-3.258
LSIZE	16.079	16.147	16.460	16.520	15.970	17.160	15.044
GROUP	0.616	0.634	0.510	0.781	0.569	0.691	0.616
Germany							
Obs.	1,749	1,117	154	296	312	77	278
LTRDINS	-3.907						
LRDINS		-4.560	-5.218	-4.900	-4.022	-4.233	-4.527
LITURN	-1.597	-1.525	-1.702	-1.532	-1.380	-1.717	-1.551
INKS	0.440	0.470	0.447	0.456	0.502	0.482	0.459
COOP	0.278	0.355	0.260	0.321	0.413	0.363	0.378
APPR	0.363	0.447	0.409	0.422	0.503	0.467	0.426
CKNOW	0.333	0.358	0.305	0.348	0.365	0.377	0.384
PCS	0.604	0.636	0.630	0.699	0.596	0.636	0.619
LHEI	-14.171	-13.930	-14.914	-14.452	-13.576	-14.098	-13.182
LEXI	-3.921	-3.274	-3.688	-2.826	-1.774	-1.793	-5.614
LSIZE	16.668	16.954	16.834	17.131	17.007	17.732	16.558
GROUP	0.453	0.471	0.416	0.476	0.477	0.623	0.446

Notes: # In the case of LITURN this relates to total research-active and product-innovative firms. * In the case of LITURN this relates to in-house research-active and product-innovative firms.

Table A3.2 (continued)

Variable	Total research-active#	In-house research-active*	*Of which:* Res.-intensive	Scale-intensive	Special. supplier	Science-based	Infor.-intensive
Norway							
Obs.	1,390	962	247	240	177	43	255
LTRDINS	−3.764						
LRDINS		−3.957	−4.668	−4.706	−3.272	−3.554	−3.109
LITURN		−1.748	−1.877	−1.997	−1.436	−2.214	−1.559
INKS	0.427	0.454	0.475	0.444	0.459	0.551	0.423
COOP	0.460	0.527	0.538	0.529	0.542	0.720	0.470
APPR	0.343	0.418	0.400	0.372	0.516	0.566	0.384
CKNOW	0.329	0.355	0.375	0.358	0.313	0.448	0.348
PCS	0.663	0.643	0.647	0.733	0.666	0.790	0.513
LHEI	−13.289	−13.269	−13.937	−13.666	−13.097	−13.706	−12.293
LEXI	−4.018	−3.483	−3.815	−3.193	−1.404	−1.041	−5.287
LSIZE	16.042	16.068	16.338	16.406	15.955	16.963	15.417
GROUP	0.725	0.737	0.781	0.717	0.734	0.837	0.698
Spain							
Obs.	3,079	1,773	434	499	338	201	301
LTRDINS	−3.881						
LRDINS		−4.532	−5.139	−4.844	−4.055	−4.314	−3.821
LITURN	−1.602	−1.577	−1.642	−1.665	−1.290	−1.927	−1.441
INKS	0.386	0.429	0.411	0.414	0.430	0.467	0.456
COOP	0.214	0.310	0.194	0.327	0.287	0.468	0.375
APPR	0.274	0.374	0.369	0.382	0.425	0.437	0.268
CKNOW	0.304	0.353	0.344	0.342	0.309	0.429	0.384
PCS	0.726	0.707	0.737	0.780	0.633	0.667	0.654
LHEI	−12.968	−13.059	−13.368	−13.297	−13.111	−13.405	−11.928
LEXI	−3.975	−3.382	−3.342	−2.872	−2.491	−2.705	−5.737
LSIZE	15.900	16.288	16.320	16.581	16.048	16.934	15.589
GROUP	0.428	0.510	0.410	0.565	0.434	0.662	0.548

Notes: # In the case of LITURN this relates to total research-active and product-innovative firms. * In the case of LITURN this relates to in-house research-active and product-innovative firms.

Table A3.2 (continued)

Variable	Total research–active#	In–house research–active*	*Of which*: Res.–intensive	Scale–intensive	Special. supplier	Science–based	Infor.–intensive
Portugal							
Obs.	746	368	112	90	45	34	87
LTRDINS	–3.932						
LRDINS		–5.503	–5.843	–5.707	–5.110	–4.992	–5.256
LITURN		–1.739	–1.779	–1.611	–1.683	–2.434	–1.539
INKS	0.349	0.392	0.402	0.409	0.436	0.421	0.329
COOP	0.273	0.399	0.419	0.322	0.444	0.529	0.379
APPR	0.216	0.309	0.318	0.289	0.385	0.294	0.283
CKNOW	0.258	0.316	0.318	0.273	0.359	0.410	0.299
PCS	0.749	0.764	0.714	0.800	0.755	0.676	0.827
LHEI	–12.371	–12.930	–13.273	–13.261	–13.159	–12.812	–12.075
LEXI	–4.091	–3.584	–2.631	–3.038	–1.979	–3.785	–6.129
LSIZE	15.370	15.841	16.117	15.364	15.868	16.081	15.869
GROUP	0.400	0.503	0.473	0.344	0.467	0.676	0.655

Notes: # In the case of LITURN this relates to total research-active and product-innovative firms. * In the case of LITURN this relates to in-house research-active and product-innovative firms.

PART TWO

Open innovation and public funding practices

4. Public intervention and innovative networking: firm-level evidence on the opening-up of the innovation process

4.1 INTRODUCTION

The main argument in favour of public intervention in R&D and innovation processes lies in the existence of market failures (Arrow, 1962). Due to these failures unintended knowledge spillovers occur and the appropriability of benefits from the innovation hinders further activity (Martin and Scott, 2000). Enterprises are then assumed to avoid R&D investment through lack of necessary incentives (Hall, 2002). Hence, supporting innovation is a major ingredient of science and technology policy, which ultimately aims to enhance entrepreneurial performance and so stimulate economic growth.

Policy-makers have at their disposal a large number of policy instruments to enhance R&D and innovation. These range from passive instruments like tax policy through fiscal incentives (Hall and Van Reenen, 2000; Klette et al., 2000; Guellec and Van Pottelsberghe, 2003) and subsidies (Klette et al., 2000; Guellec and Van Pottelsberghe, 2003) to more active involvement by performing R&D in public research organisations designed to create spillovers (Martin and Scott, 2000; OECD, 2002b). This chapter focuses on subsidies. These relate directly to the innovation projects the enterprises perform themselves (Busom, 2000).

Although the rationale behind the public funding of innovation is theoretically well grounded, the empirical literature is ambiguous in its assessment of the benefits of public funding of R&D and innovative activity (David et al., 2000, Klette et al., 2000) and there is little consensus on the effectiveness of R&D subsidies (Jaffe, 2002; Hall, 2002). This lack of consensus can be attributed to several factors. First, subsidies are non-random processes among firms, requiring resources and grant competition beforehand (Busom, 2000). Therefore, methodological problems can arise when estimating the effect of innovation policies. Among these, there is a potential endogeneity problem since public aid is not independent of other variables, and there is a risk of sample selection bias due to the existence of a distribution process at the end of which a difference arises between the firms that benefit and those that do not.

Second, much of the existing literature on the effects of public funding largely revolves around the (mostly linear) relationship between input and output identified by Kline and Rosenberg (1986), Tidd et al. (1997), Crépon et al. (1998), Klomp and Van Leeuwen (2001), and Griffith et al. (2006). The new ideas on open innovation, however, are well acquainted with the role of the business model in externalising innovation and making a profit from innovations not confined to the core business of the firm (Chesbrough, 2003a, Chesbrough and Rosenbloom, 2002; Chesbrough and Crowther, 2006; Huston and Sakkab, 2006). In this respect, many public authorities have addressed the risk taken by enterprises to engage in innovation or stimulate specific forms of innovative activity such as research cooperation by providing public funding (European Commission, 2005; Polt and Psarra, 2006).

Finally, in order to understand how R&D programs differ and how differences can be related to private R&D spending, David et al. (2000) recommended that evaluation studies use taxonomies. However, since then, little progress has been made in accounting for the impact that subsidies allocated from different R&D programs have on recipient firms (see, however, the case for Spain by Blanes and Busom, 2004).

This chapter takes these shortcomings on board and investigates the impact of public intervention, through funding programmes primarily aimed at stimulating innovative networks, on research intensity and innovative turnover at firm level. It starts from the ideas of open innovation and reverts to knowledge sourcing, engaging in cooperative arrangements and the use of benefits from innovative activities. There are two essential evaluation issues: the level of research intensity an enterprise attains as an input measure, and the level of turnover due to product innovation as an output measure.

The study relies on empirical evidence from the Third European Community Innovation Survey (CIS3) for Belgium. Public funding stems from different channels leading to variations in policy objectives. Three broad categories are examined: regional; national and European funding. Belgium is an interesting case because its science and technology policy is highly regionalised to accommodate the particular differences in institutional setting and supportive policy measures which affect entrepreneurial innovative activities. Most of the regional funding of innovation is tailored to the demands and needs especially of small and medium-sized enterprises. Funding by federal government is restricted to particular technical domains which focus on the creation of innovative networks (Capron et al., 2000; Nauwelaers, 2001). EU-funding is also directed towards the instalment of research networks rather than individual companies. Each source of public funding rests on programmes involving different agencies and imposes varying application criteria, thereby drawing attention to a possible selectivity bias in the receipt of public funding.

The chapter also highlights why the inconclusive evidence on the impact of public funding on innovation is due to the emphasis placed on the average behaviour of firms. The heterogeneity of firms in their response to receiving public funding is taken into account by discriminating between dynamic and lagging innovators by means of quantile regression analysis (Buchinsky, 1995, 1998; Ebersberger et al., 2006; Nahm, 2001; Spithoven et al., 2010).

The chapter is structured as follows: Section 4.2 briefly reviews the opening up of the innovation process, the sources of public funding, the main characteristics of recipients of public funding, and formulates the research question. Section 4.3 discusses the data, the measurement of the key variables, and specifies the structural model and the estimation strategy. In Section 4.4 the econometric evidence is presented. Section 4.5 concludes with policy implications and directions for further research.

4.2 PUBLIC FUNDING AND THE PROMOTION OF INNOVATIVE NETWORKS

Public authorities have various ways open to them to promote R&D and innovative activities. These ways range from generic measures such as organising higher education; establishing large-scale public research organisations aimed at fundamental research into future needs for industry; or providing research infrastructure such as incubators and science parks. More recently, the need to evaluate the societal benefits of public investment and the budgetary constraints in many countries has increasingly accompanied public outlays on projects. This shift coincides with the ideas of the knowledge paradox, according to which the abundance of knowledge is insufficiently commercialised. The ideas on open innovation also detect this shift (Chesbrough, 2003a). Partial public responses to this favour development projects over research programmes or stimulating the 'translation' of basic knowledge into marketable products by inducing networking activities in general and industry-science relations in particular (OECD, 2002b).

4.2.1 Open Innovation and Opportunities for Public Intervention

Besides the handling of intellectual property rights (Chesbrough, 2003b; Kline, 2003) and the provision of venture capital and spin-off creation (Chesbrough and Rosenbloom, 2002; Chesbrough, 2003c), the stimulus of cooperative arrangements with external partners is central in the debate on public intervention and open innovation (European Commission, 2005).

Cooperation on innovation involves an 'active participation in joint innovation projects with other organisations. These may either be other enterprises or non-commercial institutions. The partners need not derive immediate commercial benefit from the venture. Pure contracting out of work, where there is no active collaboration, is not regarded as cooperation. Cooperation is distinct from open information sources and acquisition of knowledge and technology in that all parties take an active part in the work' (OECD, 2005: 79–80).

Research cooperation for innovative activities has enjoyed a great deal of attention from scholars (for a review of the literature, see Fritsch and Lukas, 2001; Tether, 2002; Miotti and Sachwald, 2003) and public organisations (OECD, 2001). Two elements of research cooperation have been the subject of empirical study. First, in-house research of firms is considered a prerequisite for successful research cooperation because of the creation of so-called absorptive capacity (Cohen and Levinthal, 1990; Veugelers, 1997; Miotti and Sachwald, 2003; Negassi, 2004). Once a certain threshold of qualified knowledge workers has been established and they are well able to cognitively understand research efforts and step willingly and knowingly into joint research ventures, the impact of cooperation on research intensity is found to be positive, reflecting the need to internalise the knowledge thus created through additional in-house research efforts (Veugelers, 1997). Next, research cooperation is pursued to enhance innovative performance. This aspect has received less attention in the literature (Becker and Dietz, 2004; Negassi, 2004). This chapter picks up on both points because, due to the paradigm of open innovation, this particular exponent of knowledge-related interaction is being further explored in this new context (Laursen and Salter, 2006).

Several authors acknowledge the impact of the type and/or geographical location of the partner (Fritsch and Lukas, 2001; Tether, 2002; Miotti and Sachwald, 2003; Spithoven and Teirlinck, 2006). For the type of partner, two distinct categories are considered: inter-firm research cooperation with clients, suppliers and competitors (von Hippel, 1988; Fischer and Varga, 2002), and research cooperation with other types of organisations such as universities and government research centres – often described as industry–science relations (OECD, 2002b; Monjon and Waelbroeck, 2003; Laursen and Salter, 2004). The geographical location of the partner is related to the size of the domestic market. Innovative enterprises in small European countries tend to rely more on collaboration for innovation with partners outside the home country (Spithoven and Teirlinck, 2006).

Considerable attention has been paid to cooperative arrangements for innovation; either in the realm of inter-firm relations (see the special issue on Strategic Networks in *Strategic Management Journal* of March 2000) or in the cases of national innovation systems (OECD, 2001). Fewer papers bring in additional information on complementary external knowledge relations – incoming knowledge spillovers, appropriability conditions, and complexity of research – and firm characteristics such as sector, size, human capital, competitive pressure, etc. Due to the debates on the impact of open innovation, renewed attention is turned to the role of external knowledge relations (Laursen and Salter, 2006) and the distribution of knowledge (Coombs et al., 2003). Remarkably few studies have connected these insights to the issue of public funding aimed directly at cooperative arrangements; let alone allowed for a different type or location of external partners in their network (Tether, 2002; Polt 2001). Finally, since the literature on open innovation is to a large extent based on case studies (Chesbrough, 2003a; Chesbrough, 2003c; Chesbrough and Rosenbloom, 2002; Huston and Sakkab, 2006), it is acknowledged that not every firm adheres to the imperative to open up its innovation process (Gassmann, 2006). Thus, innovative behaviour may vary across firms as investigated by Spithoven et al. (2010) using quantile regression analysis.

The principal goal of public intervention is to bring about a mix of agents which share a common interest in performing R&D and/or to introduce innovations. To enforce this goal, the instrument of public funding is often used. But the funds publicly available to firms stem from various sources which are briefly reviewed in the next section.

4.2.2 Sources of Public Funding

Sources of public funding are to a major extent country-specific. This calls for a short positioning description of the way in which the national innovation system is organised in relation to the policy instruments and programmes designed to promote R&D and innovation (Capron et al., 2000; Nauwelaers, 2001). Four specific sources of public funding to be obtained by enterprises in Belgium are taken into account: regional government; federal government; the EU Framework Programmes; and other EU funding.

Regional government funding is by far the most frequent type of funding in Belgium (Nauwelaers, 2001). As the political responsibility for innovation policy is decentralised, administrations and public agencies in the Flemish, Walloon and Brussels regions deal with R&D subsidies. A reason for its success is that regional funding is largely inspired by a bottom-up approach and therefore tailored to the demands of the enterprises. Decentralisation makes it possible for regions to accommodate the specificities of the enterprises located on their territory (Capron et al., 2000). The general idea behind most regionally-funded programmes is to foster networking and cooperative activities, aimed particularly at the use of scientific knowledge, and therefore intended to stimulate research cooperation. It is not, therefore, aimed at enriching the R&D portfolio of the firms or lowering the threshold for innovative projects with higher risk factors.

The federal funding of R&D is in relative decline, due to state reforms which transferred in 1994 the bulk of science and technology policy to the regions (Capron et al., 2000), from 44% in total public R&D budgets in 1990 to 33% in 2000 (Nauwelaers, 2001). Federal funding is aimed at the stimulation of spillover effects. Federal R&D activities are restricted to large-scale space programs greatly benefiting private enterprises: 1 Euro invested in space research induces a 3-Euro return in industrial terms (Nauwelaers, 2001). There are also several policy initiatives related to other specific domains such as health and defence. Other federal activities pertain to fiscal measures which help lower the labour cost of researchers. Finally, a few historic remnants, such as the body of 'collective research centres', are still in place. These policy instruments make the federal government an actor in public funding, albeit moderately. Since federal government funding is also directed at infrastructural investments (for example, an advanced communication infrastructure: Belnet) and participation in international research organisations, the expected impact of this funding works through the generation of spillover effects.

European public funding, on the other hand, flows largely top-down to specific themes. The EU Framework Programmes, introduced in 1984, are the most important mechanism of public funding designed explicitly to stimulate research and technological development (European Commission, 2000; Polt and Psarra, 2006). One of the three recognised added values of funding through the EU Framework Programme can be found in pooling and leveraging resources. EU funding impacts on the scale and scope of research efforts and is able to bring together what might otherwise be fragmented R&D resources to achieve a 'critical mass' not attainable at national level. Polt and Psarra (2006) warn against exaggeration on this front: funding under the Fifth Framework Programme is only a small percentage as compared to the R&D expenditures by the participating partners. Nevertheless, the budgets of the successive Framework Programmes keep growing. The Fourth EU Framework Programme (FP) had a budget of 13,215 million ECU and spanned 1994–1998. Topics in the programme covered: ICT; industrial technologies such as new materials; environment; biotech; energy; nuclear research; transport; socio-economic research cooperation with Third countries and international organisations; and mobility of researchers such as in the Marie Curie programme. The average FP4 share-cost project has a budget of €1.160 million and on average involves 7 participants coming from 4.2 Member States, often combining universities, public research centres, SMEs and large enterprises. In the Fifth EU Framework Programme (1998–2002) the budget was €14.960 million. This Fifth Framework Programme addressed the major socio-economic challenges the EU thought itself to be facing at the time. It developed a multi-themed structure covering topics such as the quality of life; information society; sustainable growth; and energy/environment. These topics are underpinned by stressing the international dimension of research; promotion of innovation in SMEs; and improving human research potential (European Commission, 2000). The average project amounts to €1.405 million and involves 6.2 participations from 3.7 different Member States (Muldur et al., 2006).

Polt and Psarra (2006) have identified the success of the projects under the Fifth Framework Programme: the projects are of high strategic importance, tightly linked to in-house R&D, and have taken place in the core technology areas of the firm. Moreover, EU intervention has a leverage effect on private investment. It leads to 'crowding-in' (that is, it stimulates firms to invest more of their own money in R&D than they would otherwise have done. It is estimated that a 1-Euro increase in public R&D investment induced 0.93 Euro of additional private sector investment (Muldur et al., 2006). Moreover, studies on additionality demonstrate that 57% of the projects are abandoned if EU funding under the Framework Programme is not forthcoming (Polt and Psarra, 2006). Also, Framework programmes proved important to enter into or, even more the case, sustain cooperative research in technology areas which are characterised by uncertainty and high risk. EU support for R&D encourages a particular type of research project, in which private companies can collaborate with foreign partners – of which 29% are with universities; 25% with (public) research centres; 39% with enterprises (30% SMEs and 9% large companies) and 7% other (Polt and Psarra, 2006) – on a scale not possible at national level, in projects tested for excellence, and gain valuable access to complementary skills and knowledge. It is therefore reasonable to conclude that the attractiveness of EU schemes induces firms to invest more of their own funds than they would under national funding programs. Large-scale European projects enable participants to access a much wider pool of firms in their own industry than would be possible at a purely national level. This mechanism offers clear advantages to enterprises compared with national-level schemes. It broadens the scope and complexity levels of the research, promotes linkages with in-house R&D, stimulates cross-border networking and allows for a division of work according to each participant's field of specialisation (Polt and Psarra, 2006).

In addition to these programmes, which have been called into existence to fund research and innovation, complementary policy initiatives have been taken at the European level (European Commission, 2001): European Cooperation in the field of Scientific and Technical research (COST) and, if the region is eligible, the European Regional Development Funds (ERDF) might be envisaged to support innovative activities. Each of these programmes has its own policy target. And although the focus of these programmes is on other fields of policy (for example, regional policy) there is a clear impact on the private research decisions made by firms. A notable example is the programme on innovative actions aimed at facilitating interregional exchanges and collective learning through comparison and dissemination of best practices.

4.2.3 Selectivity Bias: Firm Characteristics and Public Funding

Several firm characteristics are put forward as potentially influencing the probability of receiving public funding for research. There are different opinions on the matter of firm size. Some found small firms to be more likely to obtain R&D subsidies compared to large ones (Busom, 2000 for Spain); others discovered counter evidence (Wallsten, 2000 for the US). One of the main reasons for this is the neglect of the explicit aim of the public agency to fund R&D efforts, necessitating a description of the policy goals before using econometric techniques. The finding confirms the argument made by Jaffe (2002) that it is clear that all firms have an incentive to get access to subsidies in order to reduce the private cost of doing R&D. Further, public funding showed a greater effect on private spending of large firms in comparison with small ones (Carmichael, 1981). Other studies demonstrated larger complementarities in small firms than in large ones (Gonzalez et al., 2005; Lach, 2002). Theoretically, large firms have fewer problems of information asymmetry, difficulties in accessing financial markets (Hall, 2002) and are more able to deal with the economic risk of R&D projects. As such, public funding could substitute projects for others that would have been carried out even without funding and free rider behaviour could be expected to be higher for large firms (Wallsten, 2000).

Industry is an important variable in evaluating research subsidies because it controls for differences in technological opportunities and appropriability conditions between firms (Pavitt, 1984; Klette et al., 2000;

Chesbrough and Crowther, 2006). Also, the tendency to cooperate on research differs with the technology intensity of industry (Miotti and Sachwald, 2003; Gassmann, 2006).

As seen earlier, research cooperation with outside partners is often a prerequisite for public funding. Thus, public funding is a preferred instrument to create innovative networks for projects that have positive spillover effects and hence a 'weak' appropriability regime. Appropriability is, therefore, confined to strategic measures to reap the profits from the innovation, such as secrecy, complexity of design and lead time advantage.

4.2.4 Research Question

This chapter sets out to investigate the relation between public funding of innovation and innovative input (in terms of research intensity) on the one hand, and innovative output (in terms of innovative turnover) on the other. Both the direct effects of public funding and its relation to innovative networking are examined.

Bringing in different sources of public funding – regional, federal, European – allows us to highlight the impact of the different objectives of public intervention. These different objectives are, in the case of regional funding, tailored support and regional cooperation; at federal level they focus on the stimulation of spillover effects and infrastructure, at EU level the Framework Programmes aim at the creation of critical mass and cross-border cooperative research in high risk technology areas, whereas in other EU funding research serves objectives other than S&T objectives stricto sensu.

In order to measure these direct and indirect effects appropriately and to account for the heterogeneity in firm responsiveness on public incentives, a differentiation between strong and weak innovators (measured by quantile regression) will be used.

4.3 MODEL, DATA AND ESTIMATION METHODS

4.3.1 Data

This study draws on data for Belgium gathered as part of the Third European Community Innovation Survey, CIS3, conducted in 2002 and relating to the period 1998–2000. The target population included all enterprises with ten or more employees. A survey stratified by size (small, medium, large) and economic activity (NACE) of 4,482 firms was taken, covering one third of the target population. The rate of useful answers was

45.4% and a non-response analysis revealed no systematic non-response bias in terms of innovative activity. After exclusion of 48 enterprises because of incomplete data or inconsistencies a sample of 1,987 enterprises remained, of which a majority of two thirds (1,298) claimed to be innovative. Of these innovative firms, 938 reported in-house R&D and are considered henceforth to be research-active, 724 of these had innovative turnover.

Reliance on public funding proved, purely in terms of numbers, to be relatively modest. Of the 1,298 innovative firms 255 (20%) reported funding from regional government, 65 (5%) from federal government, 53 (4%) from the EU Framework Programmes and 41 (3%) from other EU funding. The predominance of regional funding seems in line with the regionalised nature of innovation policy in Belgium (see Section 2.2). The low shares on EU funding possibly underlines the difficulty in obtaining these funds or the complexity of the EU programmes. In the case of the 938 research-active enterprises these figures are respectively 219 (23%), 55 (6%), 52 (6%) and 34 (4%), and in the case of the 724 research-active product innovators they are 181 (25%), 48 (7%), 48 (7%) and 29 (4%).

4.3.2 Model and Construction of Key Variables

The model is in the spirit of structural models (Crépon et al., 1998; Klomp and Van Leeuwen, 2001; Lööf and Heshmati, 2002, 2006; Griffith et al., 2006) that have been used to study the relation between innovation and firm performance. It concentrates on the innovation input and the innovation output equations and intends to investigate indirect funding effects via external knowledge relations, and can be presented as follows for each firm.

$$\begin{aligned}\text{LRDINS} = \text{f1}(&\text{C, IDPAV, INKS, COOP, APPR, CKNOW, LHEI, LEXI,}\\ &\text{LSIZE, GROUP, FREG, FFED, FEUF, FEUO}) \qquad (4.1)\end{aligned}$$

$$\begin{aligned}\text{LITURN} = \text{f2}(&\text{C, IDPAV, LRDINS, INKS, COOP, APPR, CKNOW, PCS,}\\ &\text{LHEI, LEXI, LSIZE, GROUP, FREG, FFED, FEUF, FEUO}) \qquad (4.2)\end{aligned}$$

The innovation input equation (4.1) explains the R&D intensity of firms, measured by the log of the share of in-house R&D expenditure in turnover, LRDINS. The focus is on in-house R&D instead of on total innovation expenditure because this is assumed to be particularly relevant to the capability of firms to use incoming knowledge spillovers, to engage in research cooperation and to appropriate innovative results. Its explanatory variables are a set of public funding variables, knowledge-related variables, measures of human capital, competitive pressure and firm size, and industry dummies depicting the technological environment.

Four sources of public funding of innovation are considered: regional governments, FREG, federal government, FFED, funding by the EU in its Framework Programmes, FEUF, and EU innovation funding other than framework aid, FEUO. All measures of public funding are binary in nature. They take a value of 1 in case funding is obtained and 0 otherwise. In the case of regional government funding, federal government funding, and EU funding other than framework aid, they relate to the period 1998–2000. The EU framework aid relates to both the Fourth and Fifth Framework Programs with respect to 1994–1998 and 1998–2002. In order to avoid potential overlap, the variable on EU funding other than framework aid is constructed by considering enterprises reporting EU funding but no framework aid, on the assumption that if they report both it concerns framework aid.

As highlighted before, because of the particular relation between funding and cooperation special attention will be paid to research cooperation, COOP. Cooperation on research (measured as a binary variable that takes value 1 in case of cooperation and 0 otherwise) is thought to stimulate R&D efforts to the extent that it raises the efficiency of the research process, and because the internalisation and commercialisation of the knowledge created through cooperation requires further research efforts by the firms concerned (Colombo and Gerrone, 1996; Tether, 2002, Monjon and Waelbroeck, 2003; Becker and Dietz, 2004; Veugelers and Cassiman, 2005). Cooperation can occur either with suppliers, clients, competitors, research labs, universities or government. In view of the importance of this variable in the analysis, estimation will be repeated while splitting it up into alternate components. Thus, binary variables will be considered depicting cooperation with suppliers, clients or competitors, COOP_SCC (or inter-firm cooperation), and cooperation with research labs, universities or government, COOP_RUG (or cooperation with organizations). A distinction will also be made between cooperation with domestic partners, COOP_DOM, and cooperation with foreign partners, COOP_FOR. This is especially relevant for European funding aimed at promoting cross-country cooperation and financing joint research projects (Luukkonen, 2001; Capron and Cincera, 2005).

Other knowledge-related variables measure the level of incoming knowledge spillovers, INKS, the complexity of knowledge, CKNOW, and appropriability of in-house knowledge, APPR. Incoming spillovers (measured in terms of the importance attached to alternative sources of information including suppliers, clients, competitors, universities, government, professional conferences, and fairs and exhibitions) are expected to affect R&D intensity positively because they increase the firm's knowledge base from which it can draw in its research process, and because the absorption of external knowledge requires additional in-house research efforts (Cohen and Levinthal, 1989, 1990).

Complexity of knowledge, CKNOW, is obtained by taking the ratio of the sum of the scores with respect to information from universities, government and professional conferences over the sum of the scores of all sources of information, as suggested by Cassiman and Veugelers (2002). It is, in other words, inferred from the structure of the incoming knowledge flows, on the assumption that this conditions the character of the firm's in-house research. It can be expected to positively affect the level of research efforts to the extent that the greater complexity of basic research requires more efforts than applied research.

Appropriability, APPR, concentrates on methods of strategic protection (including secrecy, complexity of design and lead-time advantage on competitors) rather than on methods of legal protection through patents, because the latter are highly industry-specific (Griliches, 1990; OECD, 2002b). By always giving a score of 1 if they did and 0 otherwise, an average score between 0 and 1 could be obtained to generate a measure of strategic protection. A higher capacity in appropriating in-house knowledge will tend to reduce spillovers to competitors and, thereby, increase the incentives to undertake research efforts (Spence, 1984; Teece, 1986; special issue on appropriability in *Research Policy* 2006 edited by Chesbrough et al.).

Human capital is measured by the log of the level of higher education intensity of the personnel, LHEI, which reflects the average level of qualification of the workforce. Firms will tend to spend more on research if they dispose of highly qualified workers, who are able to operate in a more complex production environment and who may more easily contribute to the innovation process itself (Redding, 1996; Frantzen, 2000). The log of export intensity, LEXI, is included as a measure of competitive pressure, since firms operating in foreign markets are likely to be subject to stronger competition. The need to remain competitive is assumed to affect R&D efforts positively.

To control for the impact of firm size on R&D intensity, two size-related variables are included. These are the log of size (measured by employment in the base year), LSIZE, and a variable reflecting group membership, GROUP. There are no a priori expectations as to their sign, since arguments have been given in the theoretical literature implying either a positive or a negative influence on R&D intensity (Cohen, 1995).

Finally, industry dummy variables based on Pavitt's taxonomy, IDPAV, are added, depicting the general technological environment in which the firm operates. These dummies are based on the taxonomy originally proposed by Pavitt (1984), and revised and elaborated by Tidd et al. (1997)

to accommodate for the growth of the services economy. This classification reflects differences in technological trajectories between groups of industries. It is based on the general characteristics of the technologies used in the production system, the qualifications needed to operate them, their sources and the monitoring strategy. Five groups of sectors are identified: resource-intensive, IDPAV1, scale-intensive, IDPAV2, specialised suppliers, IDPAV3, science-based, IDPAV4, and information-intensive, IDPAV5. The industries in this classification are included in Table A4.1 in the Appendix.

The innovation output equation (4.2) concentrates on product innovation. It can be conceived as a production function of new knowledge and it has as dependent variable the log of the proportion of new or significantly improved products (goods or services) in turnover. It was preferred over a patent-based indicator of innovative output because of the well-known shortcomings of the latter in this respect, as not all patents become innovations, the degree of patenting strongly varies between industries, and few service firms, such as software development, tend to patent (Griliches, 1990; OECD, 2002b).

The key explanatory variable in the output equation is the log of R&D intensity, LRDINS. Its influence on innovation output is not only direct, through the creation of new product-innovative ideas, but also indirect, through contribution to the generation of complementary process innovations required for the implementation of product innovation (Parisi et al., 2006). Since part of the process innovations concern production re-organisations that do not require formal research efforts, a measure of process innovation, PCS, is included as a separate variable.

Moreover, the same variables on public funding, knowledge relations, human capital, competitive pressure, size, and environmental dummies, are included in the innovation input equation. The inclusion of variables already present in the innovation input equation allows us to discover whether there is also a direct influence from these variables on innovation output, rather than merely an indirect one, through their impact on R&D intensity. The funding variables may exercise a direct influence on product innovativeness because, as put forward in the previous section, public funds may partly be used to finance aspects of the innovation process other than in-house R&D. The size-related variables will indicate whether there is a size-related research efficiency effect.

Table A4.2 in the Appendix summarises the construction and definition of all variables used and Table A4.3 their mean values.

4.3.3 Estimation Methods

The data on CIS3 are cross-sectional and refer to the period 1998–2000. The parameter estimates can, therefore, only provide a momentary impression of the long-run relations between the variables. The data do not allow introducing lagged responses.

In order to assess the impact of funding, the equations explaining research intensity and innovative turnover will be estimated, excluding and including the variables on public funding in turn. The initial aim is to obtain a general idea of the average behaviour of firms.

An equation-based estimation by ordinary least squares (OLS) will produce consistent parameter estimates provided the explanatory variables are exogenous. Here, there may be doubts in the case of some of the explanatory variables under consideration. Therefore, it may be the case that the level of research intensity, LRDINS, the knowledge-related variables – INKS, COOP, APPR, CKNOW - and the level of labour skills, LHEI, are determined simultaneously to some extent. Nor can we rule out the fact that innovative firms devote more efforts to R&D performance, information sourcing and research cooperation, perform more complex research, have a greater capacity for strategic appropriation and employ more skilled workers than less innovative firms, even in the absence of a behavioural relation between these variables. Moreover, public funding authorities may apply policies of 'picking winners' by deliberately targeting research-intensive firms, so there would be a reverse causation from research intensity to public funding in the research intensity equation. Similar arguments may hold for these respective variables in the innovative turnover equation. In this case, the level of research intensity, LRDINS, and the variable on process innovations, PCS, may also be determined in part simultaneously with the level of innovative turnover, LITURN. Furthermore, there may be feedback from level of innovative turnover, LITURN, through growth of turnover to level of R&D intensity, LRDINS. To check this, Durbin–Hausman–Wu (DHW) exogeneity tests are performed on the variables in question in the respective equations, on the assumption that the other explanatory variables are exogenous (Davidson and MacKinnon, 1993). As we see below, exogeneity is confirmed systematically in the results, so it would appear there is no need to instrument.

Following Crépon et al. (1998), recent econometric work on the relation between innovation and firm performance has devoted a great deal of attention to the issue of sample selection. After estimating the research intensity and innovative turnover equations respectively on samples of research-active and both research-active and product-innovative firms, as a check of robustness, the samples are extended and the equations always complemented with a sample selection equation, to draw inference on a larger population that also includes non-innovative firms. This sample selection equation is a probit equation explaining the probability of doing research, or of doing research and being product-innovative. When considered alongside the corresponding research intensity or innovative turnover equation, it always forms a generalised tobit model consisting of a probit and a tobit equation. Estimation by OLS ceases to be appropriate because of the binary nature of the dependent variable in the probit equation, and due to a possible correlation between its error term and the error term of the tobit equation, which will cause a sample selection bias in the parameter estimates. Thus, for each of the equations, the corresponding generalised tobit model is estimated by Maximum Likelihood (ML), using the Heckman (1979) procedure to choose the initial parameter values in the iterative procedure. The resulting standard errors are, again, heteroscedasticity-robust. This method has been shown to provide consistent and efficient estimates (Verbeek, 2000).

It would be preferable to consider all the exogenous variables in the equations on research intensity, LRDINS, and innovative turnover, LITURN, as possible explanatory variables in the probit equation. But, due to data constraints, only the variables for which the data are available over the full sample of both innovative and non-innovative firms can, in principle, be used. Apart from the intercept these are the industry dummies, IDPAV, strategic appropriability, APPR, qualified labour, LHEI, export intensity, LEXI, firm size, LSIZE, and group membership, GROUP. Since it is unlikely that firms which do not report innovative activities receive funding, a second version of the probit equation is considered, and the funding variables are included by giving them the value 0 in the case of non-innovators.

Attention then returned to the samples of research-active and both research-active and product-innovative firms, and quantile regression estimation is performed to identify systematic differences in behaviour between strong and weak innovators. A quantile regression estimates the parameters by minimising the sum of the asymmetrically-weighted absolute values of the residuals. Thus, for example, the 0.25 regression quantile estimates lend less weight to the positive residuals than to the negative ones. Therefore attention is drawn to the left-hand tail of the conditional density function of the dependent variable, and vice versa, in the case of the 0.75 quantile. Varying the quantile parameter between 0 and 1 generates all the regression quantiles, revealing the conditional distribution of the dependent variable given the explanatory variables. Quantile regression estimates have the additional advantage of being more efficient than least squares estimates when the errors deviate from normality, which is the case.

Since we cannot rule out the possibility of heteroscedasticity and, therefore, dependence between the quantile regressors and the density of the error terms, the bootstrap design method is most appropriate for estimating the standard errors (Buchinsky, 1995, 1998). Bootstrap re-sampling is performed using 1,000 repetitions. Due to space constraints, the results presented in Tables 4.4 and 4.5 below relate to the 0.25, 0.50 and 0.75 quantiles.

4.4 RESULTS AND ANALYSIS

4.4.1 Research Intensity and Innovative Turnover: Average Behaviour

Tables 4.1 to 4.3 present the estimates on the average behaviour of firms. Table 4.1 presents the baseline estimates of the equation on research intensity.

Table 4.1 Research intensity: average behaviour – baseline estimates

	OLS (i)	OLS (ii)	ML (iii)	ML (iv)	ML (v)	ML (vi)	ML (vii)	ML (viii)
Constant	–3.513	–3.577	–0.642	–3.720	–0.640	–3.744	–0.787	–3.820
	(14.7)**	(15.5)**	(5.06)**	(15.4)**	(5.07)**	(16.1)*	(5.65)**	(16.3)**
COOP	0.364	0.268		0.363		0.268		0.266
	(3.12)**	(2.30)**		(3.15)**		(2.33)**		(2.31)**
INKS	0.949	0.821		0.945		0.818		0.813
	(3.48)**	(3.01)**		(3.50)**		(3.04)**		(3.02)**
APPR	0.506	0.447	1.339	0.631	1.339	0.550	1.204	0.570
	(3.52)**	(3.09)**	(11.8)**	(4.29)**	(11.8)**	(3.75)**	(10.4)**	(3.83)**
CKNOW	0.365	0.241		0.365		0.242		0.249
	(1.72)*	(1.09)		(1.74)*		(1.11)		(1.14)
LHEI	0.129	0.116	0.117	0.143	0.117	0.128	0.108	0.131
	(2.73)**	(2.51)**	(6.12)**	(3.05)**	(6.11)**	(2.79)**	(5.57)**	(2.83)**
LEXI	0.051	0.040	0.051	0.056	0.051	0.045	0.040	0.046
	(2.18)**	(1.75)*	(4.31)**	(2.39)**	(4.33)**	(1.93)*	(3.25)**	(1.97)**
LSIZE	–0.184	–0.192	0.179	–0.167	0.178	–0.178	0.165	–0.177
	(3.50)**	(3.74)**	(6.53)**	(3.33)**	(6.54)**	(3.66)**	(5.72)**	(3.57)**
GROUP	–0.202	–0.144	–0.041	–0.205	–0.040	–0.147	0.002	–0.142
	(1.59)	(1.15)	(0.61)	(1.63)	(0.60)	(1.18)	(0.03)	(1.14)
IDPAV1	–1.282	–1.213	–0.021	–1.284	–0.020	–1.215	0.021	–1.211
	(6.92)**	(6.55)**	(0.25)	(7.00)**	(0.24)	(6.65)**	(0.25)	(6.62)**

Table 4.1 (continued)

	OLS (i)	OLS (ii)	ML (iii)	ML (iv)	ML (v)	ML (vi)	ML (vii)	ML (viii)
IDPAV2	−1.082	−1.040	0.140	−1.065	0.140	−1.026	0.108	−1.027
	(5.84)**	(5.64)**	(1.55)	(5.77)**	(1.56)	(5.61)**	(1.16)	(5.63)**
IDPAV3	−0.168	−0.202	0.250	−0.145	0.249	−0.183	0.176	−0.188
	(0.78)	(0.96)	(2.29)*	(0.68)	(2.28)*	(.88)	(1.55)	(0.01)
IDPAV4	−0.396	−0.361	0.422	−0.358	0.423	−0.330	0.407	−0.324
	(2.05)*	(1.91)	(3.26)**	(1.83)	(3.27)**	(1.73)	(3.02)**	(1.71)
FREG		0.407				0.405	0.971	0.495
		(3.13)**				(3.16)**	(8.62)**	(3.77)**
FFED		−0.080				−0.081	0.896	−0.005
		(0.38)				(0.40)	(3.95)**	(0.02)
FEUF		0.792				0.791	1.312	0.843
		(3.95)**				(4.01)**	(2.28)**	(4.41)**
FEUO		−0.198				−0.198	0.626	−0.142
		(0.56)				(0.58)	(2.09)**	(0.42)
RHO				0.115		0.095		0.138
				(2.04)*		(1.74)		(2.32)*
Obs.	938	938	1,987	938	1,987	938	1,987	938
Adj. R^2	0.202	0.219						

Notes: The dependent variable is the log of research intensity, LRDINS, except in columns (iii), (v) and (vii) where it is the probability of doing research in the total population, PRD. RHO stands for the correlation coefficient between the errors of the sample selection and tobit equations. The absolute value of the t-values is presented between brackets. These are based on heteroscedasticity-consistent estimates of the standard errors. The t tests are one-tailed or two-tailed, according to a priori expectations with respect to the parameter values. Symbols * and ** denote significance at the 5% and 1% levels.

Columns (i) and (ii) present the estimates by OLS on the sample of 938 research-active firms, respectively excluding and including the funding variables. Durbin–Hausman–Wu (DHW) tests confirm the exogeneity of incoming knowledge spillovers, INKS, research cooperation, COOP, appropriability, APPR, complexity of knowledge, CKNOW, labour skills, LHEI, and of the four funding variables, FREG, FFED, FEUF and FEUO, justifying the use of these variables as such rather than in instrumented form. This also means that, to the extent that they are statistically significant, the parameter estimates of the funding variables will not reflect a reverse causation.

As seen at the bottom of the columns (i) and (ii), including the funding variables improves somewhat the statistical fit (the adjusted R^2), confirming that there is an impact of funding on research intensity. This seems exclusively due to regional government funding, FREG, and EU framework funding, FEUF, whose parameter estimates are found positive and highly statistically significant. The results would in this case clearly point to incentive effects. Considering that only 6% of the firms in the research-active sample received EU framework funding, it appears that EU framework funding in particular is effective in stimulating R&D intensity. This is not surprising as these Programmes aim especially at projects with strategic importance and at the creation of critical mass in research (see Section 4.2.2).

Research cooperation, COOP, is highly significant in the equation excluding funding, in column (i). Column (ii) shows that including the funding variables still leaves cooperation highly significant, although the impact is somewhat reduced. This reflects that research cooperation is positively correlated with the significant funding variables, FREG and FEUF, as confirmed by the correlation patterns presented in Table A4.3 in the Appendix. These results do, therefore, suggest that there is a significant direct influence of research cooperation on research intensity, explainable by the fact that research cooperation increases the firms' knowledge base and that the absorption of the knowledge concerned always requires further in-house R&D. At the same time, there is also an indirect influence through funding, as regional and EU public authorities tend to target firms involved in research cooperation. They do so both for reasons of allocative efficiency of the funds concerned and to give a signal to firms to increase their information-sourcing efforts and to set up new cooperative networks. The deliberate targeting of firms setting up new cooperative networks tends to amplify this. In the latter case there is also, therefore, a funding influence on the amount of research cooperation, and therefore on research intensity too. Or, to put it differently, the funding policies play a part in opening up the research process. As such, firms cooperating in research are more research intensive not only because they have to be in order to absorb the knowledge in question, but because they are better able to do so as they receive more funding.

Fairly similar results are obtained for incoming knowledge spillovers, INKS. Compared to cooperation, inclusion of the funding variables shows a smaller reduction in the direct impact of incoming knowledge spillovers. Incoming knowledge spillovers clearly have a positive correlation with the significant funding variables, FREG and FEUF. As such, here too, there is a significant direct influence of incoming knowledge spillovers on research intensity, as well as an indirect influence through funding, as regional and EU public authorities tend to target firms sourcing external knowledge.

Appropriability, APPR, and complexity of knowledge, CKNOW, are also found to have a statistically significant positive impact on research intensity in column (i) of Table 4.1. This impact appears, however, far stronger in the case of appropriability. Column (ii) shows that including the funding variables slightly reduces the impact of appropriability, APPR. However, in the case of complexity of knowledge, CKNOW, there is a stronger reduction in impact, which renders this variable completely non-significant. This reflects the stronger positive correlation of CKNOW with the significant funding variables, especially FREG, as seen in Table A4.3 in the Appendix. These results suggest that appropriability mostly has a direct influence on research intensity, confirming that the capacity to appropriate the benefits of internally-developed knowledge increases incentives to undertake research efforts. In the case of complexity of knowledge the influence on research intensity appears to be weaker and, essentially, indirect in nature, due to the fact that firms undertaking complex research attract more regional public funding.

The human capital variable, LHEI, is also clearly significant, with the expected positive sign in the equation excluding funding in column (i) in Table 4.1, and including funding in column (ii), leaves its parameter estimate largely unaffected. This reflects that its correlation with funding is modest only and suggests that its impact on research intensity is mostly direct, confirming the complementarity between formal R&D and the qualification of workers. A fairly similar conclusion can be reached in the case of export orientation, LEXI, even though there is evidence here of a more important indirect influence through funding. It suggests that the stronger competitive pressure in foreign markets does, indeed, necessitate stronger research efforts and that public funding authorities act consequently.

Firm size, LSIZE, has a clearly negative impact on research intensity in the equation without funding in column (i) in Table 4.1, whereas the coefficient of the other size-related variable, group membership, GROUP, is also negative, but non-significant. Including the funding variables in column (ii) does not alter these results. This is not surprising since firm size, LSIZE, and group membership, GROUP, are only very modestly correlated with the funding variables, as seen in Table A4.3 in the Appendix. The negative impact of size can be explained by the fact that in our sample an important part of strong R&D investors are new technology-based firms. These young dynamic firms expect to grow rapidly from their innovations. They are, therefore, not greatly affected in their research decisions by their original size constraints or, for that matter, by their short-term turnover performance, and they tend to spend relatively more on R&D in relation to their size than larger, less dynamic firms.

Finally, for the industry effects, the information-intensive services dummy, IDPAV5, was excluded so that its impact is captured, among other influences, by the intercept. The estimates of the other industry dummies reflect deviations from this. The results in column (i) indicate that there is a statistically significant negative impact in the case of IDPAV1, IDPAV2 and IDPAV4. The corresponding estimates in column (ii) are much alike. In this case IDPAV4 remains only significant at a 10% level. These results suggest that general technological characteristics ensure that resource-intensive, scale-intensive and to a lesser extent science-based industries are, in Belgium, less research-intensive than the specialised supplier and information-intensive sectors.

Tables 4.2a and b concentrate again on the research-active sample and considers the effect of splitting the variable on research cooperation in the equation explaining research intensity.

Table 4.2a Research intensity: average behaviour – refining research cooperation (models i-vi)

	OLS (i)	OLS (ii)	OLS (iii)	OLS (iv)	OLS (v)	OLS (vi)
Constant	−3.499	−3.444	−3.466	−3.565	−3.532	−3.559
	(14.7)**	(14.5)**	(14.6)**	(15.5)**	(15.2)**	(15.3)**
COOP_SCC	0.453		0.328	0.384		0.366
	(3.82)**		(2.19)**	(3.28)**		(2.48)**
COOP_RUG		0.458	0.240		0.284	0.036
		(3.54)**	(1.47)		(2.12)**	(0.21)
COOP_DOM						
COOP_FOR						
INKS	0.877	0.892	0.861	0.758	0.794	0.758
	(3.23)**	(3.26)**	(3.16)**	(2.79)**	(2.91)**	(2.79)**
APPR	0.501	0.503	0.485	0.436	0.456	0.435
	(3.50)**	(3.51)**	(3.38)**	(3.04)**	(3.17)**	(3.02)**
CKNOW	0.408	0.282	0.321	0.263	0.211	0.252
	(1.94)*	(1.30)	(1.46)	(1.18)	(0.95)	(1.12)
LHEI	0.132	0.128	0.130	0.118	0.117	0.118
	(2.80)**	(2.73)**	(2.76)**	(2.56)**	(2.53)**	(2.56)**

Table 4.2a (continued)

	OLS (i)	OLS (ii)	OLS (iii)	OLS (iv)	OLS (v)	OLS (vi)
LEXI	0.051	0.048	0.050	0.041	0.039	0.040
	(2.22)**	(2.06)**	(2.13)**	(1.78)*	(1.69)*	(1.76)*
LSIZE	–0.186	–0.183	–0.190	–0.196	–0.188	–0.196
	(3.57)**	(3.51)**	(3.62)**	(3.84)**	(3.70)*	(3.84)**
GROUP	–0.204	–0.195	–0.204	–0.147	–0.141	–0.148
	(1.61)	(1.54)	(1.61)	(1.17)	(1.11)	(1.17)
IDPAV1	–1.264	–1.303	–1.274	–1.197	–1.230	–1.199
	(6.80)**	(7.09)**	(6.90)**	(6.46)**	(6.68)**	(6.48)**
IDPAV2	–1.059	–1.099	–1.072	–1.022	–1.052	–1.024
	(5.70)**	(5.99)**	(5.83)**	(5.55)**	(5.76)**	(5.60)**
IDPAV3	–0.143	–0.187	–0.160	–0.185	–0.211	–0.187
	(0.66)	(0.87)	(0.75)	(0.88)	(1.01)	(0.89)
IDPAV4	–0.376	–0.429	–0.395	–0.344	–0.384	–0.347
	(1.94)	(2.25)*	(2.06)*	(1.82)	(2.04)*	(1.84)
FREG				0.424	0.389	0.418
				(3.31)**	(2.94)**	(3.15)**
FFED				–0.088	–0.074	–0.088
				(0.43)	(0.36)	(0.43)
FEUF				0.759	0.765	0.752
				(3.78)**	(3.77)**	(3.72)**
FEUO				–0.214	–0.208	–0.218
				(0.62)	(0.59)	(0.62)
Obs.	938	938	938	938	938	938
Adj. R^2	0.206	0.203	0.207	0.219	0.218	0.222

Notes: The dependent variable is the log of research intensity, LRDINS. The absolute value of the t-values is presented between brackets. These are based on heteroscedasticity-consistent estimates of the standard errors. The t-tests are one-tailed or two-tailed, according to a priori expectations with respect to the parameter values. Symbols * and ** denote significance at the 5% and 1% levels.

Table 4.2b Research intensity: average behaviour – refining research cooperation (models vi-xii)

	OLS (vii)	OLS (viii)	OLS (ix)	OLS (x)	OLS (xi)	OLS (xii)
Constant	–3.502	–3.442	–3.440	–3.570	–3.513	–3.510
	(14.7)**	(14.5)**	(14.5)**	(15.5)**	(15.2)**	(15.2)**
COOP_SCC						
COOP_RUG						
COOP_DOM	0.338		–0.054	0.241		–0.108
	(2.81)**		(0.36)	(1.98)**		(0.71)
COOP_FOR		0.619	0.655		0.522	0.592
		(5.04)**	(4.18)**		(4.27)**	(3.80)**
INKS	0.939	0.815	0.811	0.813	0.716	0.707
	(3.45)**	(3.02)**	(3.01)**	(2.99)**	(2.65)**	(2.61)**
APPR	0.510	0.474	0.478	0.451	0.419	0.425
	(3.56)**	(3.33)**	(3.35)**	(3.13)**	(2.93)**	(2.96)**
CKNOW	0.368	0.381	0.394	0.245	0.248	0.271
	(1.73)*	(1.81)*	(1.85)*	(1.11)	(1.13)	(1.23)
LHEI	0.131	0.123	0.122	0.118	0.111	0.110
	(2.76)**	(2.56)**	(2.56)**	(2.53)**	(2.37)**	(2.35)**
LEXI	0.052	0.051	0.051	0.041	0.040	0.040
	(2.23)**	(2.17)**	(2.17)**	(1.78)*	(1.76)*	(1.74)*
LSIZE	–0.179	–0.192	–0.191	–0.188	–0.199	–0.198
	(3.43)**	(3.67)**	(3.65)**	(3.69)**	(3.90)*	(3.88)*
GROUP	–0.195	–0.232	–0.233	–0.139	–0.174	–0.176
	(1.54)	(1.84)	(1.85)	(1.10)	(1.38)	(1.40)
IDPAV1	–1.289	–1.285	–1.287	–1.217	–1.221	–1.224
	(6.94)**	(6.94)**	(6.93)**	(6.58)**	(6.59)**	(6.60)**
IDPAV2	–1.076	–1.100	–1.103	–1.035	–1.062	–1.069
	(5.79)**	(5.93)**	(5.92)**	(5.61)**	(5.76)**	(5.76)**
IDPAV3	–0.169	–0.167	–0.166	–0.202	–0.205	–0.205
	(0.78)	(0.78)	(0.78)	(0.96)	(0.99)	(0.99)
IDPAV4	–0.392	–0.426	–0.430	–0.357	–0.390	–0.399
	(2.03)*	(2.22)*	(2.23)*	(1.89)	(2.08)*	(2.12)*

Table 4.2b (continued)

	OLS (vii)	OLS (viii)	OLS (ix)	OLS (x)	OLS (xi)	OLS (xii)
FREG				0.424	0.389	0.418
				(3.31)**	(2.94)**	(3.15)**
FFED				–0.088	–0.074	–0.088
				(0.43)	(0.36)	(0.43)
FEUF				0.759	0.765	0.752
				(3.78)**	(3.77)**	(3.72)**
FEUO				–0.214	–0.208	–0.218
				(0.62)	(0.59)	(0.62)
Obs.	938	938	938	938	938	938
Adj. R^2	0.206	0.203	0.207	0.219	0.218	0.222

Notes: The dependent variable is the log of research intensity, LRDINS. The absolute value of the t-values is presented between brackets. These are based on heteroscedasticity-consistent estimates of the standard errors. The t-tests are one-tailed or two-tailed, according to a priori expectations with respect to the parameter values. Symbols * and ** denote significance at the 5% and 1% levels.

A distinction is first made between cooperation with other firms among which suppliers, clients or competitors, COOP_SCC, and industry–science linkages with research labs, universities or government, COOP_RUG. Columns (i), (ii) and (iii) of Table 4.2a look at the separate and combined effect of these variables in the research intensity equation excluding funding, and columns (iv), (v) and (vi) present the corresponding estimates in the equation including funding.

When entered separately, both inter-firm cooperation, COOP_SCC, and industry–science relations, COOP_RUG, are always found to be highly significant in the equation excluding funding, in columns (i) and (ii). The parameter estimates of the other variables are similar, except that they are lower in the case of the IDPAV4 industry dummy and complexity of research, CKNOW, in the case of industry–science cooperations, COOP_RUG, is included. This suggests, with some plausibility, that industry-science relations with research labs, universities or government predominantly take place in the science-based industries and in basic research. Combining both the COOP_SCC and COOP_RUG variables in the research intensity equation, in column (iii), reduces their positive impact and even renders COOP_RUG non-significant. This suggests the existence of strong multicollinearity between these variables: the correlation coefficient between them is indeed 0.62. This can be explained by complementarities both in terms of time perspectives (short-term versus long-term, and in terms of perspectives related to the nature of research (market oriented versus basic strategic research)).

Including the funding variables in columns (iv) to (vi) of Table 4.2a confirms these respective results. The parameter estimates of both COOP_SCC and COOP_RUG are always reduced in columns (iv) and (v) when compared to columns (i) and (ii), but this reduction is clearly stronger in the latter case. This reflects that both regional funding, FREG, and EU framework funding, FEUF, are more strongly correlated with the industry–science relations variable, COOP_RUG, as seen in Table A4.3 in the appendix. It confirms that the public funding authorities concerned tend to favour in particular firms cooperating with research labs and universities and/or setting up new cooperative networks with them. As seen in Section 4.2, both the regional funding programmes and the EU Framework Programmes are designed to expand the knowledge base of the firms and, therefore, aim to support research projects in which cooperation with universities and public research organisations is central. In the case of EU funding, cross-border cooperations in particular are envisaged, between firms and industry–science linkages.

The continuation of Table 4.2b makes a similar distinction between research cooperation with domestic partners, COOP_DOM, and research cooperation with foreign partners, COOP_FOR. The results in columns (vii) and (viii) show that cooperation with domestic and with foreign partners both have a positive impact on research intensity, but that the latter is clearly stronger. In the Belgian case this can be explained by the mere fact that not all knowledge is available in a small economy (Spithoven and Teirlinck, 2006). Multicollinearity ensures that when combined in column (ix), only cooperation with foreign partners, COOP_FOR, remains statistically significant. The correlation coefficient between COOP_DOM and COOP_FOR is 0.66. This reflects that some of the firms that cooperate do so with both domestic and foreign partners. Including funding in columns (x) to (xii) confirms these results. Again there is a reduction in the impact of the cooperation variables, which are quite evenly balanced in this case. It reflects a significant positive correlation between the FREG and FEUF variables and domestic and foreign cooperation, COOP_DOM and COOP_FOR, as seen in Table A4.3 in the Appendix. Not surprisingly, regional funding authorities appear to favour cooperation between domestic organisations in particular, whereas European Framework funding authorities have a preference for international cooperation. This is in line with the aim of the EU Framework Programmes to foster greater coherence between the European Member States and, consequently, in the domain of science and technology, make progress in founding a European Research

Area (Luukkonen, 2001; Capron and Cincera, 2005; Muldur et al., 2006; Polt and Psarra, 2006).

Table 4.3 presents the estimates of the equation on innovative turnover, LITURN.

Table 4.3 Innovative turnover: average behaviour

	OLS (i)	OLS (ii)	ML (iii)	ML (iv)	ML (v)	ML (vi)	ML (vii)	ML (viii)
Constant	–1.004	–0.972	–0.763	–1.101	–0.763	–1.074	–0.882	–1.164
	(6.19)**	(6.04)**	(6.06)**	(4.51)**	(6.06)**	(4.63)**	(6.57)**	(5.22)**
LRDINS	0.102	0.108		0.102		0.108		0.107
	(4.95)**	(5.16)**		(5.00)**		(5.22)**		(5.19)**
COOP	0.010	–0.011		0.010		0.011		0.011
	(0.12)	(0.11)		(0.12)		(0.15)		(0.14)
INKS	0.272	0.296		0.271		0.295		0.293
	(1.37)	(1.49)		(1.38)		(1.51)		(1.49)
APPR	–0.014	0.011	1.312	0.045	1.312	0.051	1.179	0.088
	(0.14)	(0.14)	(12.5)**	(0.33)	(12.5)**	(0.39)	(11.0)**	(0.72)
CKNOW	–0.147	–0.147		–0.147		–0.147		–0.144
	(0.86)	(0.85)		(0.88)		(0.86)		(0.85)
PCS	0.195	0.185		0.195		0.185		0.185
	(2.26)**	(2.15)**		(2.28)**		(2.18)**		(2.18)**
LHEI	0.004	0.007	0.150	0.012	0.150	0.016	0.142	0.021
	(0.12)	(0.23)	(6.70)**	(0.35)	(6.70)**	(0.47)	(6.46)**	(0.66)
LEXI	–0.021	–0.019	0.053	–0.019	0.053	–0.017	0.043	–0.015
	(1.49)	(1.37)	(4.37)**	(1.25)	(4.37)**	(0.05)	(3.43)**	(1.02)
LSIZE	–0.102	–0.100	0.170	–0.094	0.170	–0.092	0.154	–0.088
	(3.32)**	(3.23)**	(6.28)**	(2.82)**	(6.28)**	(2.79)**	(5.46)**	(2.78)**
GROUP	0.075	0.061	–0.033	0.073	–0.033	0.059	0.009	0.063
	(0.86)	(0.70)	(0.47)	(0.86)	(0.47)	(0.69)	(0.12)	(0.72)
IDPAV1	0.490	0.496	0.120	–0.496	–0.120	–0.502	–0.080	–0.504
	(4.30)**	(4.33)**	(1.36)	(4.38)**	(1.36)	(4.44)**	(0.91)	(4.45)**
IDPAV2	–0.359	–0.357	–0.058	–0.361	–0.058	–0.359	–0.085	–0.364
	(2.97)**	(2.92)**	(0.62)	(3.01)**	(0.62)	(2.97)**	(0.89)	(3.00)**
IDPAV3	–0.006	–0.003	0.084	–0.002	0.084	0.001	0.021	–0.006
	(0.05)	(0.028)	(0.76)	(0.02)	(0.77)	(0.00)	(0.18)	(0.05)
IDPAV4	–0.429	–0.440	0.253	–0.419	0.253	–0.429	0.230	–0.425
	(3.29)**	(3.37)**	(2.01)*	(3.22)**	(2.01)*	(3.31)**	(1.77)	(3.32)**

Table 4.3 (continued)

	OLS (i)	OLS (ii)	ML (iii)	ML (iv)	ML (v)	ML (vi)	ML (vii)	ML (viii)
FREG		–0.078				0.079	0.682	–0.027
		(0.87)				(0.90)	(6.64)**	(0.28)
FFED		0.197				0.197	0.764	0.251
		(1.29)				(1.32)	(3.77)**	(1.63)
FEUF		–0.054				–0.055	0.909	–0.014
		(0.30)				(0.32)	(3.12)**	(0.08)
FEUO		0.278				0.278	0.624	0.323
		(1.35)				(1.39)	(2.41)**	(1.61)
RHO				0.079		0.084		0.154
				(0.55)		(0.64)		(1.27)
Obs.	724	724	1,987	724	1,987	724	1,987	724
Adj. R^2	0.125	0.126						

Notes: The dependent variable is the log of innovative turnover, LITURN, except in columns (iii), (v) and (vii) where it is the probability of doing research and being product innovative in the total population, PRDINSAL. RHO stands for the correlation coefficient between the errors of the sample selection and tobit equations. The absolute value of the t-values is presented between brackets. These are based on heteroscedasticity-consistent estimates of the standard errors. The t-tests are one-tailed or two-tailed, according to a priori expectations with respect to the parameter values. Symbols * and ** denote significance at the 5% and 1% levels.

Columns (i) and (ii) show the estimates by OLS on the sample of 724 research-active product innovators. Durbin–Hausman–Wu (DHW) tests confirm the exogeneity of research intensity, LRDINS, process innovation, PCS, and also the knowledge, labour skill and funding-related variables, justifying their use as such, rather than in instrumented form. As seen at the bottom of columns (i) and (ii), inclusion of the funding variables does not improve the statistical fit (the adjusted R^2). These variables are found to be non-significant, suggesting that their influence on product innovativeness is exclusively indirect, through research intensity. Public funding is apparently used mainly to finance in-house R&D expenditures and, to the extent that it helps to finance other aspects of the innovation process, these appear less important for product innovation.

As a consequence, the inclusion of the funding variables leaves the parameter estimates of the other variables unaltered. Consider these respective parameter estimates in columns (i) and (ii) in turn. The level of research intensity, LRDINS, is by far the most significant variable, confirming that it is, indeed, the key variable explaining innovative turnover. The confirmation by the DHW test that LRDINS is exogenous in the innovative turnover equation suggests, at the same time, that its parameter estimates are not suffering an upward endogeneity bias, due to a problem of simultaneity or to feedback from the level of innovative turnover, LITURN, over the growth of turnover to the level of research intensity, LRDINS. There is also evidence of a positive impact of process innovations, PCS, on the level of innovative turnover. This confirms that production re-organisations that do not require formal research efforts also play their part in generating product innovations.

There is, however, no evidence of a direct effect of the knowledge-related variables or of human capital on product innovativeness, when controlling for research intensity. This clearly suggests that their influence on the level of product innovation is indirect, working through research intensity. Incoming spillovers, research cooperation or the ideas of workers do not appear to generate readily usable knowledge: they need first to be absorbed in the knowledge base of the firm through additional R&D efforts (Cohen and Levinthal, 1989, 1990). The lack of significance of export orientation suggests that, also for this variable, the impact is mainly indirect, through research intensity, further stressing the importance of in-house R&D efforts.

As far as the size-related variables are concerned, the impact of firm size, LSIZE, is again significant and negative, whereas group membership, GROUP, is once more non-significant, as it was in the research intensity equation. The negative influence of firm size would imply a decrease in efficiency in the innovation process as firm size increases. This can be explained by the fact that among the research-active product innovators the relatively small new technology-based firms tend to be more dynamic. They are, therefore, prepared not only to spend relatively more on R&D in relation to their size, but to do so in a more risky manner.

For industry dummies and when controlling for the impact of research intensity, the innovative output is, on average, lower for firms active in the resource-intensive, scale-intensive and science-based industries than for those operating in information-intensive and specialised supplier industries. The lower innovative output in the case of the science-based industries is somewhat surprising since these industries are also known as dynamic and hi-tech. Product innovation in industries such as biotech and pharmaceuticals is probably characterised by long and variable gestation periods (Bruneel et al., 2007).

4.4.2 Research Intensity and Innovative Turnover: Quantile Behaviour

Table 4.4 presents the quantile regression estimates of the research intensity equation on the sample of 938 research-active firms. The left-hand side of the table gives the estimates of the equation excluding funding and the right-hand side those of the equation including funding. Similar to the average estimates in Table 4.1, the statistical fit is somewhat improved by including the funding variables. This is, again, exclusively due to regional government funding, FREG, and to EU framework funding, FEUF, and appears to be so in all quantiles. This confirms that these public funding programmes are aimed at encouraging cooperative actions focused on particular research projects to create positive externalities.

Table 4.4 Research intensity: quantile behaviour

	Quantiles: no funding			Quantiles: with funding		
	0.25	0.50	0.75	0.25	0.50	0.75
Constant	–4.569	–3.450	–2.644	–4.630	–3.485	–2.694
	(12.5)**	(12.1)**	(9.86)**	(14.4)**	(11.9)**	(11.9)**
COOP	0.323	0.369	0.407	0.259	0.224	0.278
	(2.34)**	(2.54)**	(3.01)**	(1.47)	(1.54)	(2.00)**
INKS	1.017	0.998	1.010	1.042	0.757	0.937
	(2.54)**	(2.68)**	(3.45)**	(2.63)**	(1.96)**	(3.09)**
APP	0.830	0.526	0.261	0.701	0.487	0.219
	(4.26)**	(2.65)**	(1.49)	(3.33)**	(2.54)**	(1.23)
CKNOW	0.730	0.372	0.205	0.272	0.112	0.007
	(2.11)**	(1.50)	(0.76)	(0.75)	(0.44)	(0.03)
LHEI	0.154	0.069	0.088	0.127	0.058	0.128
	(2.26)**	(1.12)	(1.63)	(1.85)*	(1.01)	(2.46)**
LEXI	0.067	0.046	0.025	0.038	0.024	0.004
	(1.80)*	(1.62)	(1.07)	(1.05)	(0.83)	(0.017)
LSIZE	–0.176	–0.149	–0.134	–0.206	–0.166	–0.142
	(2.17)*	(2.50)**	(2.64)**	(2.65)**	(2.96)**	(2.93)**
GROUP	–0.456	–0.171	–0.047	–0.373	–0.086	–0.141
	(2.37)**	(0.90)	(0.31)	(2.12)*	(0.48)	(0.32)
IDPAV1	–1.186	–1.616	–1.502	–1.118	–1.463	–1.121
	(4.74)**	(6.59)**	(6.01)**	(4.75)**	(6.21)**	(4.59)**
IDPAV2	–1.015	–1.579	–1.381	–0.949	–1.479	–1.179
	(4.56)**	(6.50)**	(6.01)**	(4.24)**	(5.77)**	(5.15)**
IDPAV3	0.027	–0.239	–0.347	0.047	–0.232	–0.104
	(0.09)	(0.94)	(1.66)	(0.18)	(0.85)	(0.49)

Table 4.4 Research intensity: quantile behaviour (continued)

	Quantiles: no funding			Quantiles: with funding		
	0.25	0.50	0.75	0.25	0.50	0.75
IDPAV4	−0.160	−0.619	−0.772	−0.095	−0.568	−0.512
	(0.50)	(2.60)**	(3.15)**	(0.32)	(2.63)**	(2.11)*
FREG				0.442	0.438	0.290
				(2.24)**	(2.40)**	(1.98)**
FFED				0.259	−0.125	−0.128
				(0.81)	(0.57)	(0.56)
FEUF				1.111	0.730	0.717
				(3.78)**	(2.44)**	(3.15)**
FEUO				−0.439	−0.061	0.055
				(0.99)	(0.15)	(0.13)
Obs.	938	938	938	938	938	938
Pseudo R^2	0.111	0.122	0.153	0.125	0.136	0.167

Notes: The dependent variable is the log of research intensity, LRDINS. The t-values are based on standard errors obtained by the bootstrap design matrix method. Bootstrap re-sampling was performed using 1,000 repetitions. The t-tests are one-tailed or two-tailed, according to a priori expectations with respect to the parameter values. The pseudo R^2 is obtained as one minus the proportion of the sum of absolute weighted deviations over the raw sum of deviations. Symbols * and ** denote significance at the 5% and 1% levels.

As far as research cooperation is concerned, the estimates of the equation on exclusive funding on the left-hand side of Table 4.4 show that the parameter estimates are highly significant in all quantiles. In the case of research cooperation, COOP, both the parameter estimate and the statistical significance increase somewhat from quantile to quantile.

Similar results are found for incoming knowledge spillovers, INKS. The parameter estimate remains fairly constant, but its statistical significance increases slightly, quantile-by-quantile. This contrasts sharply with the variables related to the nature of knowledge, whose impact appears strongest in the lower quantiles. In the case of appropriability, APP, there is a very strong impact in the lowest quantile and a gradual decrease in impact, which renders this variable completely non-significant in the highest quantile. In the case of complexity of knowledge, CKNOW, there is only a significant impact in the lowest quantile.

The inclusion of funding on the right-hand side of the table has the consequence of reducing the parameter estimates for the respective knowledge-related variables, and particularly so for research cooperation, COOP, and complexity of knowledge, CKNOW, as it did with the average estimates. Research cooperation now becomes non-significant at a 5% level in the lower quantiles (although it remains so at a 10% level in a one-tailed test), and complexity of knowledge becomes completely non-significant throughout.

The equations both excluding and including funding suggest that stronger innovators in terms of research intensity have a more open innovation process than weaker ones, confirming the predictions of Chesbrough (2003a), Chesbrough and Crowther (2006) and other adherents of the 'open innovation paradigm'. Among strong R&D performers, firms cooperating in research tend to be more research-intensive than firms that do not cooperate, both because research cooperation increases the knowledge base and requires additional knowledge absorptive research effort, and also because research cooperation attracts public funding. Among weaker R&D performers, it is still the case that research cooperative firms spend relatively more on R&D than non-cooperative ones, but this appears to be mainly due to the fact that, by cooperating, they receive funding. In other words, firms that cooperate spend more on R&D if they receive funding than if they do not. This is particularly relevant in the lower quantiles, for example among R&D performers, where funding appears to be the main reason why cooperating firms spend more on R&D than non-cooperating ones. As such, the impact of public funding on the R&D spending of research cooperators and on the extent of research cooperation appears relatively strongest among weaker R&D performers. These results are confirmed when splitting the cooperation variable as in Table 4.2a and b.

The variable on human capital, LHEI, is on both sides of Table 4.4 found to be significant in at least some of the quantiles, confirming the average estimates in Table 4.1, but there is in this case no clear quantile-by-quantile pattern. In the case of export orientation, LEXI, the significance found in the average estimates is only confirmed in the lower quantile of the equation excluding funding. Including funding renders export orientation completely non-significant in all quantiles. This would suggest that it is mainly among weaker innovators that competitive pressure in foreign markets necessitates increased research efforts, and consequently that public authorities respond by providing innovation funding for the firms concerned.

The negative impact of firm size, LSIZE, on research intensity, found in the average estimates, is confirmed quantile-by-quantile. In this case, this effect is reinforced by a negative effect of group membership, GROUP, but only in the lower quantile.

Finally, the strong negative impact of the IDPAV1 and IDPAV2 industry dummies, observed in the average estimates in Table 4.1, is confirmed quantile-by-quantile on both sides of Table 4.4, and the more modest negative impact of IDPAV4 seems due to the higher quantiles. These results confirm that general technological characteristics ensure that resource-intensive, scale-intensive and, to a lesser extent, science-based industries are, in Belgium, less research-intensive than specialised supplier and information-intensive sectors.

Similarly, Table 4.5 presents the quantile regression estimates of the equation explaining innovative turnover, using the sample of 724 research-active product innovators.

Table 4.5 Innovative turnover: quantile behaviour

	Quantiles: no funding			Quantiles: with funding		
	0.25	0.50	0.75	0.25	0.50	0.75
Constant	–1.525	–0.784	–0.074	–1.565	–0.794	–0.044
	(5.74)**	(3.75)**	(0.35)	(6.39)**	(3.80)**	(0.21)
LRDINS	0.093	0.114	0.147	0.104	0.115	0.132
	(3.21)**	(4.29)**	(4.79)**	(3.33)**	(4.40)**	(4.59)**
COOP	0.124	0.005	–0.066	0.154	0.008	–0.064
	(1.09)	(0.05)	(0.59)	(1.36)	(0.07)	(0.56)
INKS	0.253	0.020	0.313	0.284	0.071	0.287
	(0.85)	(0.08)	(1.01)	(1.02)	(0.30)	(0.97)
APP	0.047	0.056	–0.050	0.086	0.077	–0.041
	(0.31)	(0.48)	(0.33)	(0.55)	(0.61)	(0.27)
BR	–0.144	–0.246	–0.292	–0.197	–0.270	–0.356
	(0.56)	(1.10)	(1.28)	(0.77)	(1.20)	(1.57)
PCS	0.256	0.127	0.187	0.278	0.128	0.216
	(1.92)*	(1.13)	(1.64)*	(1.99)**	(1.14)	(1.90)*
LHEI	–0.003	0.029	0.027	0.002	0.031	0.026
	(0.10)	(0.62)	(0.77)	(0.06)	(0.68)	(0.72)
LEXI	–0.023	–0.011	–0.044	–0.023	–0.011	–0.029
	(1.04)	(0.67)	(1.60)	(1.09)	(0.64)	(1.14)
LSIZE	–0.139	0.086	–0.124	–0.126	–0.074	–0.128
	(3.37)**	(2.74)**	(3.23)**	(2.91)**	(2.14)*	(3.32)**
GR	0.002	0.065	0.107	0.019	0.008	0.079
	(0.01)	(0.64)	(0.99)	(0.15)	(0.08)	(0.69)
IDPAV1	–0.520	–0.482	–0.400	–0.511	–0.487	–0.379
	(3.66)**	(3.67)**	(2.62)**	(3.52)**	(3.49)**	(2.27)*
IDPAV2	–0.514	–0.258	–0.116	–0.482	–0.245	–0.192
	(2.82)**	(1.73)	(0.65)	(2.61)**	(1.60)	(1.04)
IDPAV3	0.031	0.017	0.000	0.018	0.032	–0.090
	(0.19)	(0.11)	(0.00)	(0.11)	(0.20)	(0.51)

Table 4.5 (continued)

	Quantiles: no funding			Quantiles: with funding		
	0.25	0.50	0.75	0.25	0.50	0.75
IDPAV4	–0.434	–0.195	–0.345	–0.465	–0.225	–0.376
	(1.80)	(1.38)	(1.96)*	(1.88)	(1.48)	(2.17)*
FREG				–0.010	–0.136	–0.138
				(0.07)	(1.13)	(1.18)
FFED				0.195	–0.054	0.110
				(1.16)	(0.28)	(0.50)
FEUF				–0.284	0.163	0.271
				(1.11)	(0.64)	(1.38)
FEUO				0.551	–0.057	0.182
				(1.55)	(0.18)	(0.38)
Obs.	724	724	724	724	724	724
Pseudo R^2	0.070	0.086	0.091	0.075	0.088	0.098

Notes: The dependent variable is the log of innovative turnover, LITURN. The t-values are based on standard errors obtained by the bootstrap design matrix method. Bootstrap re-sampling was performed using 1,000 repetitions. The t-tests are one-tailed or two-tailed, according to a priori expectations with respect to the parameter values. The pseudo R^2 is obtained as one minus the proportion of the sum of absolute weighted deviations over the raw sum of deviations. Symbols * and ** denote significance at the 5% and 1% levels.

As in the case of the average estimates in Table 4.3, inclusion of the funding variables on the right-hand side of the table does not affect the results, as these variables are always found to be non-significant in all quantiles. This confirms that there is no direct influence of funding on product innovativeness and that its impact works exclusively through research intensity.

As in the case of the average estimates in Table 4.3, the key explanatory variable is the level of research intensity. Interestingly, its impact is seen to increase from quantile to quantile. Strong output innovators are, it appears, able to organise their research process more efficiently and, thereby, to achieve higher returns on their R&D investments. None of the other knowledge-related variables, nor worker's qualification nor export orientation, seem to have a direct impact on innovative output. Rather, as in the case of the average estimates in Table 4.3, their influence on innovative output works indirectly through their effect on research intensity.

Apart from the level of research intensity, process innovation, PCS, also appears to matter, as was the case in the average estimates in Table 4.3. Its parameter estimates show a U-shaped pattern, being highest in the lowest quantile, strongly decreasing in the median and increasing again in the highest. As more complex process innovations contributing to important product innovations are likely to require more in-house research efforts by the firm, part of their impact on innovative turnover is bound to be captured by the LRDINS variable itself. As in the case of research intensity, the negative impact of firm size, LSIZE, found in the average estimates is confirmed quantile-by-quantile. Finally, evidence of a significant negative impact of IDPAV1 is confirmed in all quantiles and appears particularly strong in the lower ones. Evidence of a significant negative impact of IDPAV2 appears to be exclusively due to the lowest quantile. This suggests that, when controlling for the influence of research intensity, not only is innovative output lower on average in the more low-tech resource-intensive and scale-intensive industries, but particularly so among weak innovators.

4.5 CONCLUSIONS

A regression analysis was presented of the relation between public funding and the level of research intensity and of innovative turnover on harmonised firm-level data from the Third Community Innovation Survey for Belgium. Four types of funding were considered: regional public funding, federal public funding, EU Framework Programme funding and other EU funding.

The initial focus was on the average behaviour of firms. Both regional public funding and EU Framework Programme funding are found to have a significant positive direct impact on research intensity. When considering the percentages of firms involved, EU framework funding appears to have the greatest impact. Differences in objectives between the types of funding could explain this. The particularly strong impact of EU framework program funding can be explained by the focus on critical mass and scale and scope. In the case of regional funding it is related to tailored support. This contrasts with federal funding aimed at the provision of infrastructure and other EU funding initially aimed at broader objectives than R&D.

This chapter paid special attention to the relation between funding and research cooperation. The latter turned out not only to have a significant positive effect on research intensity, but to correlate highly with regional and EU Framework Programme funding. The latter finding is in line with the objective to stimulate the cooperation present in both regional and European Framework Programme funding. In refining cooperative behaviour, these relations turned out to be correlated to the presence of inter-firm co-operative activities in particular. At first sight, this could be at odds with the intention of funding programmes to stimulate industry–science relations. However, the high correlation with funding due to complementarities between short-term market research and longer-term strategic research can explain this. Not surprisingly, this is also reflected in the geographical scope of the cooperation (regional funding is related more to domestic cooperation whereas EU FP funding is related more to international cooperation). This is in line with the aim of the EU Framework Programmes to foster greater coherence between the Member States. Given the positive direct effect of research cooperation on research intensity, we conclude that, at least in the Belgian case, regional and EU Framework Programme funding helps open the research process by means of cooperation and as such stimulates positively the firm's research intensity. This effect could not be found for central government funding and funding by other EU programs.

Also, other knowledge-related variables such as incoming knowledge spillovers and appropriability have a significant direct influence on research intensity. This confirms that the absorption of external knowledge requires substantial further research efforts by firms and that the capacity for appropriating in-house knowledge with respect to non-partners increases incentives to undertake R&D. The same holds true for firms undertaking more complex research.

The level of research intensity is the main variable explaining innovative turnover. Process innovations also matter. The influence of funding on innovative turnover turned out to be exclusively indirect, through research intensity. This confirms that public innovation funding is used mainly to finance in-house R&D and that, to the extent that it helps finance other aspects of the innovation process, these appear less important for product innovation. The impact of knowledge-related variables such as incoming knowledge spillovers and research cooperation is also indirect, through research intensity. This emphasises the necessity of knowledge absorption firstly, before it can be used in the innovation process.

Quantile regression estimates show that the average behaviour of firms does, however, hide substantial differences between strong and weak innovators. The impact of incoming knowledge spillovers and research cooperation on research intensity is seen to increase quantile-by-quantile in relation to that of appropriability. This suggests that firms which are more research-intensive have a more open innovation process than the less research-intensive ones. Among strong R&D performers, research co-operative firms tend to be more research-intensive than those not co-operating, both because research cooperation requires additional research effort and because it enjoys public funding. Among weaker R&D performers, it is still the case that research cooperative firms spend relatively more on R&D than non-cooperative ones, but this appears mainly due to the fact that, by cooperating, they receive public funding. As a consequence, the impact of public funding on the R&D spending of research cooperative firms and on the extent of research cooperation appears relatively strongest among weaker R&D performers. This is where the opening up of the research process is most needed.

Finally, quantile regression estimates confirm that the impact of public funding on innovative turnover is indirect, through research intensity. They show that research intensity is not only the key explanatory variable of innovative turnover, but that its impact tends to increase quantile-by-quantile. When considered in conjunction with the results on research intensity, this implicitly suggests that the greater openness of the innovation process of highly research-intensive firms allows them to achieve significant gains in product-innovative efficiency. Helping to achieve such gains in innovative efficiency is one of the main justifications for fostering research cooperation through public funding.

APPENDIX

Table A4.1 gives a description of the manner in which the variables were constructed from the raw data of the CIS3 questionnaire. Table A4.2 presents descriptive statistics of the variables with respect to the total and research-active samples, and the corresponding Pavitt industry dummies. And Table A4.3 presents the correlation patterns between the variables on funding and the other variables in the equation on research intensity.

Table A4.1 Description of the variables

Variable	Definition
Log of research intensity (LRDINS)	Natural log of the ratio of intramural R&D expenditure over turnover in the year 2000
Log of innovative turnover (LITURN)	Natural log of the % of new or significantly improved products (goods and services) in turnover in the year 2000
Research cooperation (COOP)	Cooperation = 1 if the firm collaborates either with suppliers, clients, competitors, research labs, universities or government
Incoming knowledge spillovers (INKS)	Sum of scores of the importance attached to information from suppliers, clients, competitors, universities, government, professional conferences, exhibitions (rescaled between 0 and 1)
Appropriability (APPR)	Average score of the effectiveness of strategic protection through secrecy, complexity of design and lead-time advantage
Complexity of knowledge (CKNOW)	Ratio of the sum of the scores of the importance attached to information from universities, government and professional conferences over the sum of the scores of all sources of information
Log of higher education intensity (LHEI)	Natural log of the % employees with higher education in the year 2000
Log of export intensity (LEXI)	Natural log of the proportion of exports in turnover in the year 1998
Log of size (LSIZE)	Natural log of the number of employees in the year 1998
Group affiliation (GROUP)	Group affiliation = 1 if the firm is member of a group
Process innovation (PCS)	Process innovation = 1 if the firm has introduced new or significantly improved production processes during the period 1998–2000

Table A4.1 (continued)

Variable	Definition
FREG	FUNDLOC = 1 if the firm received public funding for innovation during the period 1998–2000 from regional authorities
FFED	FUNDGMT = 1 if the firm received public funding for innovation during the period 1998–2000 from the federal government
FEUF	FEUF = 1 if the firm received funding for innovation from the 4th (1994–98) or 5th (1998–2002) EU Framework Programmes
FEUO	FEUO = 1 if the firm received funding for innovation from the EU during the period 1998–2000 and no funding from the 4th or 5th Framework Programmes
IDPAV1	IDPAV1 = 1 if the firm is classified in the 'Resource-intensive' sectors (mining, food, beverages and tobacco, wood and furniture, recycling, wholesale)
IDPAV2	IDPAV2 = 1 if the firm is classified in the 'Scale-intensive' sectors (paper and printing, rubber and plastics, non-metallic minerals, ferrous metals, non-ferrous metals, fabricated metal products, motor vehicles, other transport equipment)
IDPAV3	IDPAV3 = 1 if the firm is classified in the 'Specialised supplier' sectors (non-electrical machinery, electronics and communication, electrical machinery)
IDPAV4	IDPAV4 = 1 if the firm is classified in the 'Science-based' sectors (chemicals, pharmaceuticals, computers and office machinery, scientific instruments)
IDPAV5	IDPAV5 = 1 if the firm is classified in the 'Information-intensive' sectors (financial intermediation, telecommunication, transport services, business services, computer and related service activities, research and development, technical engineering)

Table A4.2 Descriptive statistics: mean and standard deviation

Variable	All firms	Research-active firms *	*Of which*: Res.-intensive	Scale-intensive	Special. supplier	Science-based	Inform.-intensive
Obs.	1,987	938	220	239	130	117	232
LRDINS		–4.625	–5.400	–5.194	–4.022	–4.231	–3.840
		(1.825)	(1.587)	(1.510)	(1.672)	(1.535)	(2.056)
LITURN		–1.836	–2.178	–2.044	–1.559	–2.015	–1.422
		(1.056)	(1.035)	(1.101)	(0.885)	(1.037)	(0.960)
COOP		0.361	0.304	0.359	0.423	0.393	0.366
		(0.480)	(0.461)	(0.481)	(0.495)	(0.491)	(0.483)
COOP_SCC		0.304	0.254	0.293	0.338	0.316	0.336
		(0.460)	(0.436)	(0.456)	(0.475)	(0.467)	(0.473)
COOP_RUG		0.242	0.200	0.238	0.307	0.324	0.207
		(0.428)	(0.400)	(0.427)	(0.463)	(0.470)	(0.406)
COOP_DOM		0.310	0.272	0.288	0.376	0.325	0.323
		(0.462)	(0.446)	(0.454)	(0.486)	(0.470)	(0.469)
COOP_FOR		0.271	0.214	0.284	0.300	0.342	0.262
		(0.445)	(0.411)	(0.452)	(0.460)	(0.476)	(0.441)
INKS		0.404	0.379	0.397	0.439	0.428	0.400
		(0.213)	(0.222)	(0.209)	(0.208)	(0.193)	(0.219)
APPR	0.193	0.320	0.257	0.308	0.420	0.393	0.300
	(0.322)	(0.369)	(0.352)	(0.360)	(0.405)	(0.378)	(0.353)
CKNOW		0.320	0.304	0.300	0.322	0.377	0.325
		(0.245)	(0.285)	(0.251)	(0.246)	(0.251)	(0.228)
PCS		0.730	0.745	0.799	0.646	0.700	0.706
		(0.444)	(0.436)	(0.401)	(0.480)	(0.459)	(0.456)
LHEI	–2.313	–1.913	–2.385	–2.505	–2.104	–1.522	–0.947
	(1.838)	(1.421)	(1.501)	(1.227)	(1.394)	(0.884)	(1.189)
LEXI	–3.004	–2.355	–2.404	–1.872	–1.572	–1.460	–3.697
	(2.860)	(2.734)	(2.641)	(2.409)	(2.299)	(2.238)	(3.047)
LSIZE	3.795	4.136	4.194	4.415	4.222	4.343	3.641
	(1.301)	(1.433)	(1.317)	(1.439)	(1.537)	(1.533)	(1.300)
GROUP	0.561	0.634	0.568	0.698	0.561	0.684	0.646
	(0.496)	(0.482)	(0.496)	(0.459)	(0.498)	(0.467)	(0.479)

Table A4.2 (continued)

Variable	All firms	Research-active firms *	*Of which*: Res.-intensive	Scale-intensive	Special. supplier	Science-based	Inform.-intensive
Obs.	1,987	938	220	239	130	117	232
FREG	0.128	0.233	0.150	0.238	0.392	0.290	0.189
	(0.334)	(0.423)	(0.357)	(0.427)	(0.490)	(0.456)	(0.393)
FFED	0.032	0.058	0.059	0.062	0.053	0.076	0.047
	(0.177)	(0.235)	(0.236)	(0.243)	(0.226)	0.267)	(0.212)
FEUF	0.027	0.055	0.023	0.025	0.084	0.060	0.099
	(0.161)	(0.228)	(0.149)	(0.156)	(0.279)	(0.238)	(0.299)
FEUO	0.020	0.036	0.031	0.042	0.038	0.051	0.026
	(0.142)	(0.187)	(0.175)	(0.200)	(0.193)	(0.221)	(0.159)

Notes: The standard deviation is presented between brackets beneath the mean. * In the case of LITURN this concerns research-active and product-innovative firms.

Table A4.3 Public funding by source: correlation patterns

	FREG (i)	FFED (ii)	FEUF (iii)	FEUO (iv)
LRDINS	0.192	0.004	0.206	–0.019
	(5.97)**	(0.12)	(6.44)**	(0.59)
COOP	0.209	0.057	0.186	0.080
	(6.54)**	(1.77)	(5.80)**	(2.44)**
COOP_SCC	0.129	0.062	0.184	0.070
	(3.97)**	(1.90)	(5.74)**	(2.15)*
COOP_RUG	0.318	0.060	0.266	0.104
	(10.25)**	(1.85)	(8.43)**	(3.18)**
COOP_DOM	0.218	0.058	0.180	0.104
	(6.84)**	(1.78)	(5.60)**	(3.20)**
COOP_FOR	0.167	0.051	0.229	0.074
	(5.17)**	(1.58)	(7.20)**	(2.26)*
INKS	0.172	0.052	0.155	0.017
	(5.34)**	(1.60)	(4.80)**	(0.54)
APPR	0.147	0.091	0.126	0.007
	(4.53)**	(2.78)**	(3.89)**	(0.21)
CKNOW	0.208	0.092	0.128	0.013
	(6.50)**	(2.83)**	(3.94)**	(0.40)
LHEI	0.082	–0.010	0.116	0.009
	(2.50)**	(0.32)	(3.56)**	(0.26)
LEXI	0.139	0.064	0.077	–0.026
	(4.28)**	(1.98)*	(2.37)**	(0.80)
LSIZE	0.076	0.044	0.074	0.038
	(2.33)**	(1.33)	(2.27)*	(1.17)
GROUP	–0.025	0.010	0.019	0.052
	(0.78)	(0.32)	(0.60)	(1.61)
IDPAV1	–0.109	0.001	–0.079	–0.013
	(3.36)**	(0.03)	(2.42)**	(0.40)
IDPAV2	0.007	0.010	–0.077	0.017
	(0.21)	(0.31)	(2.38)**	(0.53)
IDPAV3	0.151	–0.008	0.051	0.004
	(4.66)**	(0.25)	(1.57)	(0.15)
IDPAV4	0.051	0.029	0.007	0.030
	(1.56)	(0.90)	(0.22)	(0.93)
IDPAV5	–0.059	–0.027	0.109	–0.031
	(1.81)	(0.84)	(3.37)**	(0.97)

Table A4.3 (continued)

	FREG (i)	FFED (ii)	FEUF (iii)	FEUO (iv)
FREG		–0.052	0.219	0.095
		(1.59)	(6.86)**	(2.92)**
FFED	–0.052		0.078	0.024
	(1.59)		(2.40)**	(0.75)
FEUF	0.219	0.078		–0.047
	(6.86)**	(2.40)**		(1.44)
FEUO	0.095	0.024	–0.047	
	(2.92)**	(0.75)	(1.43)	
Obs.	938	938	938	938

Notes: The absolute value of the t-values is presented between brackets under the correlation coefficients. The t-test is two-tailed. Symbols * and ** denote significance at the 5% and 1% levels.

5. Public funding for innovation and research cooperation

5.1 INTRODUCTION

In innovation policy circles in Europe there has, in the past two decades, been a growing awareness of the importance of fostering cooperative research. This is reflected by an increasing proportion of public funding awarded by the various regional, national and EU authorities to cooperative research networks rather than individual firms. Despite this development, the bulk of the econometric evaluation literature on public funding for innovation concentrates almost exclusively on the question of whether funding actually helps increase R&D expenditure by firms or the number of patents granted.

Only a very limited number of studies have addressed the issue of whether public funding for innovation does indeed succeed in influencing the cooperative practices of firms. As stated below, these studies are not without problems. Some fail to address the problem of endogeneity caused by funding selectivity. Of those that do address the issue, Belderbos et al. (2004) make use of an instrumental variable regression analysis explaining different types of research cooperation, using lagged values of the explanatory variables as instruments. But they admit that lagged funding may still be partly endogenous in the sense that it may have been brought about by past targeting of firms involved in research cooperation, whose co-operative behaviour persists during the current period. Busom and Fernandez-Ribas (2004, 2008) and Aschhoff et al. (2006) do, for their part, present non-parametric matching analyses. But, due to the cross-sectional nature of their data with respect to a specific point in time, there are serious doubts about the exogeneity of some of their control variables – especially those related to R&D. This will bias their estimated average treatment effects. A common weakness of these respective studies is, moreover, that they always concentrate exclusively on the impact of a single measure of funding, relating to total or national funding. But since the various funding authorities apply partially different project-selection criteria, it would be interesting to discover whether there are differences in impact between different types of funding.

In this chapter we present a more systematic analysis concentrating on Belgium. Belgium has the peculiarity of a relatively strongly decentralised public funding organisation structure with distinct regional and federal levels, and is at the same time a member of the EU. Attention will therefore be drawn to the impact of funding by the regional and federal governments, as well as by the EU. Use is made of a database, constructed from the Third and Fourth Community Innovation Surveys (CIS3 and CIS4) with respect to Belgium.

A cross-section regression analysis helps first to explain the firms' respective cooperation with other firms and with public research organisations (mainly universities) by the considered types of funding and a set of control variables. The dependent variables relate to the period 2002–2004 (from CIS4), whereas the explanatory variables are instrumented by their lagged value with respect to the period 1998–2000 (from CIS3). This regression analysis is then complemented by a non-parametric matching analysis, explaining the average treatment effect on the two types of cooperation by the respective types of funding. The control variables used are again instrumented by their lagged values (from CIS3).

The chapter is organised as follows. Section 5.2 summarises the relevant findings in the theoretical and empirical literature, gives a brief account of the practice of funding in Belgium and expresses the research questions. Section 5.3 presents the empirical model, and describes the data and methods of estimation. The estimates are presented in Section 5.4, whereas Section 5.5 gives a discussion and concludes. Tables with a detailed description of the variables and with summary statistics are presented in the Appendix.

5.2 RESEARCH COOPERATION AND PUBLIC FUNDING FOR INNOVATION

5.2.1 Theoretical Background

Several hypotheses have been proposed in the industrial organisation and management literature for explaining research cooperation (for recent surveys, see Caloghirou et al., 2003; and Sena, 2004). These can be summarised as follows. First, due to its properties as a non-rival, only partially-excludable good, the generation of knowledge leads to knowledge spillovers (Arrow, 1962; Romer, 1990). A research partnership may improve learning efficiency in absorbing external knowledge, thereby increasing the impact on innovative performance of incoming spillovers. It may at the same time also provide a mechanism to reduce outgoing spillovers, by helping to internalise them. However, cooperative agreements require a sufficient degree of knowledge appropriability so as to reduce free-rider opportunities for outsiders (Kesteloot and Veugelers, 1995; Martin, 2002; Cassiman et al., 2002). Second, research cooperation may give access to intangible tacit knowledge and know-how, which does not spill over and cannot easily be contracted through market transactions (Hagedoorn, 1993; Katsoulakos and Ulph, 1998). Third, research partnerships enable exploitation of economies of scale and scope in both research and development, thereby reducing innovation costs. And, in the same vein, they help share risks (Röller et al., 1997).

Although the theoretical literature is less specific on this, the choice of different types of research partners (customers and suppliers, competitors or public research organisations) will tend to be affected by the relevance of these various factors for the firms. The nature of projects involving public research organisations as partners will generally tend to have a more risky character with a longer gestation period than is the case for mere inter-firm projects.

The theoretical rationale for public intervention in the innovation process follows directly from the character of knowledge as a non-rival, only partially-excludable good and the consequent knowledge spillovers (Arrow, 1962; Romer, 1990). These characteristics generally tend to lead to an underinvestment in innovation efforts by private, profit-motivated individual decision-makers from a social point of view, or, in other words, to market failure.

This market failure may express itself not only in an underinvestment in privately financed R&D but in an insufficient exploitation of the opportunities of innovative efficiency-enhancing research cooperation. This failure is likely to be especially important as far as the cooperation between firms and public research organisations is concerned (Martin and Scott, 2000). Although various policy instruments exist to enhance innovation performance by firms, public funding is generally viewed as most suited to the fostering of research cooperation, and in particular the creation of networks involving both firms and public research organisations such as universities (OECD, 2002b).

5.2.2 Empirical Issues

Although there is quite a lot of econometric work explaining research cooperation or the impact of public funding on innovation inputs or outputs, such as R&D effort and granted patents, only a limited number of studies have thus far investigated the influence of public funding on research cooperation. By making use of Community Innovation Survey (CIS) or other, similar survey data for European countries, some studies have presented cross-section regression analyses linking research partner choice to public funding. The main conclusion from these studies is that R&D subsidies appear to have a stronger impact on the probability of cooperating with public research organisations than with other firms. Moreover, the influence of subsidies appears to be stronger on vertical cooperation (with customers and suppliers) than on horizontal cooperation (with competitors) (see Miotti and Sachwald, 2003; Mohnen and Hoareau, 2003; Dachs et al., 2004; Abramovsky et al., 2005). Unfortunately, none of these studies faces the endogeneity problem arising from the selectivity of funding. Not only may funding be deliberately part-targeted at firms already cooperating, thereby implying a reverse causation, but, moreover, some factors that determine the likelihood of cooperation are very likely to affect the likelihood of receiving funding as well. To the extent that some of these are not observed, the funding variable will also, for this reason, be correlated with the error term, leading to biased and inconsistent estimates.

Of the few studies that address the endogeneity issue, Belderbos et al. (2004) present an instrumented variable cross-section regression analysis explaining the probability of horizontal, vertical or private–public research cooperation on Dutch CIS2 data for the period 1996–1998. Both the funding variable (total funding) and the respective control variables are instrumented by their lagged value from corresponding CIS1 data for the period 1994–1996. Lagged funding is shown to have a significant positive impact on the likelihood of vertical and private–public cooperation. The authors do, however, admit that lagged funding may still be partly endogenous, in the sense that it may have been brought about by past targeting of firms involved in research cooperation, whose cooperative behaviour persists during the period under consideration. In order to check this, they repeat estimation while restricting the sample to include only new cooperators (for example, firms that cooperate in CIS2 but not in CIS1). As a consequence, the funding variable becomes non-significant. It is, however, unclear whether this can be viewed as confirmation that lagged funding is still endogenous, or rather as a reflection of the loss in information caused by the substantial sample size reduction.

The remaining studies that have addressed the issue of endogeneity have done so by using a non-parametric matching approach. Thus, Busom and Fernandez-Ribas (2004, 2008) have estimated the average treatment effect by national funding on vertical research cooperation and on cooperation with public research organisations on a cross-section of Spanish research active manufacturing firms in 1998. Aschhoff et al. (2006) have estimated the average treatment effect by national funding on cooperation exclusively with other firms, cooperation only with public research organisations, and cooperation with both, on a cross-section of German cooperating firms in the period 1998–2003. The results suggest that there is a significant impact of funding on research cooperation and that this impact is strongest on cooperation with public research organisations or with a combination of these and other firms. The authors view this as evidence of 'behavioural additionality', in the sense that public funding does, indeed, influence the organisation of the research process of firms by stimulating cooperation and by affecting the partner choice.

However, one problem with these estimates is that, due to the cross-sectional nature of the data on a specific moment in time, there are serious doubts as to the exogeneity of the control variables under consideration. This is particularly the case with respect to the R&D variables used, such as the number of researchers or the various R&D dummies. The assumption that these are exogenous with regard to funding would imply ruling out the possibility that funding may affect the research efforts of firms, or, in other words, 'input additionality'. This contradicts the bulk of the econometric evidence on project evaluation, which suggests the existence of at least partial 'input additionality'. As a consequence, this endogeneity of the R&D related control variables will give the estimated average treatment effects an upward bias and make them inconsistent.

Making use of additional data obtained through a self-conducted survey on a subset of firms, Aschhoff et al. (2006) have also estimated probit equations explaining the probability of continued research cooperation after funding has ended. They include among the explanatory variables a dummy depicting the probability of new cooperation initiated through funding. The negative significance of the latter variable in the case of cooperation with research organisations would, according to the authors, suggest that this type of cooperative agreement is not as long-lived when it is set up through funding than when born spontaneously. Or, in other words, the 'behavioural additionality' would be more short-run in nature in this case. There are, however, two problems with these estimates. First, the new cooperation dummy is not drawn independently from the continued cooperation dummy. As a consequence, the parameter estimate of the new cooperation variable will suffer from a simultaneity bias and be inconsistent. Second, since funding tends to be highly persistent, finding firms whose project funding ended (and was not replaced) necessarily implies a vast reduction in sample size. This poses serious questions as to its representativeness.

It is clear that the econometric literature on the relation between public funding and research cooperation is in its infancy. In what follows, we will concentrate on the as-yet unresolved issue of whether funding impacts research cooperation at all.

5.2.3 Sources of Public Funding for Innovation

We consider three sources of public funding for innovation in Belgium: regional government funding, federal government funding (including institutions working on behalf of federal government) and EU funding. Public funding includes financial support in terms of grants and loans, including a subsidy element, and loan guarantees. Ordinary payments for orders of public customers are not included. Following the institutional reforms in Belgium, which transferred competency for industrial and research policy to the regions (the Flemish, the Walloon and the Brussels region), the federal government retained a coordination and stimulation role at the national level and an interface role at the European level (Capron et al., 2000). Federal funding follows a top-down logic. It is aimed mainly at stimulating research by firms in specific domains, such as health, space programs and defence (Nauwelaers, 2001).

Regional funding is largely inspired by a bottom-up approach and therefore tailored to the demands of the enterprises. Decentralisation makes it possible for regions to accommodate the specificities of enterprises located in their territory. Regionally-funded programs are increasingly aimed at fostering cooperative research networking, in order to stimulate the use of scientific knowledge by firms. Firms setting up research-cooperative partnerships with public research organisations, such as universities, and undertaking basic research are particularly favoured (Capron et al., 2000).

Despite its growing importance, EU funding still only accounts for a relatively modest proportion of total funding for innovation in Europe (Polt and Psarra, 2006). In Belgium it accounted for 1.1 % of total business R&D financing against 4.3 % for regional funding in 2002 (CFS/STAT, 2007). EU funding largely flows top-down to specific themes. The EU Framework Programs are its most important mechanism. Their main aim is to foster cross-border research cooperation, in order to increase the scale and scope of research efforts, to achieve a 'critical mass' not attainable at national level, especially in areas characterised by uncertainty and high risk. Domains that received special attention in the Fourth and Fifth Framework Programs, with respect to 1994–1998 and 1998–2002, are ICT, new materials, biotech and energy/environment (Muldur et al., 2006). The creation of international networks including both firms and public research organisations from different countries is particularly favoured.

5.2.4 Research Questions

In this chapter an attempt will be made to answer the following research questions. First, does total public funding for innovation, irrespective of the source, foster research cooperation by firms in Belgium? If so, does it affect partner choice, and more precisely the choice between cooperation with business (other firms) and cooperation with public research organisations (mainly universities)? Second, are there differences in impact on research cooperation between regional government funding, federal government funding and EU funding? If so, how can they be explained? Do they reflect the relative importance of the different sources of funding? Do differences in project selection criteria by the funding authorities affect partner choice?

5.3 DATA, EMPIRICAL MODEL AND ESTIMATION METHODS

5.3.1 Data

This chapter draws on data for Belgium gathered as part of the Third and Fourth European Community Innovation Surveys, CIS3 and CIS4, conducted in 2002 and 2006 and relating to the periods 1998–2000 and 2002–2004. The target population included all enterprises with ten or more employees. A survey stratified by size (small, medium and large) and economic activity (NACE) was made, covering about one third of the target population each time. The rate of useful answers was always slightly less than 50% and a non-response analysis revealed no systematic non-response bias. After excluding firms with incomplete data or inconsistencies, samples remained from 1,987 and 2,740 enterprises. From these, a subset of 601 firms that are present in both could be identified. This forms our main sample for investigation. Of these firms, 371 reported being innovation-active in 2002–2004, 417 did so in 1998–2000 and 314 reported being so in both.

Concentration on firms present in both periods is justified by the fact that lagged values will be used as instruments. Performing estimation on a sample that includes both firms that do report being innovative and firms that do not is justified for two reasons. First, our key variables depict probabilities of cooperation in research and of receiving public funding that gain in relevance when considered with respect to a larger population of firms than merely those reporting being innovation-active. Second, most firms can be thought of as innovation-active in some sense – not just those who officially report that they are. Quite a few new ideas emerge and experimentation takes place in an informal manner, outside the R&D department, due to qualified production workers. Nonetheless, as a check of robustness, we repeat estimation on the sub-sample of 314 firms that report themselves innovation-active in both periods.

Table 5.1 gives the frequency of research cooperation and of public funding for innovation by type in our sample.

Table 5.1 Frequency of research cooperation and public funding for innovation by type

Variable	CIS3 (1998–2000)		CIS4 (2002–2004)	
	Obs.	Per cent	Obs.	Per cent
Total	601	100.0 %	601	100.0 %
Innovative in period	417	69.4%	371	61.7%
Innovative in both periods	314	52.2%	314	52.2%
Cooperate with business	125	20.8 %	183	30.4 %
Cooperate with institutions	105	17.4 %	120	20.0 %
Total funding	127	21.1%	133	22.1%
Regional funding	98	16.3%	102	17.0 %
Federal funding	22	3.7%	43	7.2%
EU funding	37	6.1%	41	6.8%
Regional funding only	72	11.9%	63	10.5%
Federal funding only	13	2.1%	21	3.5%
EU funding only	11	1.8%	7	1.2%
Reg. and Fed. funding	3	0.0%	9	1.5%
Reg. and EU funding	20	3.3%	21	3.5%
Fed. and EU funding	3	0.0%	4	0.0%
Reg., Fed. and EU funding	3	0.0%	9	1.5%

The predominance of regional funding conforms with the highly regionalised nature of innovation policy in Belgium. The relatively low percentage of firms receiving EU funding underlines the difficulty in obtaining these funds. Note that the majority of firms that are able to do so also succeed in receiving mainly regional and to a lesser extent federal funding. This is not surprising since regional and federal funding is sometimes given to help obtain EU funding (OMC/CREST, 2007).

5.3.2 Empirical Model

The empirical analysis starts by estimating different versions of a bi-variate probit model to explain the likelihood of two types of research cooperation:

cooperation with business (other firms), COBUS, and cooperation with public research organisations, COPRO.

The dependent variables, COBUS and COPRO, depict the probabilities of engaging in research cooperation with business and with institutions. They are binary variables that take value 1 in the case of cooperation with the partners in question and 0 otherwise. Cooperation with business covers vertical cooperation (with suppliers and customers) and horizontal cooperation (with competitors). A split between vertical and horizontal cooperation was not possible due an insufficient number of observations for cooperation with competitors.

The equations are successively estimated with the following types of treatment: total funding, FTOT; regional government funding, FREG, federal government funding, FFED, and EU funding, FEU, where there may be an overlap in the sense that the same firm may receive more than one type of treatment; and, finally, again regional, federal and EU funding plus the combinations regional–federal funding, FRF, and regional–EU funding, FREU, where these respective categories are non-overlapping now. These respective variables are binary in nature. They take the value 1 in the case of type of funding or combined types of funding in question and 0 otherwise. Other possible combinations, such as federal–EU and regional–federal–EU could not be considered due to an insufficient number of observations which prevented estimation of the probit equations.

The first set of control variables is knowledge-related. Research and development activity, RDA, and patent activity, IPPAT, are binary variables that take the value 1 if the firm reports performing R&D and owning valid patents on its inventions. They are indicators of the firm's knowledge capital and reflect, in some way, its knowledge absorption capacity and innovative experience. They are expected to raise the likelihood of engaging in research cooperation. To the extent that patents are generally used to protect fairly complex innovations, we can expect the patent variable to have a particular effect on the likelihood of engaging in partnerships with public research organisations.

Other knowledge-related control variables depict the level of incoming knowledge spillovers, INKS, strategic appropriability of knowledge, APPR, and the complexity of knowledge, CKNOW. Incoming spillovers, INKS, are measured by the firm's average importance attached to various sources of information, including suppliers, customers, competitors, public research organisations, and fairs and exhibitions. They are expected to affect the likelihood of research cooperation positively to the extent that cooperation helps improve learning efficiency in absorbing the external knowledge concerned. Strategic appropriability, APPR, is measured by the average importance attached by the firm to secrecy, complexity in design and lead-time advantage on competitors. It is likely to have a positive impact on research cooperation because it reduces free-riding opportunities for outsiders to the cooperative agreement. Complexity of knowledge, CKNOW, is measured by the relative importance attached by the firm to information from public research organisations and professional conferences, as suggested by Cassiman and Veugelers (2002). It is expected to have a positive influence on cooperation because the proportion of intangible, tacit knowledge and know-how that does not easily spill over is likely to increase with the complexity of the research process. Cooperation is a way of gaining access to this knowledge. In view of their positive relation with the complexity of the research process, both strategic appropriability and complexity of knowledge are likely to affect cooperation with public research organisations in particular.

Human capital, measured by the log of higher education intensity of personnel, LHEI, is also included as a control variable, because firms disposing of highly qualified workers are better suited to the undertaking of cooperative research. As is the log of export intensity, LEXI, because firms that operate in foreign markets are likely to experience stronger competition. This may provide an incentive to increase the efficiency of innovative effort by cooperating in research.

The next set of control variables are all to some extent size-related. The most important of these is the log of size, measured by employment, LSIZE. Its impact is most likely to be positive, given that larger firms are more likely to have the means and absorptive capacity needed to engage in research cooperation. But it cannot be ruled out that small and medium-sized firms are also tempted to participate in cooperative research networks, for the very purpose of compensating for their handicaps of scale and scope. Two other size-related variables are domestic and foreign group membership, GRP_D and GRP_F. These are binary variables that take the value 1 when the firm is a member of a larger domestic or foreign group. The implied size effect may affect propensity to cooperate either way. However, in the case of foreign group membership in particular, we expect subsidiaries to be less likely to engage in cooperative research partnerships because they are able to access the knowledge in question through the parent company (Busom and Fernandez-Ribas, 2008). Furthermore, innovation hampering factors depicting risk and financial constraints, HRISK and HFIN, are also included. They are measured by the importance given by the firm to risk and financial constraints that help to prevent, delay or abandon innovation processes. Again, there are no a priori expectations as to the direction of their impact. If sufficiently important, they may prevent the firm reaching the threshold level of innovative effort necessary to engage in research cooperation. But, if not, they may, on the contrary, provide an incentive for engaging in a cooperative partnership to share risks and have access to more finance.

Finally, industry dummy variables based on Pavitt's taxonomy, IDPAV, are added depicting the general technological environment in which the firm operates, since this may affect propensity to cooperate in research. The dummies are based on the taxonomy originally proposed by Pavitt (1984) and elaborated by Tidd et al. (1997) to accommodate for the services sectors. This classification reflects differences in technological trajectories between groups of industries. The industries included in the different groups are mentioned in Table A5.1 in the Appendix.

The model can be presented as follows.

COBUS = f1(C, IDPAV, RDA, IPPAT, INKS, APPR, CKNOW, LHEI, LEXI, LSIZE, GRP_D, GRP_F, HRISK, HFIN, FTOT, FREG, FFED, FEU, FRF, FREU) (5.1)

COPRO = f2(C, IDPAV, RDA, IPPAT, INKS, APPR, CKNOW, LHEI, LEXI, LSIZE, GRP_D, GRP_F, HRISK, HFIN, FTOT, FREG, FFED, FEU, FRF, FREU) (5.2)

The models contain stochastic error terms that may be correlated as the decisions to engage in the respective types of research cooperation may be related.

5.3.3 Estimation Methods

The empirical analysis is started by estimation of the bi-variate probit model presented in equations (5.1) and (5.2) above. Estimation is performed by the method of maximum likelihood (ML), allowing for correlation between the error terms and with heteroscedasticity-robust standard errors.

Estimates are first presented, for illustrative purposes, of the model with total funding with the explanatory variables as such, in non-instrumented form (from CIS4). These are likely to suffer from endogeneity bias, not only in the case of funding due to selectivity, but also in the case of the control variables. These can be expected to be especially severe in the case of the knowledge-related variables, as they may all be expected to be correlated with un-measurable aspects of firm-specific 'innovativeness' which also affect the likelihood of cooperation. There may, moreover, be reverse causations from research cooperation to R&D effort, RD, and to incoming knowledge spillovers, INKS, as the knowledge generated through research cooperation itself requires further absorptive efforts and research cooperation leads to incoming spillovers from research partners.

Estimation is then repeated with the explanatory variables instrumented by their lagged values (from CIS3). After considering the model that includes total funding, FTOT, attention is turned successively to the model with this variable split into possibly overlapping regional, federal and EU funding, FREG, FFED and FEU, and finally to the model with mutually exclusive non-overlapping types and combinations of types of funding. However, due to an insufficient number of observations, causing perfect predictor problems, the combinations federal–EU funding and regional–federal–EU funding had to be dropped.

As explained above, there may still be some endogeneity in the case of funding as firms that attracted funding in the past because they were already cooperating may still pursue this cooperation during the current period under consideration. We therefore complement our investigation by applying a non-parametric matching procedure. The idea behind non-parametric matching is that of a controlled experiment. In order to find out what the outcome (the likelihood of undertaking a specific type of research cooperation) would have been if the treated (firms that received funding) had not received treatment, a matched control group is constructed consisting of firms with similar characteristics to those treated, only they did not receive treatment, that is, obtain funding. The non-parametric matching estimator estimates the average treatment effect on the treated (ATT) as the difference between the outcomes of the treated and the untreated from the matched control group.

In order to be valid, this method presupposes 'conditional independence', in the sense that the receipt of treatment and the potential outcome should be independent for firms with the same exogenous characteristics (Rubin, 1977). Furthermore, it requires 'common support', in the sense that all treated firms should have a counterpart in the non-treated population. If the firms in the treated and non-treated samples have no or few common characteristics, matching is not applicable.

If there would only be a single relevant common exogenous characteristic, this could be used as a control variable to condition the matching. If there are a large number of such characteristics, Rosenbaum and Rubin (1983) have shown that the matching can be conditioned on the propensity score (the probability of receiving the treatment). The matching procedure will, therefore, imply, in a first step, estimating the probit equation explaining the probability of receiving treatment (the funding concerned) and take its predicted value as an estimate of the propensity score. This is then used, in a second step, as a matching criterion. In order to obtain unbiased and consistent estimates of the ATT, the method requires that one include all the relevant control variables and that these are, indeed, exogenous with respect to both the outcome and the treatment. We will, therefore, in the matching procedure, make use of the same instrumented control variables considered in the regression analysis.

The specific matching procedure used will be 'nearest neighbour Mahalanobis matching' with 'replacement' and 'common support'. 'Nearest neighbour matching' implies one-to-one matching, in the sense that one ends up with a single non-treated twin firm for each treated one. 'Mahalanobis matching' means that we are using a hybrid matching procedure that conditions the matching not only on the propensity score of the treatment, but on a subset of the control variables used to estimate it. This lets us give more weight to the characteristics concerned (Lechner, 1999). In our case, we use the industry dummies and firm size as additional conditioning control variables, because we want to ensure that a matched control observation is in the same industry and of the same size-class as the treated firm. The term 'Mahalanobis' refers to the metric used to reduce these respective criteria to a single distance measure. 'Matching with replacement' implies that a similar observation of the untreated control group may be used repeatedly. And 'matching with common support' indicates that a common support restriction is imposed, eliminating firms with few or no characteristics in common with the treated from the potential control group. It should be noted that, as the propensity score is estimated and matching is performed with replacement, an ordinary t-statistic on mean differences is biased. We will, therefore, apply the bootstrap method for estimating the standard errors underlying the t-values used in estimating the significance of the ATT.

This matching procedure can be used for estimating the average treatment effect by total funding, FTOT, on the considered types of research cooperation. And it can also be applied for estimating the respective average treatment effects of the considered specific treatments and combinations of treatments. In the latter case, to make sure that one is matching against the untreated the total sample always has to be reduced by eliminating firms that received treatments other than the treatment concerned. In practice, due to the insufficient number of observations, the probit equations explaining the likelihood of receiving EU funding by itself, or the combinations regional–federal, federal–EU and regional–federal–EU funding could not be estimated because of perfect predictor problems. Our matching exercise will, therefore, concentrate on estimating the average treatment effect of total funding, FTOT, regional funding, FREG, federal funding, FFED, and of the combination regional–EU funding, FREU, against the untreated.

Since, as will be seen from the results presented in Table 5.5 below, both regional funding by itself, FRG, and its combination with EU funding, FREU, are found to have a significant impact, it is interesting to investigate whether there is a net impact of EU funding in case of firms receiving both types of treatment. Lechner (2001) and Gerfin and Lechner (2002) have proposed a method for handling this sort of problem in a multiple treatment framework (for another application, see also Czarnitzki et al., 2004). In order to implement it, one needs first to estimate the propensity scores of all possible treatments by estimating a multinomial probit model that explains all mutually exclusive possibilities, and then, to include the propensity scores of both treatments concerned in the Mahalanobis metric. Unfortunately, due to data limitations, we cannot estimate four of the seven probit equations involved, and hence cannot estimate the full multinomial probit model. The most we can do is to estimate a restraint version of it, allowing for three possibilities – regional funding, FREG, federal funding, FFED, and the combination regional-EU funding, FREU – besides the option of no funding at all. We will do this while reducing the total sample consequently, by excluding the firms receiving other types or combinations of treatment. The net impact of EU funding in the case of firms receiving both types of funding will then be assessed by estimating the average treatment effect of double treatment by the combination regional–EU funding, FREU, against single treatment by regional funding, FREG.

5.4 ANALYSIS

5.4.1 Regression Analysis

Table 5.2 presents the estimates of the bi-variate probit model explaining research cooperation with business and with public research organisations on the total sample of 601 firms.

Columns (i) and (ii) present the estimates of the model with total funding, FTOT, with the explanatory variables as such in non-instrumented form (from CIS4), whereas columns (iii) and (iv) present the corresponding estimates with the explanatory variables instrumented by their lagged values (from CIS3). Even though instrumenting slightly reduces its parameter estimates and t-statistics, total funding, FTOT, remains highly statistically significant in both equations explaining cooperation with business, COBUS, and cooperation with public research organisations, COPRO, and its significance is shown to be highest in the latter case. Although one cannot, at this stage, exclude some remaining endogeneity of lagged funding for the reasons explained above, this gives a first hint of a significant impact of public funding on the cooperative research behaviour of firms in Belgium.

Table 5.2 Research cooperation with business and with public research organisations by innovators and non-innovators

Dep. Var.	COBUS	COPRO	COBUS	COPRO	COBUS	COPRO	COBUS	COPRO
	No IV	No IV	IV	IV	IV	IV	IV	IV
	(i)	(ii)	(iii)	(iv)	(v)	(vi)	(vii)	(viii)
Constant	–3.071	–4.829	–2.268	–2.773	–2.259	–2.818	–2.244	–2.794
	(9.18)**	(10.3)**	(8.37)**	(8.89)**	(8.31)**	(8.81)**	(8.26)**	(8.90)**
IDPAV1	–0.104	0.479	–0.009	0.402	–0.005	0.434	–0.003	0.408
	(0.48)	(1.87)	(0.05)	(1.82)	(0.03)	(1.94)	(0.02)	(1.83)
IDPAV2	0.346	0.694	0.380	0.382	0.388	0.396	0.393	0.375
	(1.68)	(2.68)**	(2.10)*	(1.75)	(2.15)*	(1.80)	(2.17)*	(1.72)
IDPAV3	0.035	0.732	0.070	0.313	0.080	0.254	0.105	0.241
	(0.14)	(2.72)**	(0.32)	(1.40)	(0.36)	(1.10)	(0.48)	(1.05)
IDPAV4	–0.127	0.534	–0.221	0.115	–0.197	0.135	–0.177	0.104
	(0.39)	(1.39)	(0.80)	(0.39)	(0.69)	(0.44)	(0.62)	(0.33)
RDA	0.693	0.558	0.408	0.202	0.409	0.213	0.416	0.221
	(3.57)**	(2.48)**	(2.65)**	(1.06)	(2.66)**	(1.13)	(2.72)**	(1.18)
IDPAT	–0.121	0.366	0.236	0.471	0.237	0.364	0.237	0.376
	(0.68)	(1.82)*	(1.51)	(2.62)**	(1.50)	(1.97)**	(1.50)	(2.04)**
INKS	1.698	1.496	0.207	0.148	0.240	–0.038	0.244	–0.046
	(4.17)**	(3.00)**	(0.60)	(0.37)	(0.69)	(0.09)	(0.70)	(0.11)
APPR	0.232	0.196	0.177	0.549	0.189	0.631	0.197	0.638
	(1.06)	(0.87)	(0.94)	(2.46)**	(1.03)	(2.84)**	(1.06)	(2.88)**
CKNOW	0.463	3.111	0.288	0.660	0.242	0.809	0.312	0.884
	(1.10)	(6.25)**	(0.88)	(1.97)**	(0.72)	(2.15)**	(0.95)	(2.36)**
LHEI	–0.020	–0.015	0.079	0.096	0.080	0.087	0.085	0.090
	(0.39)	(0.25)	(1.71)*	(1.66)*	(1.73)*	(1.53)	(1.81)*	(1.59)
LEXI	–0.038	–0.004	–0.003	0.021	–0.003	0.028	–0.001	0.030
	(1.30)	(0.11)	(0.15)	(0.81)	(0.14)	(1.08)	(0.02)	(1.18)
LSIZE	0.254	0.133	0.303	0.268	0.299	0.266	0.297	0.268
	(4.58)**	(2.34)**	(6.28)**	(5.07)**	(6.19)**	(4.95)**	(6.04)**	(4.95)**
GRP_D	–0.242	0.484	0.021	0.051	0.029	0.114	0.029	0.094
	(1.30)	(2.23)**	(0.13)	(0.31)	(0.18)	(0.67)	(0.18)	(0.56)
GRP_F	–0.128	–0.039	–0.353	–0.431	–0.355	–0.341	–0.351	–0.347
	(0.68)	(0.17)	(2.12)*	(2.05)*	(2.11)*	(1.62)	(2.08)*	(1.66)
HRISK	–0.066	–0.051	0.065	0.042	0.036	0.117	0.026	0.130
	(0.22)	(0.14)	(0.29)	(0.17)	(0.15)	(0.49)	(0.11)	(0.54)
HFIN	–0.107	0.208	0.254	0.044	0.224	0.098	0.260	0.136
	(0.54)	(0.95)	(1.14)	(0.19)	(0.98)	(0.44)	(1.17)	(0.62)

Table 5.2 (continued)

Dep. Var.	COBUS	COPRO	COBUS	COPRO	COBUS	COPRO	COBUS	COPRO
	No IV	No IV	IV	IV	IV	IV	IV	IV
	(i)	(ii)	(iii)	(iv)	(v)	(vi)	(vii)	(viii)
FTOT	0.671	0.885	0.507	0.639				
	(4.13)**	(5.20)**	(3.28)**	(3.75)**				
FREG					0.586	0.705	0.552	0.673
					(3.55)**	(3.80)**	(3.09)**	(3.33)**
FFED					0.590	–0.435	0.620	–0.775
					(1.95)*	(1.40)	(1.66)*	(1.57)
FEU					–0.031	0.580	–0.345	0.486
					(0.13)	(2.18)**	(0.78)	(1.13)
FRF							0.052	0.293
							(0.08)	(0.47)
FREU							0.554	1.025
							(1.68)*	(2.92)**
RHO	0.904		0.867		0.890		0.895	
	(0.00)**		(0.00)**		(0.00)**		(0.00)**	
LL	–311.2		–411.6		–402.4		–402.4	

Notes: The sample has 601 observations. The absolute value of the t-values is presented between brackets. These are based on heteroscedasticity-consistent estimates of the standard errors. The t-tests are one or two-tailed, according to a priori expectations with respect to the parameter values. RHO stands for the correlation coefficient between the errors of both equations and the p-value of its Wald test is presented underneath. LL stands for the log of pseudolikelihood. * and ** denote significance at the 5% and 1% levels.

A comparison of the parameter estimates of the control variables in columns (iii) and (iv) with those in columns (i) and (ii) shows that, especially in the case of the knowledge-related variables, the endogeneity problems caused by using the variables as such may be severe. Thus, instrumenting the R&D-active variable, RDA, tends to drastically reduce its significance, especially in the case of the COPRO equation, where it becomes completely non-significant. This would suggest that there is a significant simultaneity problem and, most likely, a reverse causation, as the knowledge generated through research cooperation requires further absorptive R&D efforts. Not surprisingly, this appears to be particularly so where this knowledge is more complex in nature, as in the case of partnerships with public research organisations. The reduction in significance of the R&D variable when instrumented helps, in turn, to increase the significance of the correlated patent variable, IPPAT, which is found highly significant in the COPRO equation (iv). This is in line with a priori expectations, in the sense that patents are generally used to protect fairly complex innovations and tend, therefore, to affect the likelihood of engaging in partnerships with public research organisations in particular.

The most dramatic reduction in statistical significance among the knowledge-related control variables occurs, not surprisingly, in the case of incoming knowledge spillovers, INKS, which, from highly significant, becomes completely non-significant when instrumented. This confirms our presumption of extreme simultaneity and reverse causation, as research cooperation generates by itself spillovers from research partners. Instrumenting also allows the estimates of strategic appropriability, APPR, and complexity of knowledge, CKNOW, to take far more realistic values. Again, conforming to expectations, due to their positive relation with the complexity of the research process, both are found to affect cooperation with public research organisations in particular.

As for the remaining control variables, it is worth noting that there is a significant positive impact of firm size, LSIZE, on both types of research cooperation, which can probably be explained by the fact that larger firms are more likely to have the necessary means and absorptive capacity to engage in research cooperation. Also, conforming to expectations, foreign group membership, GRP_F, has a modest negative impact, suggesting that foreign subsidiaries are less likely to undertake cooperative research because they can access the knowledge concerned through the parent company. Finally, when the other control variables are instrumented, the industry dummies are, by and large, found to be non-significant. This suggests that the likelihood of research cooperation is mainly determined by firm-specific rather than industry-specific characteristics.

Columns (v) and (vi) in Table 5.2 replace total funding by three separate, possibly overlapping, variables: regional funding, FREG, federal funding, FFED, and EU funding, FEU. The results indicate that the strong significance of total funding in equations (iii) and (iv) is mainly due to regional funding, FREG. Federal funding, FFED, is found to be only modestly significant in the COBUS equation, whereas EU funding, FEU, appears to be only significant in the COPRO equation, albeit more clearly so. Although caution should be exercised in the interpretation, bearing the possible remaining endogeneity of the lagged funding variables in mind, these results give a clear hint of the existence of significant differences in impact between the respective types of funding. Due to the relatively heavily regionalised organisation of innovation policy in Belgium, the number of firms receiving regional public funding is, as seen from Table 5.1 above, much higher. But there is more than that. Federal funding, to the extent that it aims to stimulate research cooperation, mainly attempts to foster inter-firm partnerships. Regional public funding supports more explicitly both types of research partnerships, with the preference for public–private partnerships. The contrast is, however, greatest with EU funding, where the aim is explicitly to foster international mixed partnerships including firms and public research organisations. Our findings are, therefore, perfectly in line with what could have been expected.

Columns (vii) and (viii) in Table 5.2 repeat estimation while making use of the same three, but now mutually exclusive non-overlapping, types of funding and including, in addition, the combinations regional–federal and regional–EU funding, FRF and FREU. The results confirm the dominance of the influence of regional funding, FREG. There is again a modest significance of federal funding, FFED, in the COBUS equation. There is, however, none for EU funding by itself. The observed significance of EU funding in the COPRO equation in column (vi) in the previous model is apparently due to firms that also receive regional funding. As confirmed by the evidence presented in Table 5.1, regional and EU public funding authorities are partly targeting the same firms.

Table 5.3 repeats the entire regression analysis on the sub-sample of 314 firms that report being innovation-active in both periods, 1998–2000 and 2002–2004.

When compared with the results of the full sample in Table 5.2, there are some differences in the case of the control variables. Not surprisingly, reducing the sample to firms that systematically report being innovation-active reduces simultaneity problems due to a missing un-measurable variable, such as 'innovativeness', when the variables are not instrumented, as in columns (i) and (ii). But, as shown in Table A5.2 in the Appendix, this sample reduction implies a loss in information at the same time, since the variation in some of the variables is reduced. This is particularly the case for the R&D variable, RDA, which is now non-significant throughout.

These differences do not, however, affect the results on funding. The significance of the respective funding variables is largely unaffected – the only slight difference being that federal funding, FFED, is now merely non-significant at a 5% level (but remains so at a 10% level) in columns (v) and (vii). All the main tentative conclusions reached to this point are confirmed.

Table 5.3 Research cooperation with business and public research organisations by innovators

Dep. Var.	COBUS No IV (i)	COPRO No IV (ii)	COBUS IV (iii)	COPRO IV (iv)	COBUS IV (v)	COPRO IV (vi)	COBUS IV (vii)	COPRO IV (viii)
Constant	−2.466	−4.098	−1.882	−2.411	−1.894	−2.473	−1.892	−2.463
	(5.42)**	(8.16)**	(4.77)**	(5.61)**	(4.76)**	(5.53)**	(4.80)**	(5.63)**
IDPAV1	−0.262	0.397	−0.145	0.465	−0.146	0.505	−0.129	0.475
	(0.96)	(1.40)	(0.58)	(1.72)	(0.57)	(1.83)	(0.51)	(1.72)
IDPAV2	0.071	0.576	0.063	0.352	0.091	0.382	0.110	0.359
	(0.28)	(2.09)*	(0.24)	(1.38)	(0.35)	(1.47)	(0.43)	(1.39)
IDPAV3	−0.101	0.654	−0.162	0.340	−0.155	0.260	−0.123	0.236
	(0.36)	(2.28)*	(0.60)	(1.32)	(0.57)	(0.97)	(0.45)	(0.88)
IDPAV4	−0.386	0.487	−0.485	0.277	−0.446	0.305	−0.407	0.272
	(1.06)	(1.19)	(1.42)	(0.77)	(1.27)	(0.83)	(1.15)	(0.73)
RDA	0.369	0.395	0.273	0.047	0.297	0.058	0.312	0.068
	(1.75)*	(1.61)	(1.28)	(0.20)	(1.40)	(0.25)	(1.47)	(0.29)
IPPAT	−0.115	0.369	0.190	0.504	0.180	0.376	0.188	0.398
	(0.59)	(1.76)*	(1.05)	(2.54)**	(0.99)	(1.87)*	(1.03)	(1.98)**
INKS	0.757	0.925	0.102	−0.013	0.106	−0.224	0.125	−0.226
	(1.57)	(1.75)*	(0.22)	(0.03)	(0.23)	(0.45)	(0.27)	(0.46)
APPR	0.138	0.175	0.133	0.421	0.147	0.528	0.135	0.529
	(0.60)	(0.74)	(0.57)	(1.68)*	(0.64)	(2.13)**	(0.57)	(2.13)**
CKNOW	0.650	2.856	0.202	0.735	0.202	0.898	0.285	0.999
	(1.41)	(5.09)**	(0.49)	(1.65)*	(0.48)	(1.98)**	(0.70)	(2.20)**
LHEI	−0.020	−0.052	0.045	0.045	0.043	0.031	0.054	0.041
	(0.30)	(0.72)	(0.69)	(0.72)	(0.66)	(0.48)	(0.82)	(0.62)
LEXI	−0.029	0.007	−0.016	−0.000	−0.014	0.009	−0.012	0.010
	(0.81)	(0.18)	(0.49)	(0.00)	(0.44)	(0.26)	(0.36)	(0.31)
LSIZE	0.315	0.137	0.338	0.222	0.335	0.215	0.323	0.220
	(4.99)**	(2.35)**	(5.51)**	(3.81)**	(5.44)**	(3.59)**	(5.32)**	(3.65)**
GRP_D	−0.250	0.419	−0.115	0.156	−0.108	0.256	−0.119	0.222
	(1.15)	(1.80)	(0.55)	(0.77)	(0.52)	(1.23)	(0.57)	(1.07)
GRP_F	−0.086	−0.044	−0.250	−0.267	−0.250	−0.132	−0.246	−0.143
	(0.39)	(0.17)	(1.19)	(1.11)	(1.16)	(0.54)	(1.14)	(0.59)
HRISK	0.089	−0.108	−0.092	−0.089	−0.117	−0.000	−0.137	0.008
	(0.25)	(0.28)	(0.35)	(0.34)	(0.43)	(0.00)	(0.50)	(0.03)
HFIN	−0.107	0.054	0.328	0.036	0.285	0.109	0.329	0.184
	(0.47)	(0.23)	(1.28)	(0.15)	(1.09)	(0.45)	(1.28)	(0.77)

Table 5.3 (continued)

Dep. Var.	COBUS No IV (i)	COPRO No IV (ii)	COBUS IV (iii)	COPRO IV (iv)	COBUS IV (v)	COPRO IV (vi)	COBUS IV (vii)	COPRO IV (viii)
FTOT	0.704	0.830	0.554	0.683				
	(4.10)**	(4.69)**	(3.05)**	(3.61)**				
FREG					0.641	0.724	0.651	0.707
					(3.30)**	(3.50)**	(3.06)**	(3.16)**
FFED					0.502	–0.508	0.589	–0.836
					(1.56)	(1.53)	(1.50)	(1.62)
FEU					–0.036	0.646	–0.224	0.682
					(0.13)	(2.17)**	(0.46)	(1.35)
FRF							–0.081	0.103
							(0.12)	(0.17)
FREU							0.469	1.000
							(1.30)	(2.67)**
RHO	0.924		0.871		0.903		0.909	
	(0.00)**		(0.00)**		(0.00)**		(0.00)**	
LL	–255.5		–287.9		–279.3		–279.2	

Notes: The sample has 314 observations. The absolute value of the t-values is presented between brackets. These are based on heteroscedasticity-consistent estimates of the standard errors. The t-tests are one or two-tailed, according to a priori expectations with respect to the parameter values. RHO stands for the correlation coefficient between the errors of both equations and the p-value of its Wald test is presented underneath. LL stands for the log of pseudolikelihood. * and ** denote significance at the 5% and 1% levels.

5.4.2 Non-parametric Matching

Due to remaining doubts over the exogeneity of lagged funding we add a non-parametric matching procedure to the regression analysis. Our main interest here is to estimate the average treatment effect on the treated (ATT) of the types of funding considered on the outcome variables, research cooperation with business, COBUS, and research cooperation with public research organisations, COPRO. The matching procedure implies, in the first step, an estimation of the probit equations explaining the likelihood of receiving the treatments in question (types of funding) and taking their predicted value as an estimate of the propensity score. As explained above, we do this first for total funding, FTOT, and then repeat for its main (non-overlapping) components, regional funding, FREG, federal funding, FFED, and the combination regional-EU funding, FREU, while always reducing the sample by dropping the observations on firms that receive types or combinations of funding other than the one in question. Since the purpose is to control for common exogenous characteristics that may affect both the outcome and the treatment, the instrumented control variables in the bi-variate probit model explaining research cooperation are again used as explanatory variables (for example, the lagged values of these variables from CIS3). Table 5.4 presents the respective probit equations to explain funding.

Table 5.4 Explaining public funding for innovation

	All firms				Innovation-active firms			
Dep. Var.	FTOT	FREG	FFED	FREU	FTOT	FREG	FFED	FREU
	IV	IV	IV	IV	IV	IV	IV	IV
	(i)	(ii)	(iii)	(iv)	(v)	(vi)	(vii)	(viii)
Constant	−1.800	−2.042	−2.832	−2.900	−1.417	−1.455	−2.821	−3.051
	(6.48)**	(5.90)**	(5.24)**	(4.66)**	(3.61)**	(3.07)**	(3.74)**	(3.64)**
IDPAV1	−0.267	−0.288	−0.097	−0.328	−0.313	−0.385	−0.274	−0.096
	(1.31)	(1.18)	(0.23)	(0.63)	(1.21)	(1.26)	(0.50)	(0.17)
IDPAV2	−0.149	−0.390	0.503	0.100	−0.384	−0.494	0.059	0.360
	(0.74)	(1.51)	(1.35)	(0.20)	(1.53)	(1.60)	(0.11)	(0.64)
IDPAV3	−0.056	−0.010	0.154	0.066	−0.164	−0.022	−0.269	0.216
	(0.25)	(0.04)	(0.36)	(0.13)	(0.62)	(0.07)	(0.49)	(0.40)
IDPAV4	−0.241	−0.427	−0.234	−1.459	−0.390	−0.355	−0.664	−1.351
	(0.86)	(1.17)	(0.37)	(1.68)	(1.15)	(0.85)	(0.89)	(1.39)
RDA	0.185	0.193	−0.022	0.070	0.103	−0.032	0.195	0.229
	(1.06)	(0.91)	(0.07)	(0.17)	(0.45)	(0.12)	(0.46)	(0.41)
IPPAT	0.624	0.646	0.591	0.660	0.708	0.687	0.518	0.769
	(3.68)**	(3.07)**	(1.98)**	(2.06)**	(3.63)**	(2.94)**	(1.41)	(2.22)**
INKS	0.811	0.768	0.813	0.473	1.041	0.691	1.807	1.052
	(2.13)**	(1.68)*	(1.17)	(0.58)	(2.17)**	(1.23)	(1.89)*	(1.07)
APPR	0.415	0.134	1.126	0.526	0.336	0.120	1.152	0.200
	(2.08)**	(0.54)	(3.33)**	(1.25)	(1.43)	(0.43)	(2.71)**	(0.47)
CKNOW	0.914	0.984	0.340	1.471	1.022	1.070	0.080	1.455
	(2.62)**	(2.45)**	(0.49)	(2.00)**	(2.48)**	(2.26)**	(0.09)	(1.98)**
LHEI	0.054	0.037	0.015	0.494	0.014	0.016	−0.055	0.488
	(1.09)	(0.62)	(0.17)	(2.29)**	(1.16)	(0.20)	(0.40)	(2.07)**
LEXI	0.038	0.034	−0.046	0.090	0.038	0.018	−0.009	0.064
	(1.41)	(1.05)	(1.01)	(1.33)	(1.16)	(0.47)	(0.15)	(0.88)
LSIZE	0.137	0.107	0.092	0.191	0.082	0.038	0.063	0.143
	(2.62)**	(1.70)*	(0.80)	(1.84)*	(1.35)	(0.51)	(0.45)	(1.28)

Table 5.4 (continued)

	All firms				Innovation-active firms			
Dep. Var.	FTOT	FREG	FFED	FREU	FTOT	FREG	FFED	FREU
	IV	IV	IV	IV	IV	IV	IV	IV
	(i)	(ii)	(iii)	(iv)	(v)	(vi)	(vii)	(viii)
GRP_D	–0.019	0.129	–0.808	0.450	0.020	0.219	–1.130	0.352
	(0.11)	(0.61)	(2.07)*	(1.01)	(0.09)	(0.85)	(2.27)*	(0.73)
GRP_F	–0.501	–0.282	–0.881	0.076	–0.387	–0.082	–0.801	0.010
	(2.58)**	(1.18)	(2.19)*	(0.16)	(1.57)	(0.30)	(1.69)	(0.02)
HRISK	0.113	0.125	0.288	–0.279	–0.093	0.025	–0.044	–0.326
	(0.46)	(0.43)	(0.67)	(0.56)	(0.34)	(0.08)	(0.09)	(0.63)
HFIN	0.156	0.100	–0.207	0.718	0.120	–0.016	–0.059	0.821
	(0.66)	(0.34)	(0.45)	(1.59)	(0.46)	(0.05)	(0.11)	(1.71)*
Obs.	601	530	488	488	314	250	208	211
Pseudo R^2	0.259	0.215	0.247	0.433	0.184	0.137	0.249	0.363

Notes: In case of the equations on FREG, FFED and FREU the sample was reduced by dropping the observations on firms receiving kinds of public funding other than the funding in question. The absolute value of the t-values is presented between brackets. The t-tests are one or two-tailed, according to a priori expectations with respect to the parameter values. Symbols * and ** denote significance at the 5% and 1% levels.

Column (i) presents the estimates of the equation explaining the probability of receiving total funding, FTOT, on the total sample. Firms undertaking relatively complex innovative efforts appear to be favoured. This can be seen from the high significance of the variables on patents, IPPAT, incoming knowledge spillovers, INKS, strategic appropriability, APPR, and complexity of knowledge, CKNOW. The positive impact of size, LSIZE, indicates that larger firms are also favoured, whereas the negative impact of foreign group membership, GRP_F, suggests that the opposite is true for foreign subsidiaries.

Columns (ii), (iii) and (iv) in Table 5.4 present the corresponding estimates of the equations explaining the probability of receiving regional funding, FREG, federal funding, FFED, and the combination regional–EU funding, FREU. It is interesting to note that whereas having valid patents, PAT, is favoured by all public funding authorities, firms undertaking basic research, BR, are especially able to attract regional and EU funding, while possessing a strategic appropriation capacity, APPR, is particularly appreciated by federal funding authorities. This is perfectly in line with the different aims of the respective funding authorities mentioned above. These results are confirmed when repeating estimation on the sub-sample including only firms that report being innovation-active, in columns (v) to (viii).

Table 5.5 presents the estimates of the average treatment effect on the treated (ATT) by respective type of funding on research cooperation with business, COBUS, and with public research organisations, COPRO.

Table 5.5 Matching results: average treatment effect by public funding for innovation on research cooperation with business and with public research organisations

All Firms							
		Treatment by:					Treatment by:
Treated variable	Treatment against:	FTOT	FREG	FFED	FREU	Treatment against:	FREU
COBUS	Untreated	0.350	0.387	0.500	0.375	FREG	0.000
		(4.29)**	(3.89)**	(3.10)**	(1.87)*		(0.00)
COPRO	Untreated	0.359	0.355	0.300	0.687	FREG	0.235
		(5.63)**	(4.55)**	(1.87)*	(4.19)**		(1.76)*
Innovation-active firms							
		Treatment by:					Treatment by:
Treated variable	Treatment against:	FTOT	FREG	FFED	FREU	Treatment against:	FREU
COBUS	Untreated	0.302	0.241	0.333	0.500	FREG	0.059
		(3.76)**	(2.20)**	(1.56)	(2.50)**		(0.39)
COPRO	Untreated	0.377	0.276	0.200	0.571	FREG	0.117
		(4.83)**	(2.82)**	(0.95)	(2.89)**		(0.82)

Notes: The employed method is nearest neighbour Mahalanobis matching with replacement and common support. The Mahalanobis metric contains besides the relevant propensity score(s) the industry dummies and the size variable, LSIZE. In the case of treatment against untreated the propensity scores are obtained from the probit equations presented in Table 5.4. In the case of treatment by FREU against treatment by FREG, they are obtained from a multinomial probit model explaining the probability of these treatments as well as of treatment by FFED. The t-values are presented between brackets. The t-tests are one-tailed. The underlying standard errors were obtained by bootstrap sampling with 500 replications. Symbols * and ** denote significance at the 5% and 1% levels.

The upper part of the table presents the estimates on the total sample. The upper left quadrant presents, respectively, the effect of treatment by total funding, FTOT, by regional funding, FREG, by federal funding, FFED, and by the combination regional–EU funding, FREU, against the untreated each time.

Consider first the average treatment effect by total funding, FTOT, on the probability of research cooperation with business, COBUS. The estimated propensity score obtained from the predicted value of equation (i) in Table 5.4 is used as a matching criterion, besides the size variable, LSIZE, and the industry dummies. To ensure common support for the treated firms, 16 out of the 133 treated observations had to be dropped. In order to evaluate the quality of the matching, we re-estimated the propensity score by using only the matched sample. The covariates should be non-significant because there should be no more systematic differences in the regressors between treated and control firms. A likelihood ratio test confirms the joint non-significance of the regressors. Furthermore, t-tests on the mean differences of the control variables and of the propensity score between the treated and the untreated show that these differences disappear after matching. This confirms that the matching procedure is successful in balancing out these differences. Hence a causal interpretation of the average treatment effect is warranted. The estimate of the ATT effect by total funding, FTOT, on research cooperation with business, COBUS, presented in the upper left quadrant of Table 5.5, shows that this is highly significant. More precisely, it suggests that the probability of undertaking research cooperation with business increases by 35% for firms receiving public funding for innovation.

The estimated average treatment effect by total funding, FTOT, on the probability of research cooperation with public research organisations, COPRO, is in the same order of magnitude, but even more statistically significant. The estimates of the average treatment effects by regional funding, FREG, presented in the next column in the upper left quadrant are in the same vein. There is also evidence of a significant treatment effect by federal funding, FFED, which is in this case clearly stronger on research cooperation with business, COBUS. Also, the combination regional–EU funding is shown to have a significant impact, which is now much stronger on research cooperation with public research organisations, COPRO.

When compared to the corresponding instrumental variable regression estimates of the bi-variate probit model, presented in columns (vii) and (viii) in Table 5.2 above, these results appear remarkably coherent. The matching evidence can therefore be viewed as confirmation that a causal interpretation of the estimated impact of funding in the regression analysis is indeed warranted.

Moreover, the matching procedure allows us to test whether there is a net impact of treatment by EU funding for firms that receive regional funding at the same time. A comparison of the estimated values just considered for the

ATT effects of FREG and FREU in the upper left quadrant of Table 5.5 suggests that this may indeed be the case as far as cooperation with public research organisations, COPRO, is concerned. As stated above, in order to do so we applied a restraint version of the multiple treatment method proposed by Lechner (2001) and Gerfin and Lechner (2002). The propensity scores for treatment by the combination regional–EU funding, FREU, and for treatment by regional funding by itself, FREG, were in this case estimated from a multinomial probit model allowing for three choices, regional funding, federal funding and the combination regional–EU funding, besides the choice of no funding at all. Both the propensity scores for regional funding and for the combination regional–EU funding were used as matching criteria, besides firm size and the industry dummies. The estimates of the ATT effect by the double treatment by the combination regional–EU funding, FREU, against the single treatment by regional funding, FREG, are presented in the upper right quadrant of Table 5.5. They would suggest that whereas there is no net impact of EU funding on research cooperation with business, COBUS, at all, there is a modest net impact of EU funding on research cooperation with public research organisations, COPRO.

Finally, the lower part of Table 5.5 presents the corresponding estimates of the average treatment effects on the samples including exclusively firms reporting that they are innovation-active. All the main conclusions with respect to the treatment effects against the untreated obtained from the estimation on the total sample are confirmed. Again, as with the corresponding instrumental variable regression estimates of the bi-variate probit model in columns (vii) and (viii) in Table 5.3 above, the impact of federal funding, FFED, is now non-significant at a 5% level (but remains significant at a 10% level in the case of COBUS). A comparison of the treatments by the combination regional–EU funding with that by regional funding by itself, in the lower left quadrant of Table 5.5, would suggest that the impact of the double treatment is higher, especially in the case of cooperation with institutions, COPRO. But, as shown by the results presented in the lower right quadrant, application of the restraint multi-treatment method cannot find explicit confirmation of this. The results are in this case inconclusive.

5.5 DISCUSSION AND CONCLUSIONS

An analysis was presented of the impact of public funding for innovation on research cooperation by firms based on data from the Third and Fourth Community Innovation Surveys (CIS3 and CIS4) for Belgium. Two types of research cooperation were considered – cooperation with business and cooperation with public research organisations – and three types of public funding – regional government funding, federal government funding and EU funding – as well as different combinations. A double approach was followed to handle endogeneity problems. A regression analysis was first performed of a bi-variate probit model explaining the likelihood of research cooperation with business and with public research organisations (measured by data from CIS4) by treatment and control variables instrumented by their lagged values (from CIS3). This was then complemented by a non-parametric matching analysis estimating the average treatment effects by the considered types of funding on the two types of research cooperation, while again making use of instrumented control variables. The results were found to be remarkably coherent and allow clear answers to the considered research questions.

To the first research question, whether there is an impact of total funding, irrespective of source, on research cooperation by firms in Belgium, the answer is unambiguously affirmative. There is a significant positive impact of total funding on both types of research cooperation under consideration. Although the parameter estimates and estimated average treatment effects of total funding are always in the same order of magnitude for both types of cooperation, their statistical significance is higher in the case of cooperation with public research organisations than with business. This suggests that public funding for innovation also affects partner choice. All this provides evidence of 'behavioural additionality', in the sense that public funding affects the cooperative behaviour of firms.

These results reinforce the limited existing econometric evidence in this respect. These studies, with respect to other countries, either failed to address the endogeneity issue, as in the case of Miotti and Sachwald (2003), Mohnen and Hoareau (2003), Dachs et al. (2004) and Abramovsky et al. (2005), or, when they did address it, were only able to do so partially, thereby still obtaining inconclusive results. This was the case for Belderbos et al. (2004), who only concentrated on an instrumental variable regression analysis, still leaving doubts about a possible remaining endogeneity of lagged funding. And it was also the case for Busom and Fernandez-Ribas (2004, 2008) and Aschhoff et al. (2006), who presented non-parametric matching estimates without instrumenting their control variables.

Our second research question, whether there are differences in impact on research cooperation between different types of funding, has, to our knowledge, never been investigated econometrically before. The answer is that, in the case of Belgium, there are, indeed, substantial differences in impact between different types of funding. The impact of regional funding is by far the most important. It has an impact on both types of research cooperation, even though its significance is somewhat higher in the case of cooperation with public research organisations than with business. Federal funding has a far more modest influence, mainly on research cooperation with business, whereas EU funding mostly has an impact in combination with regional funding and this on cooperation with public research organisations. The dominance of the effect of regional funding can be explained by the relatively heavily regionalised organisation of innovation policy in Belgium, which makes the number of firms receiving regional public funding much higher.

But there is more to it. The different funding authorities have clearly different aims. Federal funding, to the extent that it aims to stimulate research cooperation, mainly attempts to foster inter-firm partnerships in specific domains. It favours firms with strategic knowledge appropriation potential through complexity of design and lead-time advantage. And it attempts to help further improve these appropriation capacities as a means of intensifying knowledge exchanges between partners and of internalising, and thereby reducing, outgoing knowledge spillovers to non-partners, in line with theoretical justifications for research cooperation given by Kesteloot and Veugelers (1995), Martin (2002), Cassiman et al. (2002) and others. This explains why the impact of federal funding is mainly on research cooperation with business and why knowledge appropriability is a major variable explaining the likelihood of receiving this type of funding.

Regionally-funded programs support more explicitly both types of research partnership under consideration, with a preference for public-private partnerships. The purpose is to foster cooperative research networking, to allow participating firms to access intangible scientific knowledge, which does not spill over and cannot easily be contracted through market transactions, in line with theoretical justifications for research cooperation given by Hagedoorn (1993), Katsoulakos and Ulph (1998) and others. Helping to exploit economies of scale and scope in both research and development and to share risks, are also reasons for regional funding authorities to stimulate research cooperation. These same reasons are also largely found in the case of the EU Framework Programs, even though these are directed towards specific domains. Here the explicit aim is to foster the creation of international networks including both firms and public research organisations from different countries. This explains why our results show that the statistical significance of the impact of regional funding is greatest on research cooperation with public research organisations, and why this is even more the case for firms receiving a combination of regional and EU funding. It also helps to explain why complexity of knowledge has a major positive influence on the likelihood of receiving these types of funding.

The preceding analysis provides some clear hints for future data improvement allowing further research. Thus, it could be useful to include in the questionnaire of future Community Innovation Surveys questions about the amount of R&D expenditure devoted to cooperative research and the actual size of the public subsidies for innovation received by the respondents. This should allow the construction of cooperation and funding intensity variables and complement the analysis of the likelihood of undertaking research cooperation with one explaining its intensity. This should allow for a more precise assessment of the extent of 'behavioural additionality' in terms of research cooperation, as even firms that were already cooperating may increase the intensity of their cooperation upon receipt of public support. Such quantitative measures of research cooperation may also enable the performance of a quantile regression analysis, to identify systematic behavioural differences between strong and weak co-operators. This should, in turn, allow an even further refinement of the assessment of the impact of public funding on research cooperation.

Finally, another possible area of data improvement and connected future research relates to the issue of persistence. The inter-temporal comparability of successive Community Innovation Surveys should be improved, by harmonising the questionnaires and by covering a sufficiently large core of systematically surveyed enterprises. Explicit questions should be asked with respect to the actual reasons for setting-up and joining research partnerships. This should allow us to identify, at the firm level, cooperative research partnerships created through, or with the support of, public funding and study their persistence relative to that of spontaneous cooperations. Special attention should be given to the persistence of public funding itself and its relationship with the actual duration of the cooperative research partnerships. This should also lead to an improvement in our assessment of the actual extent of 'behavioural additionality' of public funding in terms of research cooperation.

APPENDIX

Table A5.1 gives a description of the manner in which the variables were constructed from the raw data of the CIS3 and CIS4 questionnaires. Table A5.2 presents descriptive statistics of the variables with respect to the total sample and the sample of firms that report being innovation-active both in CIS3 and CIS4.

Table A5.1 Description of the variables

Variable	Definition
Dependent variables	
Research cooperation with business (COBUS)	COBUS = 1 if the firm collaborates either with suppliers, clients or competitors during the period 2002–2004
Research cooperation with public research organisations (COPPO)	COPRO = 1 if the firm collaborates either with research labs, universities or government during the period 2002–2004
Treatment variables	
Total funding (FTOT)	FTOT = 1 if the firm received public funding for innovation irrespective the source during the period 2002–2004 (1998–2000 when instrumented)
Regional government funding (FREG)	FREG = 1 if the firm received public funding for innovation from a Belgian regional government during the period 2002–2004 (1998–2000 when instrumented)
Federal government funding (FFED)	FFED = 1 if the firm received public funding for innovation from the Belgian federal government during the period 2002–2004 (1998–2000 when instrumented)
Regional government funding (FEU)	FEU = 1 if the firm received public funding for innovation from the EU during the period 2002–2004 (1998–2000 when instrumented)
Regional and federal government funding (FRF)	FRF = 1 if the firm received public funding from both regional and federal government during the period 2002–2004 (1998–2000 when instrumented)

Table A5.1 (continued)

Variable	Definition
Regional and federal government funding (FREU)	FREU = 1 if the firm received public funding from both a regional government and the EU during the period 2002–2004 (1998–2000 when instrumented)
Control variables	
Research and development active (RDA)	RDA = 1 if the firm performed in-house R&D during the period 2002–2004 (1998–2000 when instrumented)
Patents (IPPAT)	IPPAT = 1 if the firm had any valid patents protecting its inventions or innovations at the end of 2000
Incoming knowledge spillovers (INKS)	Average score of the importance attached to information from suppliers, clients, competitors, universities, government, professional conferences, journals, exhibitions during 2002–2004 (1998–2000 when instrumented)
Appropriability (APPR)	Average score of the effectiveness of strategic protection through secrecy, complexity of design and lead-time advantage during the period 2002–2004 (1998–2000 when instrumented)
Complexity of knowledge (CKNOW)	Ratio of the sum of the scores of the importance attached to information from universities, government, professional conferences and journals over the sum of the scores of all sources of information during the period 2002–2004 (1998–2000 when instrumented)
Log of higher education intensity (LHEI)	Natural log of the % employees with higher education in the year 2000
Log of export intensity (LEXI)	Natural log of the proportion of exports in turnover in the year 2002 (1998 when instrumented)
Log of size (LSIZE)	Natural log of the number of employees in the year 2002 (1998 when instrumented)
Domestic group affiliation (GRP_D)	GRP_D = 1 if the firm is a member of a domestic group in 2004 (2000 when instrumented)

Table A5.1 (continued)

Variable	Definition
Foreign group affiliation (GRP_F)	GRP_F = 1 if the firm is a member of a foreign group in 2004 (2000 when instrumented)
Hampering factor risks (HRISK)	Importance attached to excessive perceived economic risks hampering innovation activity during the period 1998–2000
Hampering factor finance (HFIN)	Importance attached to lack of appropriate sources of finance during the period 2002–2004 (1998–2000 when instrumented)
IDPAV1	IDPAV1 = 1 if the firm is classified in the 'Resource-intensive' sectors (mining, food, beverages and tobacco, wood and furniture, recycling, wholesale)
IDPAV2	IDPAV2 = 1 if the firm is classified in the 'Scale-intensive' sectors (paper and printing, rubber and plastics, non-metallic minerals, ferrous metals, non-ferrous metals, fabricated metal products, motor vehicles, other transport equipment)
IDPAV3	IDPAV3 = 1 if the firm is classified in the 'Specialised supplier' sectors (non-electrical machinery, electronics and communication, electrical machinery)
IDPAV4	IDPAV4 = 1 if the firm is classified in the 'Science-based' sectors (chemicals, pharmaceuticals, computers and office machinery, scientific instruments)
IDPAV5	IDPAV5 = 1 if the firm is classified in the 'Information-intensive' sectors (financial intermediation, telecommunication, transport services, business services, computer and related service activities, research and development, technical engineering)

Notes: The non-instrumented data are from CIS4 and the instrumented data from CIS3. In case of PAT, LHEI and HRISK the data are only available in CIS3, so that no instrumentation is necessary. When FRG, FFD and FEU are used in a non-overlapping way, as in columns (vii) and (viii) in Tables 5.2 and 5.3 and in the matching estimates in Table 5.5, these variables take only value 1 in case of receipt of the funding concerned and no other type of funding is received. The same also holds for the funding combination variables, FRGFD and FRGEU.

Table A5.2 Descriptive statistics: mean and standard deviation

	Total sample (601 obs.)				Innovation active in both (314 obs.)			
	CIS3 (1998–2000)		CIS4 (2002–2004)		CIS3 (1998–2000)		CIS4 (2002–2004)	
	Mean	St. dev.	Mean	St. dev.	Mean	St. dev.	Mean	St. dev.
Dependent variables								
COBUS	0.208	0.406	0.304	0.461	0.338	0.474	0.515	0.500
COPRO	0.175	0.380	0.200	0.400	0.302	0.460	0.360	0.481
Treatment variables								
FTOT	0.211	0.408	0.221	0.415	0.350	0.478	0.388	0.488
FREG	0.163	0.370	0.170	0.375	0.271	0.445	0.302	0.460
FFED	0.037	0.188	0.071	0.257	0.067	0.250	0.117	0.323
FEU	0.062	0.241	0.068	0.252	0.105	0.307	0.124	0.330
FRG*	0.120	0.325	0.105	0.307	0.191	0.394	0.188	0.391
FFD*	0.022	0.145	0.035	0.184	0.038	0.192	0.054	0.227
FEU*	0.018	0.134	0.012	0.107	0.025	0.158	0.022	0.148
FRF*	0.005	0.070	0.015	0.121	0.009	0.097	0.025	0.158
FREU*	0.033	0.180	0.035	0.184	0.061	0.239	0.063	0.244
Control variables								
RDA	0.516	0.500	0.438	0.497	0.818	0.386	0.742	0.438
IPPAT	0.183	0.387	0.183	0.387	0.283	0.451	0.283	0.451
INKS	0.280	0.254	0.291	0.278	0.436	0.198	0.487	0.194
APPR	0.250	0.351	0.350	0.415	0.376	0.376	0.605	0.384
CKNOW	0.203	0.238	0.195	0.220	0.318	0.224	0.325	0.190
LHEI	–2.223	1.695	–2.223	1.695	–1.920	1.449	–1.920	1.449
LEXI	–2.796	2.779	–2.709	2.736	–2.279	0.269	–2.018	2.495
LSIZE	4.118	1.473	4.303	1.467	4.558	1.611	4.796	1.608
GRP_D	0.275	0.447	0.266	0.442	0.322	0.468	0.321	0.467
GRP_F	0.318	0.466	0.328	0.470	0.350	0.478	0.376	0.485
HRISK	0.155	0.300	0.155	0.300	0.202	0.329	0.202	0.329
HFIN	0.166	0.318	0.404	0.377	0.216	0.350	0.484	0.367
IDPAV1	0.263	0.441	0.263	0.441	0.226	0.419	0.226	0.419
IDPAV2	0.271	0.445	0.271	0.445	0.299	0.459	0.299	0.459
IDPAV3	0.136	0.344	0.136	0.344	0.169	0.375	0.169	0.375
IDPAV4	0.070	0.255	0.070	0.255	0.086	0.281	0.086	0.281
IDPAV5	0.260	0.439	0.260	0.439	0.220	0.415	0.220	0.415

Notes: Symbol * indicates that a value of 1 for the funding or combination of types of funding concerned is not overlapping with a value of 1 for other types or combinations of funding.

6. The impact of public funding for research on private–public research cooperation

6.1 INTRODUCTION

In innovation policy circles in Europe and the OECD there has, in the past two decades, been a growing awareness of the importance of fostering co-operative research (European Commission, 2003; OECD, 2001). This is reflected by an increasing proportion of public funding awarded by the various regional, national and EU authorities to cooperative research networks rather than individual organisations. There are many types of partners eligible for research cooperation: firms (clients, suppliers, competitors); private organisations (consultants, R&D laboratories); and public organisations (universities, polytechnics, public research centres) (OECD, 2005; Tether and Tajar, 2008). The focus of this chapter is on the latter category, which is presumed to cover 'science'.

Industry–science relationships are thought to be instrumental for the transfer of knowledge and technology. The ultimate importance of industry–science relationships can be found in the stimulating effect they have (or, at least, are supposed to have) on economic development or growth and therefore welfare (European Commission, 2008).

The starting point of this chapter is the literature on behavioural additionality of public funding for research, and, more particularly, the impact of public funding on industry–science cooperation. At least three areas for improvement can be identified in most of the empirical literature related to this topic. First, there are very few studies that consider the mix of public funding for industry–science cooperation when considering the relationship between public funding and cooperation. Analyses are mostly restricted to one specific type/program of public funding (see for example, Miotti and Sachwald, 2003; Mohnen and Hoareau, 2003; Belderbos et al., 2004; Busom and Fernandez-Ribas, 2004, 2008; Clarysse et al., 2009). Second, many studies ignore the important differences in 'science' between universities on the one hand and public research centres on the other. Third, from an analytical point of view most of the studies that investigate whether public funding for innovation succeeds in influencing the cooperative practices of firms fail to address the problem of endogeneity caused by funding selectivity (initially addressed by Belderbos et al. (2004) and Aschhoff et al. (2006)).

This chapter contributes to the behavioural additionality literature by addressing these three issues at the same time. Based on micro-evidence from the biannual OECD business R&D surveys for Belgium for the years 2003 and 2005, the analysis includes public funding by means of subsidies both by regional governments and by the EU framework program, and distinguishes, among the 'science' partners, between universities and public research centres. The problem of endogeneity is dealt with by taking on board the history of public funding and the history of research cooperation to address the impact of public funding on the instalment of cooperation.

The chapter is organised as follows: Section 6.2 summarises the relevant findings in the theoretical and empirical literature, gives a brief account of the practice of funding in Belgium, and formulates the research questions. Section 6.3 describes the data and methods of estimation, and presents the empirical model based on data provided by the OECD bi-annual business R&D survey for Belgium. Section 6.4 describes the R&D dataset and Section 6.5 presents the empirical results. Section 6.6 discusses the main conclusions of the analysis and makes recommendations for future research and policy-making.

6.2 LITERATURE REVIEW, PUBLIC FUNDING OF RESEARCH COOPERATION IN BELGIUM, RESEARCH QUESTIONS

6.2.1 Literature Review

The theoretical rationale for public intervention in the innovation process follows directly from the non-rival characteristic of knowledge resulting in spillovers which generally tend to lead – from a social point of view – to an underinvestment in innovation efforts by private, profit-motivated individual decision-makers. This market failure may express itself not only in an underinvestment in privately-financed R&D, but in an insufficient exploitation of the opportunities of innovative efficiency-enhancing research cooperation. The failure is especially important where cooperation between industry and science is concerned (Martin and Scott, 2000). Since the science system is geared around publications and to a lesser extent patenting, there is a relatively larger disclosure of their R&D to the open science community (Dasgupta and David, 1994) compared to industry, in which secrecy and other protective measures more often prevent such knowledge-sharing. In the spirit of open innovation, however, the latter seems to be less the case for large hi-tech firms (Chesbrough, 2003a). Although various policy instruments exist to enhance innovation performance by firms, public funding is generally viewed as most suitable for fostering research cooperation, and especially the creation of networks involving both firms and public research organisations such as universities (OECD, 2002a). In this respect, to a large extent supported by public funding, industry–science cooperation has become a relatively cheap and low-risk source for the acquisition of specialised and new scientific and technical knowledge (Lundvall, 1992).

Questions on the efficiency and effectiveness of public funding of business R&D are of growing importance to policy-makers who seek ways to evaluate the 'additionality' of public support to business R&D, or in other words, what difference public support makes. The impact of public support for R&D has typically been approached in terms of input or output (additionality) measures only, treating the firm as a black box and not adequately capturing the impact of public intervention on the innovation process, the company rationale and attitude itself.

Behavioural additionality emerged as an observed phenomenon in early evaluations of collaborative R&D programmes when it was found that the traditional input and output formulations of additionality did not capture well the effects of programmes on firms (Buisseret et al., 1995), that is, how support interacts with and affects the strategies and capabilities of firms, and overlooked the learning effect which takes place in the short term in the company (Georghiou, 2002). The latter finding was based on an empirical observation that a common effect – during and/or after project implementation – of innovation policy goes beyond altering a stop–go decision in respect of a project and can more likely be found in a modification of the manner in which a project is carried out; that is, on a larger scale, covering a wider range of applications, or in a faster way (Georghiou, 2002).

Compared to input and output additionality, effects in terms of behavioural additionality can be expected to last beyond the period of R&D and become integrated in the firms' general capabilities (Georghiou, 2002).

Autio et al. (2008) make reference to a broader range of learning results as so called 'second-order additionality' including firm-level technological learning and innovation outcomes that result from knowledge spillovers, technology diffusion, and knowledge exchanges within communities of firms.

Clarysse et al. (2009) addressed a gap in the literature resulting from the fact that although previous studies identified the existence of behavioural additionality they did not try to explain why certain processes take place in an organisation. To do so, they performed an empirical test starting from the organisational theory-learning literature (including experiential learning, absorptive capacity and inter-organisational learning). Inspired by Huber's (1991) argument that an entity or organisation learns whether, through its processing of information, the range of its potential behaviours is changed, the authors emphasise that the idea of organisational learning suggests that the incentives for changes in firm behaviour are related both in a direct (formal) and indirect (informal) way with an internal learning process.

Related to the systems failure rationale for public intervention, Georghiou et al. (2004) raise strategy questions in terms of whether public support helps overcome a lock-in failure; gives incentives to firms to acquire new competences (skills and capabilities); and coordinates systemic innovations or builds new networks either between firms or between firms and the science base.

The focus of this chapter is on the latter dimension, that is, the creation of new networks between firms and the science base. This is only one dimension of the multi-dimensional concept of behavioural additionality (Falk, 2007). It relates to network additionality in case government support helps to create networks (OECD, 2005) and is closely linked to firm strategies in terms of knowledge acquisition. It should be noted that the distinctions between network additionality and other dimensions of behavioural additionality, such as scope, cognitive capacity, acceleration, and challenge additionality are not always clear-cut (Falk, 2007). Increasingly this knowledge acquisition is identified as a matter of managing external networks and refers to the emergence of a 'new ecology of industry' with the growth of collaborative R&D, outsourcing to specialist suppliers and universities and planned acquisition of start-up firms either on the market or through corporate venturing (Coombs and Georghiou, 2002).

The type of research partner either from industry (customers, suppliers, competitors, etc.) or from science (universities, polytechnics, public research centres) will tend to be affected by the relevance of research cooperation for firms. The nature of the projects involving science will generally tend to have a more risky character with a longer gestation period than those for mere inter-firm projects. However, empirical evidence for Belgium suggests that firms that attach more importance to publicly-available knowledge for their innovations tend to take greater advantage from cooperation with public organisations (Cassiman and Veugelers, 2002). From the perspective of science, this kind of cooperation generates additional financial resources, knowledge exchange, reduced dependency on government contracts, employment opportunities, and extra research facilities (Schibany and Schartinger, 2001). Not surprisingly, therefore, public research organisations increasingly orient their research agenda towards problem-solving research, possibly even to the detriment of the traditional generation of scientific knowledge. Yet, much of the literature aggregates the different actors in the public sector with respect to research (for example, Bacchiocchi and Montobbio, 2009).

Science has two prominent actors (OECD, 2002b): higher education institutions (universities, university research centres such as hospitals, and polytechnics – in the remainder of this chapter we will refer to 'universities' since these are the main actors in terms of research in this sector) and public research centres. The latter are usually involved in missions and often have specialised scientific domains such as microelectronics or biotechnology, and their environment has changed considerably over recent years. Some public research centres have five-yearly public agreements concerning their management. These changes can be seen in the increasing pressure these centres feel under to generate commercial income and stimulate technology transfer (Bozeman, 1994; OECD, 2002a). These centres have more business-oriented practices and experience budget constraints imposed by governments (European Commission, 2003). They are further affected by the shift to private management schemes (Boden et al., 2001). The role of universities in the innovation system as well has broadened in recent years with a growing demand for economic relevance and growing pressure to contribute more directly to the innovation systems of the national economies. The polytechnics in particular are well-versed in applied research, but universities are adapting their operations to meet industrial needs. This resulted in increasingly mission-oriented and contract-based research and the establishment of closer links with business through cooperative research, networks and exchange of information (European Commission, 2003; OECD, 2002a). Therefore, from a business perspective, cooperation with universities is aimed mainly at enabling firms to do (often pre-competitive) strategic long-term research. Compared to universities, public research centres are interesting research partners because of their more practical knowledge and ability to respond more quickly to specific firm needs (Bessant and Rush, 1995; Bruce and Morris, 1998; and Tether, 2002). For Belgium, these aspects are corroborated by the fact that, in the year 2003, 24.2% of all R&D performed by public research centres was financed by enterprises in the form of contract research; whereas this is only 11.6% in the case of R&D by the higher education institutes (Spithoven, 2007). This indicates a marked difference between public research centres and universities.

Although quite a lot of econometric work explains research cooperation or the impact of public funding on innovation inputs or outputs, such as R&D effort and granted patents, a limited number of studies have investigated the influence of public funding on research cooperation between industry and science. Making use of the Community Innovation Survey (CIS) or other similar survey data for European countries, several studies (for example, Mohnen and Hoareau, 2003; Belderbos et al., 2004) have presented cross-section regression analyses linking the type of research partner to the availability of public funding. The main conclusion from these studies is that R&D subsidies appear to have a stronger impact on the probability of cooperating with science than with industry. Moreover, the influence of subsidies appears to be stronger on vertical cooperation (with customer and suppliers) than on horizontal cooperation (with competitors) (see Miotti and Sachwald, 2003; Mohnen and Hoareau, 2003; Dachs et al., 2004; Abramovsky et al., 2005). A challenge in each of these studies is to deal with the problem of possible selectivity of funding, as well as to identify causalities between funding and cooperation. Funding may be deliberately targeted at firms already cooperating, implying reverse causation. Moreover, some factors that determine the likelihood of cooperation potentially affect the likelihood of receiving funding as well (Defazio et al., 2009). To the extent that some of these effects are not observed, the funding variable will be correlated with the error term, leading to biased and inconsistent estimates.

Based on CIS2 data for Germany, France, Spain and Ireland, Mohnen and Hoareau (2003) use a trivariate probit model with censoring to control for selection bias in cooperation with universities and public research centres. Belderbos et al. (2004) address the endogeneity issue by presenting an instrumented variable cross-section regression analysis explaining the probability of horizontal, vertical or industry–science research cooperation on Dutch CIS2 data for the period 1996–1998. Other studies that have addressed the issue of endogeneity have done so by using a non-parametric matching approach. Thus, Busom and Fernandez-Ribas (2004, 2008) have estimated the average treatment effect by national funding on vertical research cooperation and on cooperation with public research organisations on a cross-section of Spanish research-active manufacturing firms in 1998, whereas Aschhoff et al. (2006) have estimated the average treatment effect by national funding on cooperation exclusively with other firms, cooperation only with public research organisations, and cooperation with both, on a cross-section of cooperating firms in Germany in the period 1998–2003.

However, a problem with these estimates is that, due to possible selectivity bias caused by winner picking (that is, firms that are already engaged in research cooperation), data with respect to a specific moment in time raises doubts concerning the exogeneity of the control variables. To address this problem, Belderbos et al. (2004) repeated the estimation highlighted above while restricting the sample to include only new cooperators, that is, firms that cooperate in CIS2 but not in CIS1. As a consequence, the funding variable becomes non-significant. It is, however, unclear whether this can be viewed as confirmation that lagged funding is still endogenous, or as a reflection of the loss in information caused by the substantial sample size reduction. Aschhoff et al. (2006) have estimated probit equations explaining the probability of continued research cooperation after the funding has ended. They include among the explanatory variables a dummy depicting the probability of new cooperation initiated through funding. The negative significance of the latter variable in the case of cooperation with research organisations would, according to the authors, suggest that this type of cooperative agreement is less long-lived when set up through funding than it is when born spontaneously. Or, in other words, the 'behavioural additionality' involved would be more short-run in nature. There are, however, two problems with these estimates. First, the new cooperation dummy is not drawn independently from the continued cooperation dummy. As a consequence, the parameter estimate of the new cooperation variable will suffer from a simultaneity bias and be inconsistent. Second, since funding tends to be highly persistent, finding firms whose project funding ended (and was not replaced) necessarily implies a vast reduction in sample size. This poses serious questions as to its representativeness.

6.2.2 Sources of Public Funding for Innovation in Belgium

To properly identify the impact of public funding instrument(s) on research cooperation there is a need to consider the broader policy mix of different types of public funding for research (Nauwelaers, 2009). Georghiou et al. (2004) emphasise that subsidies or grants to firms form only one means to support firms. In terms of direct measure, besides grants for financing industry R&D, we can consider support for public sector research and for training and mobility. Moreover, nor can we ignore, on the supply side, services relating to information and brokerage support, and network measures. Nor, on the demand side (systemic policies, procurement, and regulation), can we ignore important policy areas for the promotion of research and innovation. Finally, we should not neglect framework conditions relating to human resources, science-based, and regulatory framework (including state aid, competition and IPR, and general fiscal environment). In principle, behavioural additionality can be expected to result (intentionally or unintentionally) from any of these measures. But for the purposes of this study the focus will be on R&D subsidies to firms.

Firms in Belgium can turn to two main public sources for subsidies to fund their R&D. Funding by regional governments is the most important source of direct funding (subsidies) for R&D in Belgium. It is largely inspired by a bottom-up approach and therefore more or less tailored to the demands of the enterprises. Decentralisation makes it possible for regions to accommodate the specificities of enterprises located on their territory. Regionally-funded programs are increasingly aimed at fostering cooperative research and innovation networking, in order to stimulate the use of scientific knowledge (by firms). Firms setting up research cooperative partnerships with science, and undertaking basic research, are particularly favoured (Capron et al., 2000). The general idea behind most regionally-funded programmes is to foster networking and cooperative activities, aimed particularly at the economic commercialisation of scientific knowledge and so aimed at stimulating research cooperation with science. Therefore, it is not aimed at enriching the R&D portfolio of firms or lowering the threshold for innovative projects with higher risk factors. It is important to note that some of the funding programs are exclusively oriented towards small and medium-sized enterprises (Soete, 2007). According to type of partner, specific measures are taken to promote industry–science linkages; be they built or not around key technologies or themes of particular regional relevance. For an overview of the existing measures in the period 2003–2005, see Nauwelaers (2004).

Subsidies from the EU framework programs account for a relatively modest proportion of total funding for innovation in Europe (Polt and Psarra, 2006). In Belgium they accounted for 1.1 % of total business R&D financing, against 4.3 % for regional funding in 2002 (CFS/STAT, 2007). EU funding largely flows top-down to specific themes and the main aim is to foster cross-border research cooperation in order to increase the scale and scope of research efforts; to achieve a 'critical mass' that is not attainable at national level, notably in areas characterised by uncertainty and high risk (Defazio et al., 2009). Domains that received special attention in the Fourth and Fifth Framework Programs, with respect to 1994–1998 and 1998–2002, are ICT, new materials, biotech and energy/environment (Muldur et al., 2006). The creation of international networks that include firms and public research organisations from different countries is especially favoured. One important reason for considering public funding via the EU framework programs together with public funding from regional governments is that, in Belgium, some regional public funding measures aim to help enterprises (particularly small and medium-sized) to prepare to apply for EU framework program funding (Nauwelaers, 2004).

6.2.3 Research Questions

The focus of this chapter is on the impact of public funding for R&D on industry–science cooperation. In the literature review, arguments are put forward to differentiate between universities and research centres among the 'science' partners. Our data pertain to two time-periods and the focus is on behavioural additionality of public funding in terms of the 'instalment' of industry–science research cooperation. This limits the analysis to firms that did not cooperate in research before. The research questions are formulated as follows: (i) does public funding for R&D increase the probability of 'installing' cooperation with universities?; and (ii) does public funding for R&D increase the probability of 'installing' cooperation with public research centres?

We can formulate a few hypotheses on the basis of the literature review. First, since regional public funding pays attention to installing industry–science research cooperation, we expect to see a positive impact of this type of funding on cooperative behaviour; that is, we expect to see a significant increase in the instalment of cooperation with universities and public research centres. Because economic commercialisation is a key concern in the case of funding by regional governments in Belgium, we expect to see a greater impact of funding on cooperation with public research centres as compared to universities. The assumption being that public research centres, in Belgium, offer more applied research results than universities. In the case of EU framework program funding we expect less in the way of instalment of research cooperation since we can expect the firms that already have to be engaged in cooperation to successfully apply for this type of funding.

6.3 DATA DESCRIPTION

Use is made of firm-level data from the bi-annual OECD business R&D survey for Belgium (CFS/STAT, 2007). The unit of analysis of the survey is firms that perform R&D on a permanent or quasi-permanent basis. A permanent or quasi-permanent R&D-active firm is any firm which declared in the OECD survey that they had: (i) at least one full-time equivalent R&D employee, and (ii) internal R&D expenditure greater than or equal to 30,000 euros. The firms are defined at the level of the smallest legal entity: that is, those having a VAT number. The list of these R&D-active firms is updated every time a new survey is launched, account taken of firms known from the past to be R&D-active and monitoring of firms that declare themselves R&D-active (for example, by means of press releases or requests for R&D grants). There is, therefore, a danger of a sample selection bias.

Two subsequent surveys were studied: the R&D survey 2004, offering results for the years 2002 and 2003; and the R&D survey 2006, offering results for 2004 and 2005. In the surveys, a firm is asked whether or not it cooperated with a certain type of partner in the respective period. Data on public funding of R&D are only available for the years 2003 and 2005. This difference in reference period/year as regards cooperation with a certain type of partner and external funding could underestimate the relationship between changes in both variables.

Some firms are excluded from the analysis: those that merged, acquired (parts of) other firms or experienced other changes with an important impact on the evolution of R&D and other enterprise variables (information on changes with an influence greater than 10% in terms of overall employment or in terms of turnover was requested). Firm-level data have been constructed to guarantee comparability over time, resulting in a database of 1,083 enterprises. Unfortunately, due to the voluntary nature of the survey there are a lot of missing values. Most information is available for internal R&D expenditure and R&D personnel. This is because, in the case of non-response on these variables, an additional survey/telephone follow-up was used to complete the data at the time the survey was conducted. Data on the public funding of internal R&D expenditure and on research cooperation were less extensively surveyed and therefore show a higher non-response rate.

To analyse the data in view of the research questions, we use information on research cooperation with universities (COUNI) and with public research centres (COPRC) and public funding for R&D. As control variables we opt for generally accepted firm- and research-related characteristics. Table 6.1 summarises the information. The control variables consider sector of activity (PAVITT); size (SME); group membership (GROUP); and export orientation (EXPORT). Involvement in research activities as opposed to development activities (RESEARCH); and the R&D employment intensity (RD_INT) and engagement in R&D outsourcing (RD_OUT) represent the research-related variables. The public funding-related variables include public regional funding (FREG), and public EU funding (FEU). As highlighted in the literature review to limit problems of endogeneity, for public funding, we also include variables indicating the 'starting' of funding (START).

The requirement of information on each of these variables for each of the two time-periods resulted in a data-set of 424 firms. Descriptive information concerning these variables is presented in Table 6.1.

Table 6.1 Explanatory variables for changes in research cooperation with a certain type of partner, and the way they are measured (N=424)

Variable	Description	Mean	St. dev.
Dependent variables			
COUNI_03	Binary variable = 1 of cooperation in research with a university in 2003	0.30	0.46
COPRC_03	Binary variable = 1 of cooperation in research with a public research centre in 2003	0.23	0.42
COUNI_05	Binary variable = 1 of cooperation in research with a university in 2005	0.28	0.45
COPRC_05	Binary variable = 1 of cooperation in research with a public research centre in 2005	0.27	0.44
Independent variables			
PAVITT	Binary variable = 1 if the firm belongs to the sector based on an extended Pavitt taxonomy - see Appendix, reference year 2003		
SME	Binary variable = 1 if the firm is a large enterprise, reference year 2003	0.78	0.42
GROUP	Binary variable = 1 if the firm is independent and does not belong to a group, reference year 2003	0.31	0.46
EXPORT	Share of export in total turnover, reference year 2003	0.58	0.36
AGE	Age of the firm in reference year 2003	24.5	22.0
RESEARCH	Share of research expenditures in internal R&D expenditures, year 2003	0.41	0.35
RD_INT	Share of R&D employment (in full-time equivalents) in average firm employment, year 2003	0.25	0.313
RD_OUT	Binary variable = 1 if the firm engages in outsourcing its R&D, year 2003	0.47	0.50

Table 6.1 (continued)

Variable	Description	Mean	St. dev.
FREG_03	Binary variable = 1 if the firm received funding from regional government in 2003	0.34	0.47
FREG_05	Binary variable = 1 if the firm received funding from regional government in 2005	0.39	0.49
FREG_START	Binary variable = 1 if the firm did not receive funding by regional government in 2003 and did receive funding by regional government in 2005	0.15	0.36
FEU_03	Binary variable = 1 if the firm received funding from EU government in 2003	0.08	0.27
FEU_05	Binary variable = 1 if the firm received funding from EU government in 2005	0.08	0.27
FEU_START	Binary variable = 1 if the firm did not receive funding by EU in 2003 and did receive funding by regional government in 2005	0.03	0.18

Compared to the firms present in the target population but for which detailed information on R&D (especially cooperation and public funding) is lacking, the 424 firms included in the analysis have a significantly larger R&D employment and are more export-oriented (see Table 6.2). If we consider only those firms out of the 424 that were not involved in cooperation with universities or public research centres in 2003 (that is, the firms that will be used in the analysis in Section 6.4 when estimating the behavioural additionality of public funding in terms of the propensity to start cooperation) and compare these firms with all other firms, no significant differences are found.

Table 6.2 Comparison of firms included in and excluded from the estimation

	Firms with data for 2003 and 2005 (N = 424)°	Firms with data for 2003 and 2005 and not cooperating with universities or public research centres in 2003 (N = 350)°°
Average employment	304.9	210.4
Average R&D employment (FTE)	28.1*	16.9
R&D intensity	0.21	0.19
Export share	0.58*	0.57
Research share in total R&D	0.44	0.43

Notes: R&D intensity is measured as the ratio of R&D employment to average employment; * significant difference between group means (at 5% level); ° compared to firms for which not all data for the analysis is present (N = 659); °° compared to all other firms (N = 733).

Comparison of the 424 firms for which we have a complete data set with the 659 firms for which not all information is available reveals that as far as firm size, industrial sector (which is based on the Pavitt taxonomy and reprised in the Appendix) and group membership is concerned, services and small firms prove under-represented in the analysis. Compared to the overall odds (defined as the frequency of one category of a variable compared to the frequency of another) of inclusion versus non-inclusion (0.65), the odds for inclusion versus non-inclusion for science-based firms is, however, 1.02. For specialised suppliers the odds amount to 0.98. Therefore, science-based and specialised suppliers in the manufacturing industry are over-represented in this sample. Specialised suppliers in services (0.35) and information-intensive services (0.45) are under-represented. The odds of inclusion versus non-inclusion in terms of firm size is 0.82 for large (250 and more employees); 0.76 for medium-sized (50 to 249 employees); and 0.54 for small firms (fewer than 50 employees). This means that within the category 'SME' included in the analysis, the smallest firms are under-represented. No significant differences are found in terms of whether or not a firm belongs to a group. For the 350 firms with no cooperation in the year 2003, no significant differences are found in terms of size and group membership. The differences in terms of sector of activity are less significant. However, compared to the overall odds of inclusion versus non-inclusion (0.48), science-based firms (0.61) and specialised suppliers (0.69) remain over-represented, whereas specialised suppliers in services (0.32) and information-intensive services (0.39) remain under-represented.

In the year 2005, 120 firms were involved in research cooperation with universities, and 114 firms in cooperation with public research centres. Between 2003 and 2005, 38 firms (out of 424, or 9.0%) started research cooperation with universities, and 52 (12.3%) did so with public research centres. It has to be noted that in the analysis, public research centres include several top research centres such as IMEC, the renowned Interuniversity Centre on Microelectronics, and private non-profit organisations involved in R&D activities such as certain foundations (for example, on Cancer treatment). Belgium also has specialised collective research centres that perform R&D and related R&D activities on behalf of industry and are governed and mainly financed by industry but receive a large share of public funding (Spithoven, 2007). These organisations are all grouped under the umbrella of public research centres.

In terms of public funding by means of subsidies by regional governments and the EU framework program, 33.5% (resp. 38.9%) of the firms received public funding by regional governments in 2003 (resp. 2005). For EU funding this amounted to 8.0% (resp. 7.8%). In terms of initiation of funding, 15.1% (resp. 3.3%) of the firms received funding in 2005 without having already received funding in 2003 for funding by regional governments (resp. EU framework programs).

The data presented above suggest a high persistence both in terms of cooperation and in terms of funding by public governments. Based on the correlation matrix presented in Table 6.3, it turns out that cooperation with both universities and public research centres, as well as regional and EU funding, are all heavily correlated and this both within a time period and over time. This justifies the focus in this chapter on the instalment of research cooperation and clearly highlights the shortcomings of the bulk of literature using lagged values or matching to explain industry–science cooperation. The strong correlation between cooperation with universities and with public research centres indicates that both types of cooperation need to be considered in relation to each other.

Table 6.3 Tetrachoric correlation matrix including cooperation and public funding for research (N=424)

	COUNI_ 03	COPRC_ 03	COUNI_ 05	COPRC_ 05	FREG_ 03	FREG_ 05	FEU_ 03
COUNI_03	-						
COPRC_03	0.80**	-					
COUNI_05	0.76**	0.61**	-				
COPRC_05	0.55**	0.67**	0.84**	-			
FREG_03	0.37**	0.41**	0.25*	0.30**	-		
FREG_05	0.43**	0.43**	0.39**	0.53**	0.68**	-	
FEU_03	0.51**	0.44**	0.42**	0.37**	0.49**	0.38**	-
FEU_05	0.38**	0.34*	0.36*	0.31*	0.33*	0.37*	0.83*

Note: The symbols *, ** indicate significance of correlation at 1% and 0.1% level.

6.4 ESTIMATING THE INSTALMENT AND PERSISTENCE OF COOPERATION

Following Belderbos et al. (2004) we can, to a large extent, avoid problems of endogeneity between funding and cooperation behaviour by taking as our starting point firms that did not cooperate before. This means, in this study, that we look for firms that did not cooperate with universities or public research centres in the period 2002–2003. Even if it is assumed that in this case no reverse causation occurred due to funding being targeted at firms that were already cooperating, unobserved factors might affect both the likelihood of starting cooperation and of starting to receive funding (although it should be noted that the control variables can partly reduce this problem). Moreover, a degree of reverse causation cannot be excluded in this case, since funding authorities might pick out firms that start cooperating.

The omission of firms that had already cooperated reduced the sample to 350 (see above). A comparison in this group of 350 firms with and without funding allows us to identify behavioural additionality in the sense that a distinction can be made between funding-related cooperation and non-funding-related cooperation: that is, the counterfactual situation in which there was no public intervention. Since cooperation with universities and/or public research centres are not mutually exclusive activities, we use a bi-variate probit model to estimate propensity to cooperate with both types of partners (in a model covering two periods: 2002–2003 and 2004–2005).

In relation to the research questions, the empirical model investigates the results of funding in terms of instalment of industry–science cooperation.

Taking into account the highly significant correlation between cooperation with universities and cooperation with public research centres the bivariate probit model can be denoted as follows.

$$\text{COUNI\_05} = \text{f1(C, PAVITT, SME, AGE, GROUP, EXPORT, RESEARCH, RD\_INT, RD\_OUT, FREG\_START, FEU\_START)} \quad (5.1)$$

$$\text{COPRC\_05} = \text{f2(C, PAVITT, SME, AGE, GROUP, EXPORT,RESEARCH, RD\_INT, RD\_OUT, FREG\_START, FEU\_START)} \quad (5.2)$$

The variable FREG_START is given the value 1 if there was no regional funding in 2003 (FREG_03 = 0) and if the firm received regional funding in 2005 (FREG_05 = 1). The same goes for variable FEU_START, that is, it is given the value 1 if the firm received no European funding in 2003 (FEU_03 = 0) and did receive European funding in 2005 (FEU_05 = 1). The stochastic error terms may be correlated since the decision to cooperate with the two types of science partners may be interdependent.

First, we present the 'model without funding'. The significance of RHO justifies a bivariate probit approach. For the set of (quasi-) permanent R&D-active enterprises, the sector, firm size, age or export orientation do not seem to determine the propensity to engage in research collaboration with science partners. The exception of the highly significant negative effect of firms belonging to the information-intensive sector on propensity to cooperate with public research centres can be explained by the fact that the only centre in this case – IMEC, the renowned Interuniversitary Centre for Microelectronics – is a hi-tech centre conducting fundamental research which caters far less for SMEs in the software sector that require applied research. IMEC is, moreover, especially active in international research which is also not fully captured in our sample, which only considers firms located in Belgium. Also, the research-versus-development orientation of the R&D activities has no significant influence on research cooperation. On the contrary, the firms' internal R&D employment intensity (R&D personnel on total firm employment) positively influences the propensity to start engaging in cooperation with universities (but not with public research centres). It is interesting to note that it is mainly the linkages with other firms, either through belonging to a group or through outsourcing R&D to third parties, that significantly influence in a positive way the propensity to cooperate both with universities and with public research centres. This can be related to higher needs in terms of absorptive capacity which could be strengthened by means of collaboration with a science partner.

Table 6.4 Bi-variate probit model relating funding of R&D to formal R&D cooperation

	Model without funding		Model with funding	
	COUNI_start	COPRC_start	COUNI_start	COPRC_start
Constant	−1.99 (−4.00)***	−1.28 (−3.38)***	−2.01 (−4.16)***	-1.54 (-4.00)
PAV_LI	0.04 (0.09)	0.20 (0.72)	0.03 (0.10)	0.30 (1.03)
PAV_SB	−0.38 (−1.22)	−0.16 (−0.51)	−0.37 (−1.12)	0.03 (0.09)
PAV_II	−0.70 (−1.74)	−0.83 (−2.17)*	−0.67 (−1.66)	-0.84 (-2.30)*
PAV_RI	−0.32 (−0.96)	−0.44 (−1.55)	−0.30 (−0.91)	-0.39 (-1.32)
PAV_SI	0.17 (0.59)	−0.17 (−0.61)	0.19 (.64)	0.07 (0.24)
PAV_SSS	−0.65 (−1.32)	−0.15 (−0.39)	−0.68 (−1.34)	-0.01 (-0.02)
SME	0.12 (0.44)	−0.10 (−0.42)	0.13 (0.47)	-0.11 (-0.49)
AGE	0.01 (1.81)	0.00 (0.51)	0.01 (1.82)	0.00 (0.60)
GROUP	−0.51 (−2.36)*	−0.68 (−3.04)**	−0.52 (−2.35)*	-0.63 (-2.83)**
EXPORT	0.20 (.68)	0.32 (1.23)	0.17 (0.60)	0.27 (1.05)
RESEARCH	0.33 (1.33)	0.17 (0.71)	0.33 (1.37)	0.24 (0.96)
RD_INT	0.82 (2.57)**	0.42 (1.21)	0.78 (2.35)*	0.31 (0.90)
RD_OUT	0.51 (2.77)**	0.49 (2.78)**	0.51 (2.78)**	0.48 (2.60)**
FREG _START			0.13 (.44)	0.90 (3.87)***
FEU_START			0.30 (.62)	0.47 (1.16)
RHO	0.70 ***		0.74***	
Wald chi² test	63.87 ***		93.16 ***	
Obs.	350		350	

Notes: Reference categories: PAV_SS (specialised suppliers in manufacturing industry); firms that did not receive external funding for their internal R&D. The z-values appear between brackets (two-tailed tests); and the symbols *, **, *** denote the significance at 5%, 1% and 0.1% level.

These results are robust if we include public funding by means of R&D subsidies by regional governments and EU framework programs (model with funding). Receipt of regional funding for the 'first' time (FREG_START), turns out to have a positive effect on the instalment of industry–science cooperation, but only with public research centres. This can be related to the focus on economic commercialisation of research funded by the regional governments in Belgium. EU framework program funding has been found not to influence the instalment of research cooperation with science partners. This can be due to the fact that to apply for funding under the EU framework programs, firms already had to be successfully engaged in cooperation. In this respect, the absence of influence could be related to a selection bias towards firms already engaged in cooperation.

To facilitate the interpretation of the coefficients of the bivariate probit model we calculate the marginal effects estimates (Greene, 1996). Table 6.5 shows the change in probability of cooperating with universities and public research centres for an infinitesimal change in the dependent variables at reference (mean) points. This implies that the coefficients we discuss consider cooperation with both universities and public research centres. This discussion is warranted since their tetrachoric correlation in Table 6.3 was 0.80 in 2003 and 0.84 in 2005.

Table 6.5 Marginal effects after the bivariate probit model (N = 350)

	Model without funding		Model with funding	
	Marginal effects (dy/dx)	z–value	Marginal effects (dy/dx)	z–value
Predicted probability of COUNI and COPRC	0.42		0.43	
PAV_LI	0.011	0.38	0.014	0.45
PAV_SB	–0.023	–1.40	–0.018	–0.92
PAV_II	–0.046	–3.38***	–0.046	–3.22***
PAV_RI	–0.028	–1.78*	–0.026	–1.59
PAV_SI	0.002	0.07	0.014	0.50
PAV_SSS	–0.031	–1.74*	–0.031	–1.58
SME	0.003	0.16	0.003	0.15
AGE	0.001	1.70*	0.001	1.62
GROUP	–0.047	–2.85***	–0.046	–2.84***
EXPORT	0.023	1.13	0.020	0.97
RESEARCH	0.025	1.31	0.028	1.44
RD_INT	0.062	2.56**	0.055	2.08**
RD_OUT	0.053	2.76***	0.053	2.71***
FREG _START			0.044	1.08
FEU_START			0.047	0.79

Note: The symbols *, **, *** denote the significance at 10%, 5% and 1% level.

In the case of the model without funding, the predicted probability of cooperating with universities and public research centres is 4% for independent small firms active in the specialised supplier sectors and engaging in R&D outsourcing and which are, on average, 24 years, have a research share of 39%, an export share of 57% and an R&D employment intensity of 23%. Group membership increases the probability of cooperating by 4.7% compared to independent firms. Being active in an information-intensive sector reduces this probability by 4.6% compared to firms active in the specialised supplier sector. Outsourcing R&D increases the probability of cooperating by 5.3% more than firms performing their entire R&D in-house. For an increase of 1% in the share of R&D employment intensity, holding other covariates at the reference points, there is an increase in the predicted probability of cooperating of 6.2%.

The marginal effects of firms active in information-intensive-activities (–4.6%); having group membership (–4.6%); engaging in R&D outsourcing (5.3%) are fairly similar to those in the model including the variables on funding, which would indicate that our findings are robust. A 1% increase in R&D employment intensity also increases by 5.5% the probability of cooperating with universities and public research centres. More important are the non-significant effects that firms have on the probability of cooperating both with universities and public research centres when starting to receive funding from authorities, although both estimates have positive signs.

Therefore, the analysis of marginal effects empirically corroborates the need to discuss the effects of cooperating with universities or public research centres separately, as in Table 6.4.

6.5 CONCLUSIONS

An analysis was presented of the impact of public R&D funding on research cooperation by firms based on data from the 2004 and 2006 OECD bi-annual business R&D surveys for Belgium. The analysis focuses on the creation of firm behavioural additionality for public R&D funding by governments in terms of the instalment of firms' cooperation with universities and public research centres. It takes account of funding in the form of subsidies from regional governments and from EU framework programs. Over two time-periods, the starting point for the analysis is those firms that were not engaged in industry–science cooperation in the first period. We cannot rule out a partial bias in the results towards science-based firms and specialised suppliers, whereas firms in services are slightly under-represented.

In terms of the instalment of research cooperation with science partners, funding by regional governments turned out to have a positive effect on the instalment of industry–science cooperation, but only with public research centres. This can be related to the focus on economic commercialisation of research funded by the regional governments in Belgium. EU framework program funding has been found not to influence the instalment of research cooperation with science partners. This might be explained by a selection bias towards firms already engaged in cooperation. Besides funding, engagement in R&D outsourcing activities in particular had a highly positive impact on research cooperation in terms of cooperation with both universities and public research centres. The opposite was found for firms that are not part of a group. Internal R&D employment intensity (share of R&D personnel in total firm employment) turned out to be an important factor for the set-up of industry–science cooperation, but only in the case of cooperation with universities.

The descriptive analysis that relates research cooperation to public funding showed that both regional and EU funding are biased towards firms that are already engaged in cooperation and that a positive correlation exists between cooperation with universities and public research centres and the propensity to cooperate is persistent over time. This implies that the use of lagged public funding to explain research cooperation has serious shortcomings as a measure of behavioural additionality through the public funding of research.

The preceding analysis offers several hints for improving future data for the purposes of research. One such area of data improvement and ensuing future research on the issue of instalment of cooperation is that of improving the inter-temporal comparability of successive R&D surveys by further harmonising the questionnaires and covering a sufficiently large core of systematically surveyed enterprises. Explicit questions should be asked with regard to the actual reasons for setting-up and joining research partnerships. This should enable us to identify, at firm level, those cooperative research partnerships that were created through or with the support of public funding relative to those created through cooperation that emerges more 'spontaneously'.

Since regional and EU public funding seem to be targeted at firms that are already cooperating and do not favour the instalment of new cooperation with science partners, this kind of more detailed information (compared to the binary questions that were actually posed on engagement in cooperation) would allow a more correct evaluation of the additionality of public funding in terms of research cooperation.

Another aspect that cannot be ignored is that there may, in the innovation system, be limits to increasing the propensity to collaborate on innovation. To name just one: cooperation carries additional costs in the way of coordination activities. The analysis in this chapter did not cover effects that might occur outside the funded firms, such as an increase in the system's innovation or networking capacity.

From a policy-making viewpoint, when dealing with (questions on) behavioural additionality it is important to understand that effects may (or may not) be intentional on the part of either the policy-maker or funding recipient. Also, it is of interest to assess the persistence of the effects, given that behavioural additionality effects can be expected to last beyond the period of R&D and become integrated in the firms' general capabilities (Georghiou, 2002). Innovation policies from a system perspective place a heavy emphasis on the formation and promotion of the resulting networks. However, it is important to assess the values of the linkages, because they may be a cost as opposed to a benefit at one extreme. Since persistence is detected once a firm engages in cooperation, it is important, for regional and EU public funding alike, to persuade more firms that do not engage in this sort of activity (instead of selecting firms already involved in industry–science cooperation - that is, winner picking). Once funding has succeeded in encouraging firms to engage in industry–science cooperation, it is interesting to discover whether and how these partnerships are maintained (sustained technology alliances), that is, what is the capacity of a company to retain learning in its routines? From a policy perspective, this relates to the question of whether we need reinforcement policies for learning.

Also, this chapter has looked at R&D subsidies for firms. However, in several countries – as in Belgium – the same firms are likely to have benefited from innovation support services provided by public agencies and possibly from demand-side policies and targeted changes in framework conditions. In this chapter a first attempt was made by considering R&D subsidies provided by different government levels (that is, regional and EU level). In order to more adequately assess the additionality of policy measures for R&D, the necessity to take into account the policy mix as a whole (the complementarity of the different programs) and assesses the performance of the innovation system in an international perspective can been highlighted. For this reason, further research can contribute by analysing the effects of R&D subsidies in the context of other measures experienced by firms (and of other influences that act on the aspects of behaviour that are of interest).

APPENDIX

Table A6.1 Sector classification based on Pavitt taxonomy

Variable	Definition	Sources of innovation	Learning regime *
PAV_LI	PAV_LI = 1 if the firm is classified in the 'labour-intensive' sectors (fabricated metal products, textile and clothing, construction, other manufacturing)	Specialised suppliers	LBU
PAV_SB	PAV_SB = 1 if the firm is classified in the 'science-based' sectors (biotech, computers, office machinery, scientific instruments, pharmaceuticals, aerospace)	Universities and research centres	LBS - LBD
PAV_II	PAV_II = 1 if the firm is classified in the 'Information-intensive' sectors (computer and related activities, telecommunications, research and development, technical engineering)	Universities and research centres	LBS - LBD
PAV_RI	PAV_RI = 1 if the firm is classified in the 'resource-intensive' sectors (food, beverages and tobacco, gas, water, electricity, non-metallic mineral products, wood and furniture, non-ferrous metals, recycling, petroleum refining)	Specialised suppliers	LBU
PAV_SI	PAV_SI = 1 if the firm is classified in the 'scale-intensive' sectors (other transport equipment, rubber and plastic products, chemicals, motor vehicles, ferrous metals, shipbuilding, paper printing)	Production and specialised suppliers	LBD - LBU
PAV_SSS	PAV_SSS = 1 if the firm is classified in the 'specialised suppliers services' sectors (financial intermediation, logistic services, business services, wholesale, posts)	Science-based firms/customers	LBI - LBD-LBS
PAV_SS	PAV_SS = 1 if the firm is classified in the 'specialised suppliers' sectors (non-electrical machinery, electronics-communications, electrical machinery)	Science-based firms/customers	LBI - LBD-LBS

Sources: Pavitt (1984), Tidd et al. (1997).

*Notes:** LBS: Learning by searching; LBD: Learning by doing; LBU: Learning by using; LBI: Learning by interacting.

PART THREE

A closer look at open innovation: R&D outsourcing

7. The impact of internal R&D, strategic appropriability and cooperation on R&D outsourcing

7.1 INTRODUCTION

In the past two decades there has been a remarkable increase in the R&D outsourcing activities of firms in developed market economies. This is observed through the substantial increase in the number of R&D partnerships, through interactions between partners and through the significant growth of firm expenditure on external R&D. This has all taken place in a context of rapidly developing, elaborate R&D networks that combine different kinds of partnership, from straight licensing, consulting and arm's-length pure subcontracting, through more profound long-term types of subcontracting, to various forms of explicit R&D cooperation. These partnerships involve different types of partners, such as competing firms, firms connected by vertical supplier–customer relations, and public research institutions such as universities and public research centres. And they tend increasingly to become international in nature, involving partners from different countries (Roijakkers et al., 1994; OECD, 2008b).

These developments are somewhat puzzling in the light of transaction cost economics. According to this, the development of technology market transactions through subcontracting, and of lasting collaborative R&D agreements, is hampered by factors such as asset specificity, the uncertainty of the outcome of the R&D process, and information asymmetries between buyer and supplier (subcontractor), or between research partners, leading to free-riding behaviour (Williamson, 1981, 1985; Walker and Weber, 1984; Milgrom and Roberts, 1987; Pisano, 1990). In the same vein, papers in the field of industrial organisation have focused on the difficulties in reaching an adequate level of appropriability, through legal protection and secrecy, to allow persistent cooperative partnerships in R&D (Shapiro and Willig, 1990; Kesteloot and Veugelers, 1995).

The perception of the above-mentioned tendencies has, however, led to a gradual emergence of new views on the purpose of R&D outsourcing and on its relation to in-house R&D and appropriation. Thus, it has been argued that firms' awareness of the presence of substantial network externalities fosters the establishment of trust-based long-term inter-organisational R&D relations (Dyer and Singh, 1998; Zaheer et al., 1998). The necessity of absorbing the knowledge obtained through research partnership has made in-house and external R&D increasingly complementary in nature (Kamien and Zang, 2000). And, in such a more open innovation context, the appropriability required to allow R&D partnership tends to occur in a more informal manner, through complexity of design, rather than through the deliberate actions of legal protection or secrecy (Chesbrough et al., 2006).

However, the related econometric literature is still in its infancy. Some regression analyses have provided evidence that being in-house R&D-active increases the probability of cooperating in R&D (Becker and Dietz, 2004; Belderbos et al., 2004; Fontana et al., 2004). But this may vastly underestimate the actual effect of in-house R&D on R&D outsourcing intensity. There is also some scattered evidence that strategic appropriability has a positive impact on the probability of R&D cooperation (Cassiman and Veugelers, 2002; Abramovsky et al., 2005). But the measures of strategic appropriability used are aggregated. They do not allow identification of the impact of their component parts relating to secrecy, complexity of design and lead-time advantage. And it is, again, unclear what the actual impact is on R&D outsourcing intensity. The issue as to whether firms involved in formal cooperation have, on average, a different intensity of R&D outsourcing than other firms has, until now, never been investigated. This issue is, however, of great interest in view of the preference of public funding authorities for this type of partnership, especially when industry–science relations are involved.

This chapter attempts to clarify these issues by estimating the impact of in-house R&D, appropriability and cooperation on the R&D outsourcing intensity of firms in Belgium. A cross-section regression analysis is performed on data from the Third and Fourth European Community Innovation Surveys for Belgium, for 1998–2000 and 2002–2004, and from the Belgian R&D Surveys, for 2003 and 2005. The dependent variable is always obtained from the most recent of these surveys, whereas the explanatory variables are instrumented by lagged values from the corresponding previous ones. The sample selection problem is handled by a generalised tobit approach.

The chapter is organised as follows. Section 7.2 summarises the relevant findings in the theoretical and empirical literature and expresses the research questions. Section 7.3 presents the empirical relations and describes the data and methods of estimation. The estimates are presented in Section 7.4, whereas Section 7.5 gives a discussion and concludes. Tables with a detailed description of the variables and with summary statistics are presented in the Appendix.

7.2 R&D OUTSOURCING: THEORY, EMPIRICS AND RESEARCH HYPOTHESES

7.2.1 Theory

The most influential formal explanations of the outsourcing of economic activities by firms have been offered by transaction cost economics and contract theory. According to these, outsourcing occurs under conditions of low transaction costs. This will be the case when asset specificity, the frequency of transactions and uncertainty are low (Williamson, 1981, 1985; Walker and Weber, 1984). This may be difficult in the case of technology. The tacit components of technological knowledge tend to make it specific. The uncertainty of the outcome of the R&D process and the information differentials between buyer and supplier (subcontractor) will create serious monitoring problems (Milgrom and Roberts, 1987). External R&D expenditure will fall prey to moral hazard, as the subcontractor may be tempted to free-ride and sell the knowledge concerned to others. Technology transactions should, therefore, be confined to standardised technologies, so that the outsourcers do not feel threatened by their partners' opportunism. In the case of more complex specific technologies, hierarchical internalisation within the firm should be preferred as an organisational structure.

A complementary explanation of outsourcing focuses on the capabilities of the firm. According to this, the firm will outsource non core activities in which it is not particularly specialised, but may still be important as support (Kogut and Zander, 1992; Barney, 1999; Quinn, 2000). Technology outsourcing should, therefore, concern relatively standardised technological knowledge which is non-core to the firm's activities.

Close R&D cooperation has also been explained within a transaction cost economics logic, as a hybrid form of organisation between hierarchical internal transactions within the firm and arm's-length pure subcontracting transactions in the market place (Pisano, 1990). Just like own innovative activity within the firm, close R&D collaboration with others limits transaction costs, but it has the advantage that it allows access to the specialist knowledge of research partners. The uncertain outcome of the R&D process and information asymmetries between research partners will, however, make reaching such collaborative R&D agreements rather difficult and they will form a permanent threat to their survival.

Papers in the field of industrial organisation have especially focused on the relation between the appropriation capacity of the results of the innovation process and R&D cooperation. On the one hand, information sharing in cooperative agreements allows cooperating firms to internalise incoming knowledge spillovers more efficiently than non-cooperating firms (D'Aspremont and Jacquemin, 1988; Kamien et al., 1992). On the other hand, imperfect appropriability increases the incentive of firms to free-ride on their partners' R&D investments and it fosters free-riding by outsiders to the cooperative partnership (Shapiro and Willig, 1990; Kesteloot and Veugelers, 1995). These models concentrate on horizontal cooperation with competitors, and they think of appropriability mainly in terms of deliberate actions of legal protection and of strategic protection, through patenting and through secrecy.

A growing awareness of the increasingly open nature of the innovation process has, however, led to new developing views on the role and nature of R&D outsourcing and its relation to in-house R&D effort, and on the issue of knowledge appropriability in the innovation process. Thus, the notion of absorptive capacity, originally introduced by Cohen and Levinthal (1989), was further developed by Kamien and Zang (2000) in order to explain why in-house R&D is increasingly aimed at creating the prior knowledge necessary to effectively absorb external ideas. This concerns not only direct incoming knowledge spillovers, but also knowledge obtained through research partnership. In-house R&D effort and R&D outsourcing become major complements rather than minor substitutes, as suggested by earlier conventional wisdom.

Relational rent and trust arguments have been put forward to explain how establishing long-term trusting inter-organisational relations, including buyer–supplier relations, can provide benefits similar to internal hierarchies, without the cost and knowledge disadvantages associated with in-house innovation effort on its own (Dyer and Singh, 1998; Zaheer et al., 1998). The establishment of such trust-based relationships is fostered by awareness of the presence of substantial network externalities. Long-term subcontracting agreements, which mostly involve an element of research collaboration aimed at facilitating knowledge absorption, then become a feasible option, besides closer forms of R&D cooperation such as R&D joint ventures. Vertical R&D outsourcing partnerships with clients and suppliers, whether through long-term subcontracting or through explicit cooperative agreements, play an important part in this. Partnerships with lead customers help to reduce the risk associated with market introduction (von Hippel, 1988; Tether, 2002), whereas partnerships with suppliers may be particularly important for reducing R&D costs (Atallah, 2002). In such a more open innovation context, appropriation with respect to non-partners occurs in a more informal manner, mainly through complexity of design, rather than through deliberate actions of legal protection and strategic protection, through patenting and through secrecy (Chesbrough et al., 2006).

Industry–science relations between firms and universities and public research centres also play an increasing part in the rapidly developing elaborate R&D networks (Laursen and Salter, 2004; Perkmann and Walsh, 2007). Whereas vertical inter-firm partnerships are aimed more at applied research and especially at the development part of R&D, industry–science relations are geared towards basic research. To allow effective absorption of the knowledge concerned, firms need their own basic research capability, which can, itself, be built through cooperative agreements with the public research institutions (Fontana et al., 2004). Due to their more generic nature, research projects with public research institutions tend to involve fewer appropriation issues as compared to the more commercially sensitive partnerships with competitors (Veugelers and Cassiman, 2005).

Finally, some scholars have attempted to explain the increasingly elaborate nature of R&D networks, in which different types of knowledge interactions, such as through informal contacts, licensing, consulting, outsourcing through contract research and through various forms of formal cooperative research partnerships, are combined with one another. Each type of interaction has its unique advantages and shortcomings. An adequate combination will, therefore, tend to improve flexibility and innovative efficiency (Perkmann and Walsh, 2007; Aschhoff and Sofka, 2008).

7.2.2 Empirical Framework

The econometric work explaining R&D outsourcing intensity is very limited. Most related econometric studies concentrate on R&D cooperation. Firm level cross-section regression analyses are presented explaining the probability of cooperating in R&D. Firm size is invariably found to have a strong positive effect on the probability of R&D cooperation. Also, being in-house R&D-active most often has a significant positive impact, albeit less important. And this is also the case for measures of legal protection or strategic appropriability (Bayona et al., 2001; Cassiman and Veugelers, 2002; Tether, 2002; Miotti and Sachwald, 2003; Becker and Dietz, 2004; Belderbos et al., 2004; Dachs et al., 2004; Fontana et al., 2004; Abramovsky et al., 2005; Dhondt-Peltrault and Pfister, 2007). These findings are interpreted as a confirmation, to some extent, of the necessity of the presence of a significant knowledge absorptive capacity and appropriability in order to allow R&D cooperation.

These studies do, however, suffer from a series of shortcomings. Due to data constraints, most of them are unable to cope adequately with the endogeneity problem caused by the simultaneous determination of dependent and explanatory variables. The two-step estimation procedures they use rely on an arbitrary choice of the supposedly endogenous and exogenous variables. And the instruments thus constructed are invariably found to be highly collinear with the included exogenous variables, leading to very imprecise parameter estimates. Only studies using data sets with at least two points in time, such as Belderbos et al. (2004) and Dhondt-Peltrault and Pfister (2007), were able to reduce this problem by instrumenting the explanatory variables by lagged values. Furthermore, the impact of being in-house R&D-active on the probability of cooperating in R&D may vastly underestimate the actual effect of in-house R&D intensity on cooperative R&D outsourcing intensity (the depth of cooperation). Knowledge appropriation capacity is, for its part, most often captured exclusively by patent-based measures of legal protection. These may be deceptive due to the different propensities to patent between industries (Griliches, 1990). When included at all, the measures of strategic appropriability used tend to be aggregated. They do not allow identification of the separate impact of their component parts relating to very distinct factors, such as secrecy, complexity of design or lead-time advantage.

Some of these studies have also included dummy variables depicting the receipt of public funding for innovation. Public funding is always found to have a significant positive impact on the probability of R&D cooperation, especially on cooperation with public research institutions such as universities and research centres (Miotti and Sachwald, 2003; Belderbos et al., 2004; Dachs et al., 2004; Abramovsky et al., 2005). But even when lagged values are used as instruments, this may not suffice in this case, since public authorities may have deliberately targeted firms that were already co-operating in the past and whose cooperation persists during the current period. Non-parametric methods were, therefore, used by Aschhoff et al. (2006) and Busom and Fernandez-Ribas (2008) estimating the average treatment effect of public funding on the probability of R&D cooperation. Though not without problems themselves, especially when the control variables used are not fully exogenous, these estimates tend to confirm the existence of a causal influence of public funding for innovation on co-operative R&D behaviour.

As far as the explanation of R&D outsourcing intensity itself is concerned, some authors, such as Dhondt-Peltrault and Pfister (2007), measure R&D outsourcing intensity by the number of interactions with R&D partners. They find that this is positively affected by the level of in-house R&D intensity, measured either by the ratio of in-house R&D expenditure over turnover or by the share of R&D employees in total employment, while instrumenting their explanatory variables by lagged values. The measure of R&D outsourcing intensity used has, however, the drawback that it cannot capture the actual importance of the R&D interactions in question.

A more appropriate manner of measuring R&D outsourcing intensity would be to use the ratio of expenditure on external R&D over turnover. We are, however, unaware of studies that have attempted to explain R&D outsourcing intensity by doing so. Closest to this is probably the regression analysis by Mol (2005) to explain total outsourcing intensity, measured by the sum of external production and R&D expenditure over turnover, on a cross-section of average industry-level data with respect to a set of Dutch manufacturing industries. Both equations to explain the level of total outsourcing intensity in 1994 and its change between 1994 and 1998 are estimated while instrumenting the explanatory variables by lagged values. The results show that, whereas in the early 1990s the level of total outsourcing intensity is negatively affected by in-house R&D intensity, its change during the 1990s is positively affected. This would, according to the author, suggest that in the beginning of the period-production and innovation were still mainly in-house activities of the firm, and that during this period profound organisational changes have been taken place, whereby the production and R&D departments have been opening up. Expenditures on production and R&D outsourcing have augmented and an increasing part of in-house R&D activity has been directed at external knowledge absorption, making it complementary to R&D outsourcing.

Although interesting, these estimates suffer from a series of serious shortcomings. First, by only considering total outsourcing intensity, it is not possible to identify the possibly different impact of the explanatory variables on its separate production and R&D components. Second, by aggregating the data at the industry level, one cannot distinguish between specific firm- and industry-level effects. Thirdl by reducing the sample to in-house R&D-active industries, a sample selection problem is created which will lead to an underestimation of the impact of in-house R&D intensity. And finally, major control variables relating to knowledge appropriability and public funding for innovation are omitted. This may lead to severely biased and inconsistent estimates, to the extent that these omitted variables are correlated with the included variables, which is likely in the case of in-house R&D intensity.

7.2.3 Research Questions

This chapter attempts to answer the following research questions. The first research question asks whether in-house R&D has an intense negative effect on the R&D outsourcing intensity of firms in Belgium, conforming to conventional views on the innovation process. Or does it, on the contrary, have a positive effect on R&D outsourcing intensity, as suggested by more recent views emphasising the importance of knowledge absorptive capacity?

The second research question asks whether appropriability positively affects the R&D outsourcing intensity of firms. If so, is it mainly through deliberate actions, through patenting and secrecy, as emphasised in earlier papers in the field of industrial organisation? Or does it occur in a more informal manner, mainly through complexity of design, in line with more recent views on open innovation?

The third research question asks whether formal cooperation implies, on average, a higher R&D outsourcing intensity. If so, is this particularly the case for cooperation with public research institutions such as universities and research centres? And is there a net effect after controlling for complexity of knowledge and public funding for innovation?

7.3 DATA, EMPIRICAL RELATIONS AND METHODS OF ESTIMATION

7.3.1 Data

This study draws on two datasets for Belgium. The first set is gathered as part of the Third and Fourth European Community Innovation Surveys, CIS3 and CIS4, conducted in 2002 and 2006 and relating to the periods 1998–2000 and 2002–2004. The target population included all enterprises with ten or more employees. A survey stratified by size (small, medium and large) and economic activity (NACE) was made, covering about one third of the target population each time. The rate of useful answers was always slightly less than 50% and a non-response analysis revealed no systematic non-response bias. After exclusion of firms with incomplete data or inconsistencies, samples of 1,987 and 2,740 enterprises remained. In these, a subset of 601 firms could be identified that are present in both. Of these, 102 reported R&D outsourcing in 2002–2004 and in-house R&D in both 2002–2004 and 1998–2000.

The second dataset is gathered as part of the Belgian R&D surveys for 2003 and 2005. The focus of the R&D survey is on enterprises that perform R&D on a permanent or quasi-permanent basis. The data relate to the reference years 2003 (survey 2004) and 2005 (survey 2006). After exclusion of firms that merged, acquired (parts of) other firms or experienced other changes with an important impact on the evolution of R&D and other enterprise variables, 424 firms remained for which the data required for the analysis are available for both years. Of these 424 firms, 211 reported R&D outsourcing in 2005.

Table 7.1 gives the frequency of firms reporting in-house R&D, R&D outsourcing, making use of legal and strategic methods of appropriation and cooperating in research in our total samples under consideration.

Table 7.1 Frequency of in-house R&D, R&D outsourcing, appropriability and R&D cooperation

Databases: CIS4 and CIS3	Number	Per cent
Total observations	601	100.0 %
In-house R&D-active in 2002–2004	263	43.7%
In-house R&D-active in both 2002–2004 and 1998–2000	206	34.3%
R&D outsourcing in 2002–2004	143	23.8%
R&D outsourcing in 2002–2004 and in-house R&D-active in both 2002–2004 and 1998–2000	102	17.0%
Patenting in 2000	110	18.3%
Strategic appropriability in 2002–2004	284	47.3%
Strategic appropriability through secrecy in 2002–2004	207	34.4%
Strategic appropriability through complexity of design in 2002–2004	154	25.6%
Strategic appropriability through lead-time advantage in 2002–2004	270	44.9%
Cooperate in R&D in 2002–2004	205	34.1%
Cooperate in R&D with business in 2002–2004	183	30.4 %
Cooperate in R&D with organisations in 2002–2004	120	20.0 %

Databases: 2005 and 2003 R&D surveys	Number	Per cent
In-house R&D-active in both 2005 and 2003	424	100.0%
R&D outsourcing in 2005	211	50.0%
Cooperate in R&D in 2005	188	44.3%
Cooperate in R&D with business in 2005	186	43.8%
Cooperate in R&D with organisations in 2005	150	35.4%

Both databases suggest that about half the number of firms reporting permanent in-house R&D do also outsource R&D.

7.3.2 Empirical Relations

Estimates will be presented of equations explaining the level of R&D outsourcing intensity. The explanatory variables can be subdivided into predictor variables, emphasised by the research questions, and control variables, that may also affect R&D outsourcing intensity. Use will be made of both data sets described above. The detailed description of the construction of the variables and the precise data sources are presented in Tables A7.1 and A7.2 of the Appendix.

The baseline equation when using CIS data is as follows. The dependent variable is the level of R&D outsourcing intensity, LOUTI. This is measured by the natural log of the ratio of expenditure on external R&D over turnover. The predictor variables relate to in-house R&D intensity and knowledge appropriability. In-house R&D intensity, LRDINS, is measured by the natural log of the ratio of expenditure on in-house R&D over turnover. Its impact on R&D outsourcing intensity can be expected to be either negative or positive, depending on the organisation of the innovation process. Evidence in this respect will provide a first hint in answering the first research question. Two measures of appropriability are included. Patenting, IPPAT, relates to legal protection. It is a binary variable taking the value 1 if the firm reports having valid patents on its inventions. Average strategic appropriability, APPR, is measured by the average importance attached by the firm to appropriation through secrecy, complexity of design and lead-time advantage on competitors. The impact of IPPAT and APPR on R&D outsourcing intensity can, in principle, be expected to be positive and the extent of this will provide a first hint in answering the second research question.

The first two control variables are also knowledge-related. Incoming knowledge spillovers, INKS, are measured by the average importance the firm attaches to various sources of information, including suppliers, customers, competitors, public research organisations and fairs and exhibitions. They may affect the level of R&D outsourcing intensity positively, to the extent that R&D partnerships help to improve learning efficiency in absorbing the external knowledge concerned. But one cannot rule out that some frequent recipients of incoming knowledge may prefer to attempt direct absorption of this knowledge through in-house R&D efforts, so that the expected impact on R&D outsourcing intensity is unclear. Complexity of knowledge, CKNOW, is measured by the relative importance that the firm attaches to information from public research organisations and professional conferences, as suggested by Cassiman and Veugelers (2002). It is expected to have a positive influence on R&D outsourcing intensity as the proportion of intangible tacit knowledge and know-how that does not easily spill over is likely to increase with the complexity of the research process. R&D outsourcing, especially through cooperation, is a way to get access to this knowledge.

Human capital, measured by the log of higher education intensity of the personnel, LHEI, is also included as a control variable, because firms disposing of highly qualified workers are better suited to the undertaking of R&D partnerships. As is the log of export intensity, LEXI, because firms operating in foreign markets are likely to face stronger competition. This may provide an incentive for increasing the efficiency of innovative effort by setting up R&D partnerships.

The next set of control variables are all to some extent size-related. The most important of these is the log of firm size, measured by employment, LSIZE. Its impact is most likely to be positive, because larger firms are more likely to have the necessary means and absorptive capacity to engage in R&D partnerships. But it cannot be ruled out that small and medium-sized firms are also tempted to participate in R&D networks, for the very purpose of compensating for their handicaps of scale and scope. Two other size-related variables are domestic and foreign group membership, GRP_D and GRP_F. These are binary variables that take the value 1 when the firm is a member of a larger domestic or foreign group. The implied size effect may affect the propensity to engage in R&D partnerships either way. In the case of foreign group membership, it could be argued that subsidiaries are less likely to engage in R&D partnerships with non-group members, if they can access the knowledge concerned through the parent company (Busom and Fernandez-Ribas, 2008). Accessing this knowledge may itself, however, require an R&D partnership with the parent company and other group members, so that expectations concerning the impact of group membership remain unclear.

Furthermore, innovation-hampering factors depicting risk and financial constraints, HRISK and HFIN, are also included. They are measured by the importance given by the firm to risk and financial constraints that help to prevent, delay or abandon innovation processes. Again, there are no a priori expectations as to the direction of their impact. If sufficiently important, they may prevent the firm reaching the threshold level of innovative effort necessary to engage in R&D partnerships. But, if not, they may, on the contrary, provide an incentive for engaging in such partnerships to share risks and have access to more finance.

One way of overcoming financial constraints is to receive public funding for innovation. This is captured by a binary variable, FUND, taking the value 1 if the firm receives such funding, irrespective of the public authority concerned. Due to the heavily regionalised organisation of Belgian innovation policy, public funding from regional government is far more important than that from the federal government (Capron et al., 2000). Some firms also receive innovation funding from the EU, especially through its Framework Programs. Despite its growing importance, this is still modest, however, compared to regional funding (CFS/STAT, 2007).

Finally, industry dummy variables based on Pavitt's taxonomy, IDPAV, are added, depicting the general technological environment in which the firm operates, since this may affect the propensity to outsource R&D. The dummies are based on the taxonomy originally proposed by Pavitt (1984) and elaborated by Tidd et al. (1997) to accommodate for the services sectors. This classification reflects differences in technological trajectories between groups of industries. The industries included in the different groups are mentioned in Table A7.1 in the appendix.

One shortcoming of the CIS data is that they only allow the construction of a binary variable for funding, FUND, reflecting receipt by the firm of public funding for innovation or otherwise. They do not provide data on the actual amount of funding concerned and do not, therefore, enable us to obtain a measure of funding intensity, LFUNDI, which may be far more appropriate in explaining R&D outsourcing intensity. The R&D surveys in Belgium do provide such data on the amount of public funding for innovation, but in this case no reliable data are available on firm turnover, so that the normalisation used to obtain the intensity variables has to be based on employment, as described in Table A7.2 in the Appendix. Thus LFUNDI is calculated as the natural log of the ratio of total public funding for innovation over employment. Since data are available on R&D employment, preference is given here to a measure of the log of in-house R&D intensity, LRDINS, based on the ratio of R&D employment over total employment. There are, however, no data available to allow us to construct measures of knowledge appropriability, incoming spillovers, complexity of knowledge, higher education intensity, and hampering risk and finance constraints, so that these variables have to be dropped.

Considering the estimates of the above-mentioned equations on the respective data-sets as a whole allows us to answer the first research question, on the impact of in-house R&D intensity on R&D outsourcing intensity. To allow a more precise answer to the second research question, on the impact of the various modes of knowledge appropriation, the baseline equation used for the CIS data is then adapted by splitting the average measure of strategic appropriability into its component parts. The average level of strategic appropriability, APPR, is thereby replaced by variables depicting the importance to the firm of strategic appropriability through secrecy, APPR_SE, complexity of design, APPR_CD, and lead-time advantage on competitors, APPR_LT. Finally, to answer the third research question, on the impact of R&D cooperation on the level of R&D

outsourcing intensity, the baseline equation is extended to include a binary variable, COOP, that takes the value 1 when the firm is involved in formal R&D cooperation. This is then, in turn, split into comparable variables depicting whether the firm cooperates with business (other firms), COBUS, and/or with public research organisations (universities or research centres), COPRC.

7.3.3 Methods of Estimation

A straight cross-section regression estimation of the baseline equation by means of ordinary least squares on a sample of firms that are outsourcing R&D and in-house R&D-active poses major problems. First, entering the explanatory variables as such, measured with respect to the same point in time as the dependent variable, will most likely lead to biased and inconsistent estimates, due to the simultaneous determination of dependent and explanatory variables. The endogeneity bias that this involves can be expected to be particularly severe in the case of the knowledge-related variables, as they may all be correlated with unmeasurable aspects of 'innovativeness', which also affect R&D outsourcing intensity. Second, drawing statistical inference on a specific sub-population of firms that are both R&D outsourcing and permanent in-house R&D-active may be of interest in itself, but could be highly misleading when assessing the actual impact of the explanatory variables on the level of R&D outsourcing intensity. No account is taken of the sample selection involved. Also, this may lead to biased and inconsistent estimates. The selection bias that this involves may be severe, to the extent that the behaviour of firms outsourcing R&D differs significantly from those that do not.

To avoid the endogeneity problem, or at least reduce it considerably, the dependent variable relates to 2004 (from CIS4) or to 2005 (from the R&D Survey for 2005), whereas the explanatory variables are instrumented by their lagged values from earlier corresponding surveys relating to the period 1998–2000 (from CIS3) or to 2003 (from the R&D survey for 2003). To cope with the sample selection problem, the sample is extended to include firms that are not outsourcing R&D and the R&D outsourcing intensity equations under consideration are always complemented by a sample selection equation to draw inference on a larger population. In the case of estimation on the CIS data, the larger population concerns all firms, including those not reporting themselves as in-house R&D-active or innovative, whereas in the case of estimation on the R&D Survey data it concerns all in-house R&D-active firms. These sample selection equations are probit equations explaining the probability of outsourcing R&D and being permanently in-house R&D-active in the total population, OUTRD, or the probability of outsourcing R&D in the in-house R&D-active population, OUT. They have a binary-dependent variable that takes the value 1 in case of positive R&D outsourcing and permanent in-house R&D, or in case of positive R&D outsourcing; and the value 0 otherwise. When considered together with the corresponding R&D outsourcing intensity equation, they always form a generalised tobit model. The sample selection bias caused by the correlation between the error terms of the probit and corresponding outsourcing intensity equation is dealt with by applying a system estimation by the method of maximum likelihood. The Heckman (1979) two-step procedure is used to choose the initial parameter values in the iterative process. In order to cope with the possible problem of heteroscedasticity, use is made of (White) heteroscedasticity-robust standard errors. This method has been shown to provide consistent and efficient estimates (Verbeek, 2000).

It should be noted that, although the same explanatory variables may enter both the sample selection and the corresponding R&D outsourcing intensity equations, proper identification requires that there should be at least some difference. Since it is difficult to think of reasons why most of the variables considered to explain R&D outsourcing intensity, LOUTI, may not at the same time affect the probability of outsourcing R&D and of being in-house R&D-active in the total population, OUTRD, or of outsourcing R&D in the in-house R&D-active population, OUT, they are always entered in both the sample selection and the corresponding R&D outsourcing equations. In the case of estimation on the CIS data, an obvious difference is obtained by only including the log of in-house R&D intensity, LRDINS, in the R&D outsourcing intensity equation, but not in the corresponding sample selection equation, since it only relates to firms that are in-house R&D-active. In the case of estimation on the R&D Survey data, a difference is created by including the funding dummy, FUND, in the sample selection equation and the log of funding intensity, LFUNDI, instead in the R&D outsourcing intensity equation, where it is more appropriate.

7.4 ESTIMATES AND ANALYSIS

Table 7.2 presents the estimates of the baseline estimates of the generalised tobit models on the respective considered data-sets.

Table 7.2 Explaining R&D outsourcing intensity

	Database: CIS4 All firms in 2004		Database: R&D survey R&D-active firms in 2005	
	OUTRD (i)	LOUTI (ii)	OUT (iii)	LOUTI (iv)
LRDINS		0.290	0.172	0.783
		(2.52)**	(2.97)**	(5.96)**
IPPAT	0.402	0.600		
	(2.34)**	(1.65)*		
APPR	0.794	1.523		
	(3.58)**	(2.82)**		
INKS	0.861	–0.661		
	(2.19)**	(0.66)		
CKNOW	0.789	2.685		
	(2.02)**	(2.94)**		
LHEI	0.131	0.290		
	(2.15)**	(1.43)		
LEXI	–0.030	–0.122	–0.022	–0.032
	(1.10)	(1.93)	(0.63)	(0.48)
LSIZE	0.321	–0.004	0.166	0.156
	(5.29)**	(0.03)	(3.45)**	(1.83)
GRP_D	0.063	0.277	0.121	0.139
	(0.30)	(0.46)	(0.84)	(0.55)
GRP_F	–0.335	0.671	0.266	0.603
	(1.47)	(1.22)	(1.51)	(2.01)*
HRISK	–0.314	–0.929		
	(1.09)	(1.37)		
HFIN	–0.533	0.387		
	(1.63)	(0.56)		
FUND	0.641	0.337	0.315	
	(3.94)**	(0.89)	(2.42)**	
LFUNDI				0.065
				(2.04)**

Table 7.2 (continued)

	Database: CIS4 All firms in 2004		Database: R&D survey R&D-active firms in 2005	
	OUTRD (i)	LOUTI (ii)	OUT (iii)	LOUTI (iv)
IDPAV1	0.272	-0.242	0.202	-0.009
	(1.21)	(0.46)	(1.07)	(0.03)
IDPAV2	0.248	-0.145	-0.532	0.546
	(1.14)	(0.26)	(2.59)**	(1.49)
IDPAV3	0.172	-0.170	0.240	-0.088
	(0.68)	(0.31)	(1.20)	(0.26)
IDPAV4	0.609	-0.469	0.046	1.339
	(2.08)*	(0.57)	(0.21)	(3.38)**
Constant	-3.343	-7.861	-0.511	0.128
	(10.3)**	(3.77)**	(2.18)*	(0.26)
RHO	0.795		0.806	
	(0.01)**		(0.00)**	
LL	-345.447		-637.209	
Obs.	601		424	

Notes: The estimation is by the method of maximum likelihood. The explanatory variables are instrumented by their lagged values from the period 1998–2000 from CIS3 in columns (i) and (ii) and by their lagged value from 2003 from the previous Belgian R&D Survey in columns (iii) and (iv). The absolute value of the t-values is presented between brackets. These are based on heteroscedasticity-consistent estimates of the standard errors. The t-tests are one or two-tailed, according to a priori expectations with respect to the parameter values. RHO stands for the correlation coefficient between the errors of the sample selection and tobit equations and the p-value of its Wald test is presented underneath. LL stands for the log of pseudolikelihood. Symbols * and ** denote significance at the 5% and 1% levels.

Columns (i) and (ii) present the estimates of the respective sample selection and R&D outsourcing intensity equations on the CIS data, and columns (iii) and (iv) the corresponding estimates of these equations on R&D Survey data.

Consider first the estimates on CIS data, in columns (i) and (ii). In this case, the sample selection equation, in column (i), explains the probability of both outsourcing R&D and being permanently in-house R&D-active in the total population of firms, OUTRD. The correlation of its error term with that of the corresponding R&D outsourcing equation, in column (ii), RHO, is presented at the bottom of column (i), below the parameter estimates. It is seen to be highly statistically significant, suggesting the presence of a substantial sample selection effect.

Consider the parameter estimates in turn. In-house R&D, LRDINS, is seen to have a highly significant positive impact on R&D outsourcing intensity. This provides the first hint in answering the first research question. This result would conform to the views that emphasise the importance of knowledge absorption and the consequent complementary nature of in-house R&D and R&D outsourcing.

As far as the other predictor variables are concerned, column (i) in Table 7.2 shows that both patenting, PAT, and average strategic appropriability, APPR, are seen to have a significant positive effect on the probability of R&D outsourcing and being permanently in-house R&D-active, OUTRD. Their actual impact on R&D outsourcing intensity, in column (ii), is also found to be positive and statistically significant. The impact of strategic appropriability, APPR, is, however, always stronger, especially in the case of outsourcing intensity. This provides the first hint that strategic appropriability matters more to R&D outsourcing intensity than legal protection.

In the case of the control variables, complexity of knowledge, CKNOW, is seen as having a significant positive impact on the probability of R&D outsourcing and being permanently in-house R&D-active, OUTRD, and even more so on R&D outsourcing intensity, LOUTI. This reflects the fact that R&D partnerships are increasingly used to access intangible knowledge produced by complex generic research.

Furthermore, column (i) in Table 7.2 shows that some control variables, such as incoming knowledge spillovers, INKS, higher education intensity, LHEI, firm size, LSIZE, and the funding dummy, FUND, always have a significant positive impact on the probability of R&D outsourcing and being permanently in-house R&D-active, OUTRD, but column (ii) shows that they have, it would seem, no impact on R&D outsourcing intensity. This may implicitly suggest that their impact is mainly on the probability of being permanently in-house R&D-active rather than on the probability of R&D outsourcing. In the case of firm size, LSIZE, it may also reflect the existence of a non-linearity in the relationship, as a threshold level of size may be necessary for in-house R&D and R&D outsourcing. Beyond this level, further size increase may lead only to a proportional increase in expenditure on in-house and external R&D. In the case of public funding for innovation, the explanation could simply be that a binary dummy variable, such as FUND, may only be appropriate in explaining the probability of R&D outsourcing and being permanently in-house R&D-active, OUTRD, but not the level of R&D outsourcing intensity, LOUTI.

In the case of the estimates on R&D survey data, the sample selection equation in column (iii) of Table 7.2 explains the probability of outsourcing R&D in the permanently in-house R&D-active population, OUT. The correlation of its error term with that of the corresponding R&D outsourcing intensity equation in column (iv), RHO, is also found to be highly statistically significant here, suggesting again the existence of a substantial sample selection effect. In this case in-house R&D intensity, LRDINS, could be entered in both the sample selection and outsourcing intensity equations and it is always found to be highly statistically significant. This confirms the conclusions reached on the basis of the estimates on the CIS data, emphasising the importance of knowledge absorption.

As far as the control variables are concerned, it is especially interesting to note that now there is not only a positive, highly significant impact on the funding dummy, FUND, on the probability of outsourcing R&D in the permanently in-house active population, OUT, as seen in column (iii), but also of funding intensity, LFUNDI, on outsourcing intensity, LOUTI, as seen in column (iv). The inclusion, due to data constraints, of the funding dummy in the equation to explain outsourcing intensity when performing estimation on the CIS data clearly failed to identify this effect.

Furthermore, evidence that firm size apparently impacts positively on the probability of outsourcing R&D in the permanently in-house R&D-active population, OUT, but not on R&D outsourcing intensity, LOUTI, may suggest that even among in-house R&D-active firms there exists a threshold level of firm size necessary for outsourcing R&D. Beyond this level, further size increase may lead only to a proportional increase in expenditure on external R&D. Finally, there is evidence of a highly significant positive impact of the 'science-based' industries dummy, IDPAV4, on R&D outsourcing intensity, LOUTI, in column (iv), while there was none in the corresponding estimates on the CIS data, in column (ii). This can possibly be explained by an upward bias caused by the exclusion, due to data constraints, of the variable on complexity of knowledge, BR, as the outsourcing of complex research tends to take place especially in major 'science-based' industries, such as chemicals and pharmaceuticals.

The baseline estimates on the CIS data suggest that average strategic appropriability, APPR, has apparently a significant positive impact on the level of outsourcing intensity. But it is unclear to which of its components this is due. Table 7.3 clarifies this issue.

Table 7.3 Explaining the R&D outsourcing intensity: splitting strategic appropriability

	Database CIS4 in 2004							
	OUTRD	LOUTI	OUTRD	LOUTI	OUTRD	LOUTI	OUTRD	LOUTI
	(i)	(ii)	(iii)	(iv)	(v)	(vi)	(vii)	(viii)
LRDINS		0.301		0.314		0.290		0.310
		(2.71)**		(2.79)**		(2.52)**		(2.65)**
IPPAT	0.476	0.793	0.441	0.702	0.447	0.740	0.450	0.699
	(2.73)**	(2.07)**	(2.63)**	(1.88)*	(2.65)**	(1.99)**	(2.52)**	(1.76)*
APPR_SE	0.205	0.675	0.249	0.753				
	(1.06)	(1.27)	(1.49)	(1.90)*				
APPR_CD	0.371	0.762			0.384	0.819		
	(1.88)*	(1.89)*			(2.17)**	(2.08)**		
APPR_LT	–0.188	–0.398					0.033	0.172
	(0.98)	(0.91)					(0.19)	(0.50)
INKS	1.026	–0.447	1.079	–0.348	1.015	–0.386	1.171	–0.112
	(2.67)**	(0.41)	(2.81)**	(0.33)	(2.69)**	(0.36)	(3.05)**	(0.10)
CKNOW	0.712	2.739	0.721	2.790	0.748	2.563	0.704	2.537
	(1.86)*	(2.97)**	(1.90)*	(3.01)**	(1.97)**	(2.67)**	(1.83)*	(2.67)**
LHEI	0.143	0.284	0.137	0.265	0.156	0.351	0.157	0.349
	(2.34)**	(1.40)	(2.25)**	(1.28)	(2.53)**	(1.74)*	(2.53)**	(1.67)*
LEXI	–0.020	–0.115	–0.017	–0.109	–0.019	–0.108	–0.013	–0.094
	(0.77)	(1.86)	(0.67)	(1.86)	(0.73)	(1.79)	(0.51)	(1.84)
LSIZE	0.320	–0.010	0.312	–0.025	0.322	–0.002	0.318	–0.007
	(5.25)**	(0.07)	(5.15)**	(0.17)	(5.28)**	(0.01)	(5.22)**	(0.05)
GRP_D	0.112	0.486	0.132	0.619	0.130	0.445	0.162	0.607
	(0.55)	(0.79)	(0.65)	(1.02)	(0.64)	(0.70)	(0.80)	(0.95)
GRP_F	–0.307	0.782	–0.280	0.904	–0.288	0.770	–0.257	0.908
	(1.38)	(1.37)	(1.26)	(1.58)	(1.30)	(1.35)	(1.17)	(1.56)
HRISK	–0.274	–1.030	–0.331	–1.164	–0.250	–0.806	–0.290	–0.939
	(0.97)	(1.51)	(1.21)	(1.76)	(0.88)	(1.17)	(1.06)	(1.39)
HFIN	–0.461	0.593	–0.412	0.758	–0.470	0.452	–0.406	0.664
	(1.50)	(0.85)	(1.35)	(1.12)	(1.52)	(0.65)	(1.34)	(0.99)
FUND	0.698	0.482	0.681	0.469	0.692	0.475	0.690	0.479
	(4.29)**	(1.22)	(4.23)**	(1.20)	(4.30)**	(1.19)	(4.27)**	(1.17)

Table 7.3 (continued)

	Database CIS4 in 2004							
	OUTRD	LOUTI	OUTRD	LOUTI	OUTRD	LOUTI	OUTRD	LOUTI
	(i)	(ii)	(iii)	(iv)	(v)	(vi)	(vii)	(viii)
IDPAV1	0.308	0.017	0.268	–0.070	0.296	–0.105	0.252	–0.227
	(1.41)	(0.03)	(1.22)	(0.13)	(1.37)	(0.18)	(1.16)	(0.42)
IDPAV2	0.267	–0.073	0.236	–0.084	0.280	–0.005	0.252	–0.008
	(1.28)	(0.13)	(1.13)	(0.15)	(1.32)	(0.01)	(1.19)	(0.02)
IDPAV3	0.259	0.065	0.230	0.024	0.257	0.091	0.247	0.067
	(1.04)	(0.12)	(0.93)	(0.05)	(1.03)	(0.16)	(1.00)	(0.12)
IDPAV4	0.645	–0.257	0.607	–0.287	0.677	–0.231	0.650	–0.231
	(2.21)*	(0.29)	(2.07)*	(0.33)	(2.32)**	(0.27)	(2.21)*	(0.26)
Constant	–3.263	–8.144	–3.238	–8.197	–3.245	–7.779	–3.192	–7.836
	(10.6)**	(4.06)**	(10.8)*	(4.31)**	(10.6)*	(3.54)**	(10.7)*	(3.66)**
RHO	0.825		0.839		0.802		0.818	
	(0.01)**		(0.00)**		(0.02)**		(0.01)**	
LL	–348.894		–		–		–	
			351.245		350.175		353.123	
Obs.	601		601		601		601	

Notes: The estimation is by the method of maximum likelihood. The explanatory variables are instrumented by their lagged values from the period 1998–2000 from CIS3. The absolute value of the t-values is presented between brackets. These are based on heteroscedasticity-consistent estimates of the standard errors. The t-tests are one or two-tailed, according to a priori expectations with respect to the parameter values. RHO stands for the correlation coefficient between the errors of the sample selection and tobit equations and the p-value of its Wald test is presented underneath. LL stands for the log of pseudolikelihood. Symbols * and ** denote significance at the 5% and 1% levels.

Columns (i) and (ii) present the estimates of the generalised tobit model on the CIS data while replacing the average level of strategic appropriability, APPR, by variables depicting the importance to the firm of strategic appropriability through secrecy, APPR_SE, complexity of design, APPR_CD, and lead-time advantage with respect to competitors, APPR_LT. Although one has to be careful with the interpretation due to the multicollinearity between these variables and between some of them, especially those relating to secrecy and lead-time advantage, and the patent variable, the estimates suggest that the impact of appropriability through complexity of design, APPR_CD, clearly dominates that of the other strategic appropriability variables. This is confirmed when entering the respective strategic appropriability variables separately in turn, in columns

(iii) to (viii). There is in this case only a modest statistically significant effect of strategic appropriability through secrecy, APPR_SE, on R&D outsourcing intensity, in column (iv), a strong statistically significant effect of complexity of design, APPR_CD, both on the probability of R&D outsourcing and being permanently in-house R&D-active and on R&D outsourcing intensity, in columns (v) and (vi), and no significant effect at all of lead-time advantage with respect to competitors, APPR_LT.

A final issue relates to the impact of formal R&D cooperation on the level of R&D outsourcing intensity. Does it imply, on average, a higher R&D outsourcing intensity? And if so, does the type of partner in cooperation matter? Tables 7.4 and 7.5 clarify this issue.

Table 7.4 R&D outsourcing intensity: the impact of R&D cooperation

	Database CIS4 in 2004							
	OUTRD	LOUTI	OUTRD	LOUTI	OUTRD	LOUTI	OUTRD	LOUTI
	(i)	(ii)	(iii)	(iv)	(v)	(vi)	(vii)	(viii)
LRDINS		0.286		0.299		0.264		0.302
		(2.71)**		(2.76)**		(2.26)**		(2.87)**
IPPAT	0.377	0.533	0.376	0.372	0.367	0.511	0.384	0.383
	(2.18)**	(1.53)	(2.14)**	(1.10)	(2.11)**	(1.48)	(2.20)**	(1.13)
APPR	0.777	1.419	0.782	1.409	0.795	1.468	0.786	1.402
	(3.49)**	(2.77)**	(3.48)**	(2.69)**	(3.57)**	(2.82)**	(3.50)**	(2.69)**
COOP	0.216	0.824						
	(1.32)	(2.37)**						
COBUS			0.284	0.216	0.286	0.654		
			(1.47)	(0.53)	(1.76)*	(1.91)*		
COPRC			–0.005	0.760			0.184	0.903
			(0.02)	(1.74)*			(0.96)	(2.48)**
INKS	0.827	–0.727	0.811	–0.790	0.791	–0.748	0.841	–0.762
	(2.09)**	(0.76)	(2.05)**	(0.82)	(1.98)**	(0.78)	(2.15)**	(0.79)
CKNOW	0.738	2.550	0.730	2.507	0.736	2.564	0.739	2.494
	(1.89)*	(2.93)**	(1.86)*	(2.93)**	(1.89)*	(2.86)**	(1.87)*	(2.91)**
LHEI	0.125	0.278	0.118	0.281	0.118	0.254	0.124	0.281
	(2.06)**	(1.42)	(1.98)**	(1.39)	(2.00)**	(1.26)	(2.03)**	(1.41)
LEXI	–0.029	–0.100	–0.030	–0.098	–0.029	–0.105	–0.030	–0.097
	(1.07)	(1.92)	(1.06)	(1.87)	(1.06)	(1.93)	(1.09)	(1.86)
LSIZE	0.313	–0.063	0.308	–0.074	0.307	–0.053	0.312	–0.070
	(5.08)**	(0.47)	(5.02)**	(0.54)	(5.01)**	(0.38)	(5.08)**	(0.51)

Table 7.4 (continued)

	Database CIS4 in 2004							
	OUTRD	LOUTI	OUTRD	LOUTI	OUTRD	LOUTI	OUTRD	LOUTI
	(i)	(ii)	(iii)	(iv)	(v)	(vi)	(vii)	(viii)
GRP_D	0.029	0.259	0.050	0.162	0.052	0.231	0.046	0.182
	(0.14)	(0.42)	(0.24)	(0.28)	(0.25)	(0.39)	(0.22)	(0.31)
GRP_F	−0.342	0.745	−0.324	0.672	−0.324	0.722	−0.340	0.680
	(1.48)	(1.25)	(1.41)	(1.17)	(1.41)	(1.26)	(1.48)	(1.20)
HRISK	−0.328	−1.052	−0.347	−1.137	−0.337	−1.006	−0.338	−1.133
	(1.16)	(1.69)	(1.54)	(1.91)	(1.17)	(1.54)	(1.17)	(1.90)
HFIN	−0.524	0.462	−0.501	0.546	−0.518	0.425	−0.510	0.537
	(1.60)	(0.70)	(1.54)	(0.86)	(1.59)	(0.63)	(1.55)	(0.84)
FUND	0.588	0.091	0.612	−0.069	0.604	0.209	0.583	−0.080
	(3.45)**	(0.24)	(3.42)**	(0.17)	(3.63)**	(0.56)	(3.28)**	(0.20)
IDPAV1	0.275	−0.303	0.291	−0.395	0.289	−0.322	0.267	−0.419
	(1.21)	(0.63)	(1.28)	(0.77)	(1.28)	(0.63)	(1.18)	(0.83)
IDPAV2	0.257	−0.164	0.263	−0.207	0.266	−0.192	0.251	−0.227
	(1.17)	(0.31)	(1.19)	(0.39)	(1.20)	(0.35)	(1.14)	(0.43)
IDPAV3	0.171	−0.153	0.173	−0.233	0.177	−0.178	0.168	−0.247
	(0.67)	(0.30)	(0.67)	(0.47)	(0.68)	(0.34)	(0.65)	(0.49)
IDPAV4	0.639	−0.452	0.636	−0.513	0.636	−0.503	0.652	−0.537
	(2.18)*	(0.57)	(2.16)*	(0.67)	(2.18)*	(0.62)	(2.11)*	(0.70)
Concept	−3.332	−7.713	−3.346	−7.122	−3.338	−7.742	−3.313	−7.095
	(10.3)**	(4.04)**	(10.2)**	(3.65)**	(10.4)**	(3.99)**	(10.1)**	(3.65)**
RHO	0.806		0.763		0.784		0.764	
	(0.01)**		(0.01)**		(0.01)**		(0.01)**	
LL	−342.716		−341.720		−343.307		−342.597	
Obs.	601		601		601		601	

Notes: The estimation is by the method of maximum likelihood. The explanatory variables are instrumented by their lagged values from the period 1998–2000 from CIS3. The absolute value of the t-values is presented between brackets. These are based on heteroscedasticity-consistent estimates of the standard errors. The t-tests are one or two-tailed, according to a priori expectations with respect to the parameter values. RHO stands for the correlation coefficient between the errors of the sample selection and tobit equations and the p-value of its Wald test is presented underneath. LL stands for the log of pseudolikelihood. Symbols * and ** denote significance at the 5% and 1% levels.

Table 7.5 R&D outsourcing intensity: the impact of R&D cooperation

	Database R&D survey in 2005							
	OUTRD	LOUTI	OUTRD	LOUTI	OUTRD	LOUTI	OUTRD	LOUTI
	(i)	(ii)	(iii)	(iv)	(v)	(vi)	(vii)	(viii)
LRDINS	0.146	0.733	0.144	0.743	0.163	0.760	0.141	0.739
	(2.50)**	(5.71)**	(2.48)**	(5.85)**	(2.80)**	(5.80)**	(2.43)**	(5.88)**
COOP	0.393	0.627						
	(2.90)**	(2.46)**						
COBUS			–0.178	–0.001	0.128	0.396		
			(1.17)	(0.00)	(0.96)	(1.67)*		
COPRC			0.591	0.771			0.488	0.749
			(3.67)**	(2.84)**			(3.49)**	(2.93)**
LEXI	–0.029	–0.035	–0.021	–0.026	–0.022	–0.036	–0.023	–0.023
	(0.85)	(0.55)	(0.61)	(0.41)	(0.65)	(0.55)	(0.67)	(0.36)
LSIZE	0.136	0.100	0.141	0.098	0.157	0.126	0.133	0.096
	(2.73)**	(1.15)	(2.81)**	(1.11)	(3.16)**	(1.39)	(2.68)**	(1.12)
GRP_D	0.143	0.128	0.092	0.084	0.123	0.136	0.112	0.074
	(0.98)	(0.52)	(0.63)	(0.34)	(0.86)	(0.55)	(0.77)	(0.30)
GRP_F	0.260	0.560	0.232	0.502	0.274	0.599	0.239	0.503
	(1.48)	(1.88)	(1.31)	(1.69)	(1.56)	(1.98)*	(1.36)	(1.72)
FUND	0.237		0.204		0.289		0.193	
	(1.77)*		(1.51)		(2.19)**		(1.42)	
LFUNDI		0.047		0.040		0.058		0.037
		(1.53)		(1.28)		(1.85)*		(1.19)
IDPAV1	0.145	–0.095	0.168	–0.074	0.186	–0.054	0.158	–0.084
	(0.76)	(0.30)	(0.89)	(0.23)	(0.99)	(0.17)	(0.83)	(0.26)
IDPAV2	–0.546	0.561	–0.496	0.595	–0.536	0.519	–0.516	0.626
	(2.66)**	(1.54)	(2.39)**	(1.63)	(2.60)**	(1.40)	(2.50)**	(1.76)
IDPAV3	0.182	–0.158	0.238	–0.112	0.221	–0.149	0.213	–0.117
	(0.89)	(0.47)	(1.18)	(0.33)	(1.10)	(0.44)	(1.06)	(0.35)
IDPAV4	0.042	1.347	0.032	1.300	0.048	1.317	0.024	1.318
	(0.19)	(3.41)**	(0.14)	(3.28)**	(0.21)	(3.27)**	(0.11)	(3.36)**
Constant	–0.580	–0.060	–0.534	–0.078	–0.537	–0.005	–0.548	–0.033
	(2.47)**	(0.12)	(2.27)*	(0.16)	(2.28)*	(0.01)	(2.34)*	(0.07)
RHO	0.795		0.819		0.817		0.819	
	(0.00)**		(0.00)**		(0.00)**		(0.00)**	
LL	–632.16		–629.51		–635.84		–630.29	
Obs.	424		424		424		424	

Notes: See Table 7.4. The explanatory variables are instrumented by their lagged values from the period 2003 from the previous Belgian R&D Survey.

Consider first the estimates on the CIS data in Table 7.4. Columns (i) and (ii) present the estimates of the baseline generalised tobit model extended with a binary variable, COOP, that takes the value 1 when the firm is involved in explicit research cooperation. The estimates show that there is a clear positive statistically significant impact on the level of R&D outsourcing intensity, in column (ii), suggesting that firms formally co-operating in R&D tend, on average, to spend more on external R&D. Does the type of cooperative partner matter? To find out, the aggregate cooperation variable is split up into comparable variables depicting whether the firm cooperates with business (other firms), COBUS, and/or with public research organisations (universities or research centres), COPRC. Columns (iii) and (iv) present the estimates with both variables included. Once again, one should be careful with the interpretation due to the multicollinearity between these variables, as a substantial number of firms cooperate both with business and with organisations. The results show that there is only a modest statistically significant impact of R&D cooperation with public research organisations, COPRC. Entering these variables separately in turn shows a modest statistically significant impact on R&D outsourcing intensity of cooperation with business, COBUS, in column (vi), and a highly significant impact of cooperation with institutions, COPRC, in column (viii). This would suggest that they are, in essence, firms co-operating with scientific institutions, such as universities and research centres, that spend relatively more on external R&D than firms involved in other kinds of R&D partnerships.

A comparison of the results in Table 7.4 with the corresponding baseline estimates in columns (i) and (ii) of Table 7.2, shows, moreover, that including variables on R&D cooperation does not affect the parameter estimates of the other explanatory variables. This is especially interesting in the case of complexity of knowledge, CKNOW, since this variable is known to have a positive impact on the probability of cooperating with institutions, and the more basic character of research can, at the same time, be thought of as a reason why firms involved in this type of cooperation tend to spend more on external R&D. This result suggests that there is clearly more to it.

Another possible reason for explaining why firms cooperating with scientific institutions tend to spend more on external R&D is that they are favoured by public funding authorities. The CIS-based estimates in Table 7.4 are, however, ill-suited to an investigation of this issue, since they include only a binary funding variable, FUND, which cannot properly capture the impact of public funding on the intensity of R&D outsourcing. The estimates on R&D survey data, presented in Table 7.5, are more appropriate for this purpose. Here too, the estimates show that there is a statistically significant positive impact of R&D cooperation on R&D

outsourcing intensity, and that this is due mainly to R&D cooperation with organisations, COPRC. Including total R&D cooperation, COOP, or R&D cooperation with organisations, COPRC, in the R&D outsourcing intensity equation in columns (ii), (iv) and (viii), always renders the funding intensity variable, LFUNDI, non-significant, whereas it was found to be highly significant in the corresponding baseline equation in column (iv) in Table 7.2. These results implicitly confirm that public funding authorities favour cooperative industry-science relations. By doing so, they allow the firms concerned to raise their level of external R&D expenditure and, thereby, their outsourcing intensity.

However, it is not possible to ascribe entirely to funding, and unaccounted by the other control variables, the above-average level of the R&D outsourcing intensity of firms cooperating with public research organisations. The statistical significance of total cooperation and cooperation with organisations, COOP and COPRC, in columns (ii), (iv) and (viii) of Table 7.5, is clearly higher than that of funding intensity, LFUNDI, in the corresponding baseline equation in column (iv) of Table 7.2. This renders the latter variable completely non-significant and improves the statistical fit of the regression, measured by the log of pseudolikelihood, LL, when these variables on cooperation are included. Or, put differently, part of the explanation of the above-average level of R&D outsourcing intensity of firms cooperating with public research organisations is to be found elsewhere.

7.5 DISCUSSION AND CONCLUSIONS

A cross-section regression analysis was presented of the impact of in-house R&D, knowledge appropriability and cooperation on the R&D outsourcing intensity of firms in Belgium. Estimation was performed both on data from the Third and Fourth European Community Innovation Surveys for Belgium, for 1998–2000 and 2002–2004, and on data from the R&D Surveys for 2003 and 2005. To reduce the endogeneity problem, the dependent variable was measured by data from the most recent of these surveys, whereas the explanatory variables were instrumented by lagged values from the respective corresponding previous ones. To handle the problem of sample selection, estimates were presented of generalised tobit models on extended samples, including firms not reporting R&D outsourcing.

To the first research question, with respect to the impact of in-house R&D, the answer is unambiguous. At present, in-house R&D intensity has a strong positive impact on the R&D outsourcing intensity of firms in

Belgium. This confirms the importance of creating prior knowledge to effectively absorb the knowledge obtained through research partnership, as emphasised by Kamien and Zang (2000) and others. In-house R&D and R&D outsourcing have, in the process, become major complements to one another, rather than minor substitutes, as implied by earlier conventional economic theoretical wisdom.

This finding strengthens existing scattered related evidence in this respect, showing that being in-house R&D-active increases the probability of cooperating in R&D (Becker and Dietz, 2004; Belderbos et al., 2004; Fontana et al., 2004). It also improves on the study by Dhondt-Peltrault and Pfister (2007), who could only find a modest positive impact of R&D intensity on R&D outsourcing intensity, while measuring this by the number of partnership interactions rather than by the actual expenditure on external R&D over turnover or employment.

To the second research question, with respect to the impact of appropriability, the answer is equally clear. Appropriability has a positive impact on the R&D outsourcing intensity of firms. The impact of strategic appropriability is, on average, far more important than that of legal protection through patenting. This is mainly due to complexity of design, rather than to secrecy or lead-time advantage on competitors. This evidence tends to confirm the fact that, in an increasingly open innovation context, appropriability occurs in a more informal manner, rather than through deliberate actions of legal protection or secrecy. Complexity of design is hereby of paramount importance, contrary to lead-time advantage on competitors, which is often too brief to affect R&D partnership behaviour.

These findings clearly add to the related econometric literature, which has, until now, only analysed the impact of appropriability on the probability of R&D cooperation. If considered at all, the measures of strategic appropriability used were always composite indicators (Cassiman and Veugelers, 2002; Veugelers and Cassiman, 2005; Abramovsky et al., 2005). This does not allow us to identify the apparently very different impact of its component parts.

To the third research question, with respect to the impact of the type of partnership, the answer is that R&D cooperation has a positive impact on the R&D outsourcing intensity of firms. This suggests that firms that are formally cooperating tend to spend relatively more on external R&D. Furthermore, the results show that this is mainly due to firms cooperating with public research institutions, such as universities and research centres.

Including a variable on R&D cooperation with institutions does not, interestingly, significantly reduce the impact of the variable on complexity of knowledge, suggesting that there is a net effect of this type of cooperation, which cannot be explained in this way. Also, other kinds of industry–science

R&D partnerships, such as subcontracting to scientific research institutions without explicit cooperation, tend to involve more basic research than outsourcing to business.

Including a variable on R&D cooperation with institutions does, however, strongly reduce the impact of funding intensity, which becomes statistically non-significant as a result. This confirms implicitly that public funding authorities favour cooperative industry–science relations (Capron et al., 2000). By doing so, they allow the firms concerned to raise their R&D outsourcing intensity. This strengthens earlier findings in the empirical literature, which have shown that the receipt of public funding for innovation increases the firms' probability of cooperation in R&D, especially with public research institutions (Belderbos et al., 2004; Aschhoff et al., 2006; Busom and Fernandez-Ribas, 2008).

Still, the evidence that the impact of R&D cooperation with institutions is clearly stronger than that of funding intensity suggests that there is more to it. Part of the explanation could be that the network externalities obtainable through R&D partnership are greatest in elaborate networks involving industry–science relations, and that explicit cooperation allows more profound interactions, and therefore involvement, than looser forms of R&D partnership. But these are clearly conjectures that need further investigation.

APPENDIX

Table A7.1 gives a description of the manner in which the variables were constructed from the raw data of the CIS3 and CIS4 questionnaires and Table A7.2 does so for the variables constructed from the data from the 2003 and 2005 Belgian R&D Surveys. Tables A7.3 and A7.4 present corresponding descriptive statistics of the variables concerned.

Table A7.1 Description of the variables measured by data from the CIS4 (2002–2004) and CIS3 (1998–2000) databases

Variable	Definition
Dependent variables	
Probability of R&D outsourcing and doing permanent in-house R&D (OUTRD)	OUTRD = 1 if the firm outsources R&D during 2002–2004 and performs in-house R&D during 1998–2000 and 2002–2004
Log of outsourcing intensity (LOUTI)	Natural log of the ratio of expenditure on external R&D over turnover in 2004
Independent Variables	
Log of in-house R&D intensity (LRDINS)	Natural log of the ratio of expenditure on in-house R&D over turnover in 2004 (2000 when instrumented)
Patents (IPPAT)	IPPAT = 1 if the firm had any valid patents protecting its inventions or innovations at the end of 2000
Strategic appropriability (APPR)	Average score on the effectiveness of strategic protection through secrecy, complexity of design and lead-time advantage during 2002–2004 (1998–2000 when instrumented)
Strategic appropriability through secrecy (APPR_SE)	Score on the effectiveness of strategic protection through secrecy during 2002 2004 (1998–2000 when instrumented)
Strategic appropriability through complexity of design (APPR_CD)	Score on the effectiveness of strategic protection through complexity of design during 2002–2004 (1998–2000 when instrumented)
Strategic appropriability through lead-time Advantage (APPR_LT)	Score on the effectiveness of strategic protection through lead-time advantage during 2002–2004 (1998–2000 when instrumented)
R&D cooperation (COOP)	COOP = 1 if the firm collaborates either with suppliers, clients, competitors, research labs, universities or government during 2002–2004

Table A7.1 (continued)

Variable	Definition
R&D cooperation with business (COBUS)	COBUS = 1 if the firm collaborates either with suppliers, clients or competitors during 2002–2004
R&D cooperation with public research institutions (COPRC)	COPRC = 1 if the firm collaborates either with research labs, universities or government during 2002–2004
Control variables	
Incoming knowledge spillovers (INKS)	Average score on the importance attached to information from suppliers, clients, competitors, universities, government, professional conferences, journals, exhibitions during 2002–2004 (1998–2000 when instrumented)
Complexity of knowledge (CKNOW)	Ratio of the sum of the scores of the importance attached to information from universities, government, professional conferences and journals over the sum of the scores of all sources of information during 2002–2004 (1998–2000 when instrumented)
Log of higher education intensity (LHEI)	Natural log of the % employees with higher education in the year 2000
Log of export intensity (LEXI)	Natural log of the proportion of exports in turnover in the year 2002 (1998 when instrumented)
Log of size (LSIZE)	Natural log of the number of employees in the year 2002 (1998 when instrumented)
Domestic group affiliation (GRP_D)	Domestic group affiliation = 1 if the firm is a member of a domestic group in 2004 (2000 when instrumented)
Foreign group affiliation (GRP_F)	Foreign group affiliation = 1 if the firm is a member of a foreign group in 2004 (2000 when instrumented)
Hampering risks (HRISK)	Importance attached to excessive perceived economic risks hampering innovation activity during 1998–2000
Hampering finance (HFIN)	Importance attached to lack of appropriate sources of finance during 2002–2004 (1998–2000 when instrumented)

Table A7.1 (continued)

Variable	Definition
Public funding (FUND)	FUND = 1 if the firm received public funding for innovation irrespective of source during 2002–2004 (1998–2000 when instrumented)
IDPAV1	IDPAV1 = 1 if the firm is classified in the 'resource-intensive' sectors (mining, food, beverages and tobacco, wood and furniture, recycling, wholesale)
IDPAV2	IDPAV2 = 1 if the firm is classified in the 'scale-intensive' sectors (paper and printing, rubber and plastics, non-metallic minerals, ferrous metals, non-ferrous metals, fabricated metal products, motor vehicles, other transport equipment)
IDPAV3	IDPAV3 = 1 if the firm is classified in the 'specialised supplier' sectors (non-electrical machinery, electronics and communication, electrical machinery)
IDPAV4	IDPAV4 = 1 if the firm is classified in the 'science-based' sectors (chemicals, pharmaceuticals, computers and office machinery, scientific instruments)
IDPAV5	IDPAV5 = 1 if the firm is classified in the 'information-intensive services' sectors (financial intermediation, telecommunication, transport services, business services, computer and related service activities, research and development, technical engineering)

Notes: The non-instrumented data are from CIS4 and the instrumented data from CIS3. In case of IPPAT, LEDUC and HRISK the data are only available in CIS3, so that no instrumentation is necessary.

Table A7.2 Description of the variables measured by data from the Belgian 2005 and 2003 R&D survey databases

Variable	Definition
Dependent variables	
Log of outsourcing intensity (LOUTI)	Natural log of the ratio of expenditure on external R&D over employment in 2005
Probability of R&D outsourcing (OUT)	OUT = 1 if the firm outsources R&D during the period 2005 in the population of in-house R&D-active firms in 2005 and 2003
Independent variables	
Log of in-house research and development intensity (LRDINS)	Natural log of the ratio of R&D employment over total employment in 2005 (2003 when instrumented)
R&D cooperation (COOP)	COOP = 1 if the firm collaborates either with suppliers, clients, competitors, research labs, universities or government in 2005 (2003 when instrumented)
R&D cooperation with business (COBUS)	COBUS = 1 if the firm collaborates either with suppliers, clients or competitors in 2005 (2003 when instrumented)
R&D cooperation with public research institutions (COPRC)	COPRC = 1 if the firm collaborates either with research labs, universities or government in 2005 (2003 when instrumented)
Log of export intensity (LEXI)	Natural log of the proportion of exports in turnover in the year 2005 (2003 when instrumented)
Log of size (LSIZE)	Natural log of the number of employees in the year 2005 (2003 when instrumented)
Domestic group affiliation (GRP_D)	Domestic group affiliation = 1 if the firm is a member of a domestic group in 2005 (2003 when instrumented)
Foreign group affiliation (GRP_F)	Foreign group affiliation = 1 if the firm is a member of a foreign group in 2005 (2003 when instrumented)
Public funding (FUND)	FUND = 1 if the firm received public funding for innovation irrespective of source during 2005 (2003 when instrumented)
Log of public funding intensity (LFUNDI)	Natural log of the ratio of total public funding for innovation irrespective of source over employment in 2005 (2003 when instrumented)

Notes: The outsourcing and in-house R&D intensity variables were in this case obtained by normalizing by employment rather than by turnover, because of greater data reliability. The same holds for the measure of funding intensity. In the sample concerned some of the observations of the funding intensity variables had a zero value. To allow taking their natural log, these were replaced by a 0.001 value. The industry dummies correspond to those described in Table A7.1.

Table A7.3 Descriptive statistics: mean and standard deviation (variables measured by data from the CIS databases; 601 observations)

	CIS4 (2002–2004)		CIS3 (1998–2000)	
	Mean	St. dev.	Mean	St. dev.
Dependent variables				
OUTRD	0.256	0.438	0.258	0.438
LOUTI #	–6.019	1.782	–6.049	2.093
Independent variables				
LRDINS *	–4.078	1.898	–4.242	1.867
IPPAT	0.183	0.387	0.183	0.387
APPR	0.350	0.415	0.250	0.351
APPR_SE	0.344	0.476	0.309	0.463
APPR_CD	0.256	0.437	0.146	0.354
APPR_LT	0.449	0.497	0.286	0.452
COOP	0.341	0.474	0.247	0.432
COBUS	0.304	0.460	0.208	0.406
COPRC	0.199	0.400	0.174	0.380
Control variables				
INKS	0.291	0.278	0.280	0.254
CKNOW	0.195	0.220	0.203	0.238
LEDUC	–2.223	1.695	–2.223	1.695
LEXI	–2.709	2.736	–2.796	2.779
LSIZE	4.303	1.467	4.118	1.473
GRP_D	0.266	0.442	0.275	0.447
GRP_F	0.328	0.470	0.318	0.466
HRISK	0.155	0.300	0.155	0.300
HFIN	0.404	0.377	0.166	0.318
FUND	0.221	0.415	0.211	0.408
IDPAV1	0.263	0.441	0.263	0.441
IDPAV2	0.271	0.445	0.271	0.445
IDPAV3	0.136	0.344	0.136	0.344
IDPAV4	0.070	0.255	0.070	0.255
IDPAV5	0.260	0.439	0.260	0.439

Notes: * indicates that in the case of the log of in-house R&D intensity, LRDINS, the sample consists exclusively of firms that were in-house R&D-active in both periods, and # indicates that in the case of the log of outsourcing intensity, LOUTI, it consists of firms that were not only in-house R&D-active in both periods but also outsourcing R&D in the period concerned.

Table A7.4 Descriptive statistics: mean and standard deviation (variables measured by data from the Belgian R&D survey databases; 424 observations)

	R&D survey (2005)		R&D survey (2003)	
	Mean	St. dev.	Mean	St. dev.
Dependent variables				
OUT	0.497	0.500	0.469	0.499
LOUTI	–3.296	3.901	–3.492	3.946
Independent variables				
LRDINS	–2.372	1.466	–2.367	1.580
COOP	0.443	0.497	0.443	0.497
COBUS	0.443	0.497	0.438	0.496
COPRC	0.349	0.477	0.354	0.478
Control variables				
LEXI	–1.230	1.934	–1.278	1.981
LSIZE	4.162	1.741	4.116	1.815
GRP_D	0.304	0.461	0.304	0.461
GRP_F	0.203	0.403	0.203	0.403
FUND	0.415	0.493	0.370	0.483
LFUNDI	–3.816	3.881	–4.170	3.767
IDPAV1	0.255	0.436	0.255	0.436
IDPAV2	0.205	0.404	0.205	0.404
IDPAV3	0.177	0.382	0.177	0.382
IDPAV4	0.153	0.360	0.153	0.360
IDPAV5	0.210	0.408	0.210	0.408

Note: The sample consists of firms that were in-house R&D-active in both periods.

8. Managing R&D outsourcing and the impact on firms' R&D employment

8.1 INTRODUCTION

Outsourcing activities are fuelling a heated debate among economists and the general public (Chesbrough, 2003a; Howells et al., 2008). Not that the purchase of goods and services from suppliers is anything new, but it is feared that the surge in outsourcing (off-shoring) displaces domestic employment. Due to the rising level of education among most of the populations of the developed world, the outsourcing of low-skilled labour has not yet set alarm bells ringing. However, with the increasing division of labour at firm level, the availability of specialist knowledge outside the firm's walls, the need to remain competitive through innovation and the swift development of emerging economies such as India and China, it is believed that R&D outsourcing could potentially affect highly-skilled employees and displace R&D jobs.

This chapter focuses on the impact on R&D employment intensity of firm-level decisions to start, increase, decrease or stop R&D outsourcing. This topic relates to the important move in strategic R&D management in the 1990s towards an increasing intensification of companies' dependence on external sources of technology, including a trend towards more outsourcing (Coombs et al., 2003; Edler et al., 2002). Quinn and Hilmer (1995) attributed this tendency to the corporate strategies of firms to subcontract non-core or peripheral tasks. The higher reliance on external sources is considered part of a fundamental shift in the way companies generate new ideas and bring them to market, as emphasised by the concept of 'open innovation' (Chesbrough, 2003a).

The empirical literature on R&D outsourcing has shown great interest in the determinants for R&D outsourcing (Mol, 2005), in the topic of competition between R&D outsourcing with internal R&D activities stressing the question of complementarity versus substitution effects (Veugelers, 1997), and in the search for the 'optimal' level or balance between in-house and outsourced R&D (Belderbos et al., 2004). These analyses have largely considered R&D outsourcing budgets.

To our knowledge, empirical evidence on the effects of the outsourcing of R&D on the composition of skills in the total labour force of the firm, and more precisely the relative importance of R&D workers (that is, the internal R&D employment intensity) remains a blank spot in the literature. Studying this topic is in line with research investigating the impact of technological change on employment in terms of changes in the quality of employment. The literature on open innovation rightfully acknowledges that firms can never have all the expertise in-house (Chesbrough, 2003a); but also acknowledges that for their core activities the best brains work better for the firm in enabling it to incorporate the external ideas into profitable innovation (Quinn, 2000; Cohen and Levinthal, 1990). These arguments draw attention to the effects of outsourcing on labour composition at the firm level.

This chapter contributes to the empirical literature on R&D outsourcing in several ways. First, the topic is approached in a dynamic framework, using a panel of permanent R&D-active firms. In contrast to extant empirical studies, the R&D budgets involved in outsourcing are complemented with the strategic decision of the firm to start, to increase, to decrease or to stop R&D outsourcing. Second, the effects of R&D outsourcing on the dynamics of internal R&D employment intensity (measured in terms of the evolution in the share of R&D employment in total employment at the firm level) are the central issue in this study. R&D employment, more than R&D budgets, is a vital element of the knowledge base of the firm and as such cannot be made redundant without careful consideration (Quinn, 2000). Third, R&D outsourcing is refined by taking the type and location of the contractor on board. The type of contractor varies between other (non-affiliated) firms, universities, and public and private research centres (von Hippel, 1988; Tether, 2002). Outsourcing to universities is likely to be associated more with basic research and the resulting path from invention to commercialisation might be longer than in the case of applied research by suppliers directed towards troubleshooting. Consequently, the choice of partner is not without relevance. Next, the location is taken into account because geographical proximity exerts an impact on the interaction possibilities (Teirlinck and Spithoven, 2008), because of the labour cost argument put forward in the offshoring literature (Sorenson et al., 2006; Cusmano et al., 2008), and because of cultural factors (Hofstede, 1991). A distinction can be made between outsourced R&D performed within the region, the country, and abroad (OECD, 2002a).

The chapter is organised as follows: Section 8.2 illustrates the theoretical background on R&D outsourcing and relates this to internal R&D

employment intensity. The main drivers of R&D outsourcing are identified as well as the relationship between outsourcing and internal R&D employment intensity. Next, the importance of the type and geographical location of contractors is discussed. From this literature review the main research questions are derived. Section 8.3 describes the nature of the data-set, the sample and the method used. Next, in Section 8.4, the data analysis is carried out and the results are discussed. The concluding section focuses on the main findings, offers some implications for practitioners and policy, and identifies a few limitations of the analysis.

8.2 THEORETICAL BACKGROUND

8.2.1 Firm Motives for R&D Outsourcing

From a firm's point of view, the decision as to whether to perform R&D activities internally or outsource them can be approached from a cost perspective (Quinn, 2000) or from an external technology sourcing-based view (in line with the ideas of distributed or open innovation, Chesbrough, 2003a).

From a cost perspective, firms have to weigh the costs of outsourcing a given activity against the cost of conducting it in-house. Williamson's (1975) ideas on transaction costs include both the direct costs of the transaction and the possible opportunity costs of inferior governance decisions (Holmstrom and Milgrom, 1994). With low uncertainty, reduced asset specificity and low frequency of iteration, firms will prefer a market transaction to internalising the activity (Williamson, 1985). However, as is often the case in R&D projects, transaction cost economics underline the risks of outsourcing activities based on specific assets or under conditions of uncertainty.

From a technology sourcing-based view, a distinction can be made according to whether or not the outsourced activities belong to the core activities of the firm. On the one hand, outsourcing non-core activities allows firms to increase managerial attention and resource allocation to those tasks they do best (Gilley and Rasheed, 2000; Narula, 2001). On the other hand, R&D outsourcing of more core-related activities facilitates access to fresh knowledge and new technology and helps benefit from complementary capabilities. However, the appropriation of the results of outsourced R&D is an important element in this type of activity (Teece, 1986; Chesbrough, 2003a). R&D aimed at the provision of enhanced information transfer and coordination across activities within the knowledge production system (implying second- and higher-order interactions among a set of value-chain activities) faces greater appropriation concerns than R&D directly tied to the ability to identify, isolate, or solve a specific set of independent technical problems (Cassiman et al., 2005). The necessity of absorptive capacity, suggesting that the impact of internal and external technology on firm performance is not independent, supports this argument (Cohen and Levinthal, 1990).

Empirical evidence reveals that R&D outsourcing is driven by the lack of in-house R&D and technical expertise, followed by the reduction of development time and time to market (Howells et al., 2004, 2008). Reducing development costs ranked only in third place, followed at some distance by the reduction of risk (see also Mehta and Peters, 2007) and uncertainty through sharing R&D with another (set of) partner(s). These drivers should be seen in conjunction with a greater codification of R&D processes (Granstrand et al., 1992; Narula, 2001). A main driver for R&D outsourcing includes the set-up of strategic research projects where partner firms or institutions possess complementary technological capabilities not available in-house (for example, Pisano, 1990, Chesbrough, 2003a).

These empirical findings are not fully in line with the normative logic that technology-outsourcing alliances allow the outsourcing firms to further specialise in their domain of core competence without being distracted by non-core activities (see, for example, Coombs, 1996). However, it does confirm the finding that 'peripheral' knowledge (that is, specialised knowledge in the domain of outsourced activities) complements control in technology outsourcing alliances (Brusoni et al., 2001; Mehta and Peters, 2007; Tiwana and Keil, 2007).

The potential downside of R&D outsourcing is that this kind of activity may induce an element of external technology dependence (Swan and Allred, 2003), increase organisational cost (Jacobides and Billinger, 2006) and weaken the integration between production and R&D (Teece, 1986). Another concern is the possible deterioration of internal innovation capacity and loss of comparative advantage. Outsourcing R&D also means that other parties are provided with information about the firm and its products and this reduces both the exclusivity of the resulting new knowledge and the ability of the firm to appropriate the results of its (outsourced) R&D (Arrow, 1962).

According to Antras (2005), the transfer of new technology should usually be conducted internally, while that of technology between five and ten years old can be completed through outsourcing. This confirms earlier findings by Mansfield and Romeo (1980). Davidson and McFetridge (1984) find that the probability of outsourcing is lower the more radical the technology and the larger the R&D department of a firm. Other studies indicate that, in general, outsourcing is more frequent the more codifiable and less complex the technology is – with complexity being measured through the amount of R&D necessary to produce a good (Wilson, 1977; Kogut and Zander, 1993). Similarly, Trefler (2005) argues that outsourcing is the appropriate way for a sufficiently routine project that can be fully scoped or described, while projects difficult to describe from the outset should be conducted in-house. When the task outsourced is specific, there are fewer risks involved, since contract design, enforcement and monitoring are relatively easy.

Finally, it has to be recognised that the use of external sources is also influenced by more generic elements. A first element relates to the technological opportunities in a given industry (Klevorick et al., 1995). The openness of individual firms to external sources of knowledge is subject to a degree of firm-level decision, but many of these decisions are shaped by the search activities of other firms in their industry and by the nature of the environment in which they operate (Cohen and Levinthal, 1990). Laursen and Salter (2006) found that the higher the technological opportunities offered by the industry in which firms are operating, the higher the degree of openness to external sources of innovation. A second element relates to firm size. In this respect, and in support of Chesbrough (2003a), Laursen and Salter (2006) found larger firms more likely to be open to external knowledge due to the higher availability of resources and innovations which are more complex in nature.

8.2.2 R&D Outsourcing and Internal R&D Employment Intensity

It is widely accepted that access to external technology, in combination with the presence of internal R&D necessary to absorb external knowledge (Cohen and Levinthal, 1990) allows firms to improve the performance of internal R&D efforts (Cassiman and Veugelers, 2006). In addition, internal R&D enables firms to translate external knowledge into innovation opportunities and outsourced R&D facilitates spillovers from outside the network to the firm's specific knowledge base (Audretsch et al., 1996).

In line with these arguments, several empirical studies found a positive relationship between the share of external R&D in total R&D and productivity (Bönte, 2003; Beneito, 2006; Griffith et al., 2004). Firms with high levels of absorptive capacity were found likely to be more open to external R&D because retaining skills and expert staff enables them to access wider networks (Rosenberg, 1990). Knowledge-intensive small firms were less likely to be involved in this kind of activity because they often rely on the exploitation of new ideas and, given the danger they face from leakage of their ideas, they limit the nature and scope of their external interaction (Laursen and Salter, 2006).

Evolutionary economics provides a different perspective on the effects of outsourcing knowledge-intensive or strategic activities, by stressing the impact of changing firm boundaries on the dynamic capabilities of the firm, and hence on its ability to integrate, build and reconfigure internal and external competencies to address changing environments (Teece et al., 1997). In this respect outsourcing is seen as part of a knowledge-searching strategy in which external players contribute to the division of labour in research tasks and complement in-house R&D (Arora and Gambardella, 1994a; Chesbrough, 2003a).

However, empirical evidence does not always support the idea of complementarity between internal and external R&D. Based on a cross-section of Dutch manufacturing firms, Audretsch et al. (1996) reported a substitution effect of external R&D activity in low and medium technology industries (the reverse was true in high-tech industries). Cassiman and Veugelers (2006) found that firms combining internal and external R&D strategies for innovation show superior innovative performance. It also turns out that large, resourceful firms have a higher propensity to combine both strategies (Veugelers and Cassiman, 1999). Piga and Vivarelli (2004) modelled the decision to outsource R&D jointly with the decision on internal R&D and found that firms having larger internal R&D expenditures and pursuing broader objectives related to both process and product innovation have a greater propensity to engage in R&D outsourcing.

In contrast with the interest in the topic of complementarity of – or substitution by – R&D outsourcing, to our knowledge, empirical evidence on the effects of the outsourcing of R&D on the composition of skills in the total labour force of the firm, and more precisely the relative importance of R&D (that is, the internal R&D employment intensity) remains a blank spot in the literature. Studying this topic is in line with studies investigating the impact of technological change on employment in terms of changes in the quality of employment. Relying on evidence in this broader context, Kahn (2000) reveals that technological standardisation – which drives the growth of outsourcing – reduces the importance of firm-specific knowledge, as standardisation reduces the specificity of internal skills and induces convergence of firm-specific skills to general skills over time. Pianta (2004) studied – at macro level – the impact of technological innovation on the quality of jobs by looking at the relative importance of skilled and unskilled jobs and found the diffusion of technologies to have a strong skill bias effect in favour of skilled jobs for educated workers. Tarabusi (1997) also found a trend towards skill upgrading, but revealed that the effects on qualifications depend on the importance of the nature of technology and the context in which technological changes are embedded. The latter finding is in line with the management literature (see, for example, Tushman and Anderson, 1997;

Piva et al., 2003; Chesbrough, 2003a) highlighting the importance of the organisational context for changes in the skill composition of employment.

This argument can be extended to R&D outsourcing at the firm level by investigating the impact of firms' decisions in terms of technology sourcing (by means of R&D outsourcing) on the relative importance of R&D skills in total firm employment. This relates to the question of whether (and if so, in which direction) firm level decisions in terms of R&D outsourcing change the relative importance of R&D labour in total employment.

8.2.3 Type and Geographical Location of the Contractor

As technology markets evolve, the number of contractors grows and this eventually becomes a specialised autonomous industry (Howells et al., 2004). Unfortunately, little attention has been paid to the type of contractor and their location. The selection of a certain type of contractor is dependent on the supply on the technology market on the one hand (Howells et al., 2004), and the nature of the R&D project or service that is outsourced on the other.

Agreements with universities are formed whenever risk is not an important obstacle to innovation, and typically serve to share costs. These agreements are found to be complementary to other innovative activities such as performing internal R&D, sourcing public information and cooperative agreements with suppliers and customers (Veugelers and Cassiman, 2005). Public research organisations and private research centres are partners that provide basic research as well as applied R&D for direct use. The same applies for polytechnics and other higher education institutes. Cassiman et al. (2005) find that in the case where project knowledge has high strategic value, internal development or simple contracting of some specific activities is preferred. Projects contracted out to universities show a significantly higher strategic importance. The authors argue that both the governance form and the partner's characteristics may mitigate the risk of leakage. Outsourcing R&D to other firms tends to happen in a more competitive market environment.

Besides the type of contractor, the geographical location of the contractor also needs to be taken into account. The topic of geographical location is related to the aspect of proximity and is found to be important, especially for the decision to outsource core activities (Sorenson et al., 2006; Cusmano et al., 2008). The main argument put forward in favour of local outsourcing is that proximity facilitates (the coordination of) knowledge exchange and trust by enhancing personal contacts (Love and Roper, 2001) which is important to create stable and prolonged relationships for transferring complex knowledge and enhancing innovation (Sorenson et al., 2006). Since knowledge and technology transfer activities are in general localised phenomena (Audretsch and Feldman, 1986; Jaffe, 1986), the regional level is an initially appropriate geographical entity to be taken into account.

On the other hand, the increasingly global dispersion of R&D (specialisation) creates a need for firms to access distant knowledge sources and to tap into global networks. This has been facilitated by advances in communication technologies, developments in management and business processes, and the increasing openness of economies (Mol, 2005). Also, expansion (opportunities) in different (especially emerging) markets creates the need to adapt R&D to the market strategy of a given host country. In this respect R&D outsourcing is enabled by a large pool of 'local' qualified human capital. Again, the governance and appropriability of the knowledge provided by outsourced R&D tends to become more difficult with the complexity of the knowledge transferred (as put forward by the global value chain literature, Gereffi et al., 2005).

Between the local and the global context in which R&D outsourcing takes place, institutional (including, for example, tax policy, social security) and cultural values and norms can be supposed to play a decisive role in the search process for contractors (Hofstede, 1991; Schwartz, 1999). As such, national specificities remain important.

8.2.4 Research Questions

This chapter contributes to the debate on R&D outsourcing in different ways. First, it examines the impact of firm-level decisions with regard to outsourcing of R&D on internal R&D employment intensity. These firm-level decisions take qualitative values: a firm can start with R&D outsourcing; increase its on-going outsourcing activities; decrease them; or stop outsourcing.

Second, the partners to whom the activities are outsourced are differentiated by type (universities; (quasi-) public research centres; other firms) and by geographical location (regional; national (but outside the firm's region); or abroad).

In order to assess the impact of firms' strategies in terms of R&D outsourcing on changes in the internal employment intensity of R&D workers, two research questions are formulated. First, how do corporate decisions to start, increase, decrease, or stop R&D outsourcing influence internal R&D employment intensity? Second, does the influence on the internal R&D employment intensity of corporate decisions to start, increase, decrease, or stop R&D outsourcing vary across the type and geographical location of the contractor?

8.3 MODEL AND DATA

8.3.1 The Model

Starting from the ideas put forward in the literature and based on the research questions, the basic model for the analysis in first differences (except for dummy variables) is as follows;

$$\Delta RDI_{it} = \sum_{t=2}^{T} \alpha_{\alpha t} D_t + \sum_{p=2}^{P} \alpha_p P_i + \sum_{c=1}^{C} \alpha_{lc} \Delta X_{ict} + \beta_0 \Delta RD_{it}^{START} + \beta_1 \Delta RD_{it}^{INCREASE}$$
$$+ \beta_2 \Delta RD_{it}^{DECREASE} + \beta_3 \Delta RD_{it}^{STOP} + \varepsilon_{it} \quad \varepsilon_{it} \sim N(0, \sigma^2) \qquad (8.1)$$

Notes: i = firm; t = year (2001; 2003 and 2005); Δ denotes first differences (for example $\Delta RDI_{it} = RDI_{it} - RDI_{it-1}$). D_t: Year dummies; P_i: Dummies denoting the Pavitt group to which a firm belongs (p = 1...P: number of Pavitt categories); X_{ict}: control variables (see text, c = 1...C: number of control variables).

The internal R&D employment intensity (RDI) is measured as the ratio of R&D personnel to total employment, both measured in full-time equivalents. Year dummies (D_t) are included because R&D intensity tends to change over time depending on general economic conditions (for example, business cycle). As the basic model is expressed in terms of changes over time, and as detailed information of R&D data is available only for the odd years (see data description in Section 8.3.2), three time-periods can be considered: 1999–2001; 2001–2003 and 2003–2005. Firms' internal R&D intensity clearly was on the rise at the beginning of the period under consideration (+ 1.3 per cent point between 1999 and 2001 in terms of total business enterprise R&D personnel per thousand employees in industry), fell back sharply in the period 2001–2003, and stagnated in the period 2003–2005 at a level close to the 1999 level (OECD, 2007).

As suggested by the literature review, a number of potential determinants of internal R&D intensity have been included as control variables. First, cross-sector differences in technological opportunities are captured by dummies denoting the Pavitt group to which a given firm belongs (P_i). The taxonomy proposed by Pavitt (1984) distinguished five different industry types for the manufacturing sector: science-based, scale-intensive, labour-intensive, resource-intensive and specialised suppliers. Tidd et al. (1997) extended the taxonomy to include the services sector, adding an information-intensive group as well as specialised suppliers in services.

All other control variables (X_{ict}) are continuous and are included in first differences in the estimation, except for group affiliation which is classified by three dummies denoting whether the firm is independent, belongs to a Belgian group or to a foreign group respectively (based on country of origin of the ultimate controller). Average employment is considered for firm size. Other firm characteristics included are: exports (average share of turnover realised outside the home country); the share of research expenditures in the total R&D budget (see, for example, Howells et al., 2008); the ratio of R&D that is outsourced to total R&D expenditures; and the share of R&D personnel in total employment at the beginning of the period under consideration (that is, t–1).

The variables of interest in this chapter are those related to firms' decisions to start, increase, decrease or stop outsourcing R&D. These binary choice variables were multiplied by the absolute change in the ratio of outsourced R&D to total R&D. This allows us to consider the magnitude of the changes. As highlighted before, these corporate decisions in R&D outsourcing will be further specified to account for the type of contractor and its geographical location.

8.3.2 Data Availability and Variable Description

The data on firms located in Belgium used in this analysis are provided by the Federal Cooperation Commission, CFS/STAT, which gathers this information on the basis of the biannual OECD R&D business survey. The results of four subsequent surveys were taken into account: R&D survey 2000 (results for 1998 and 1999); R&D survey 2002 (results for 2000 and 2001); R&D survey 2004 (results for 2002 and 2003); and R&D survey 2006 (results for 2004 and 2005). Detailed information on the variable 'outsourced R&D by type of contractor and geographical location' is only available for the odd years (1999; 2001; 2003; and 2005). Those years are used in the analysis.

The focus of the survey is on enterprises (defined at the level of the smallest legal entity – that is, those having a VAT number) that perform R&D on a permanent or quasi-permanent basis. All firms that are known to perform R&D are included. An update is made every time a new survey is launched, taking into account both the firms known from the past to be R&D-active and a monitoring of firms declaring themselves R&D-active, for example, by means of press releases or request for R&D grants (for a more detailed overview see: CFS/STAT, 2001). For the analysis in this chapter, a permanent or quasi-permanent R&D-active firm is any firm which declared in the OECD survey: (i) at least one full-time equivalent R&D employee in at least one year during the period 1999–2005; and (ii) internal R&D expenditures greater than or equal to 30,000 euros in at least one year during the same period.

During the period 1999–2005 firms went bankrupt, merged, acquired (parts of) other firms or experienced other changes with an important impact on the evolution of R&D and other enterprise variables. Therefore, information on changes with an impact larger than 10% (negative or positive) in terms of overall employment or in terms of turnover was requested in each of the four surveys. In the case of mergers, acquisitions or sale of a part of a company, information was collected on other enterprises involved in the operation and the different enterprises involved were taken together as one entity for the entire period. Firms that went bankrupt are excluded from the analysis. This way firm-level data have been constructed to guarantee comparability over time, resulting in a database with 1,971 enterprises as a starting point for the analysis. There are three time cohorts: 1999–2001; 2001–2003; and 2003–2005, and so 5,913 potential observations. Unfortunately, due to the voluntary nature of the survey there are a lot of missing values. More information is available for key variables on internal R&D activities and for R&D personnel, because an additional short survey/telephone follow-up and other sources are used to gather these data. Data on the evolution of outsourced R&D expenditures were available for 1,378 observations. However, in order to estimate the model as described in section 3.1, only 1,161 observations for 761 enterprises could be taken into account for a model not differentiating between type (firms outside the group – universities – public and private research centres) and geographical location (regional, national, abroad) of the contractor. Including the latter distinction, the need for more detailed information on outsourcing R&D left a dataset of 553 observations for 384 enterprises.

The comparison in Table 8.2 of the observations (for the total of the three time-periods) included in the analysis with the ones pertaining to the population of (quasi-) permanent R&D-active enterprises in Belgium but not included in the analysis, reveals a selection bias towards large, less export-oriented, and more R&D-intensive enterprises. However, these firms have a lower share of outsourced R&D in the total (internal and outsourced) R&D budget. The findings seem to support the negative relation between outsourcing and firm size found by Coase (1937), but contrasts the findings of Laursen and Salter (2006) for innovative activities. In terms of change in R&D employment intensity (the dependent variable in the analysis in Section 8.4), no significant differences are found between both groups of observations.

Table 8.1 Frequency of R&D outsourcing

Target population: R&D-active firms				
Evolutions: 1999–2001/ 2001–2003/2003–2005		5,913 observations (1,971 firms)		
Full model				
Panel estimation	1,161 observations (761 firms)	Of which:		
		1999–2001: 260 observations	2001–2003: 498 observations	2003–2005: 403 observations
Start	170			
Increase	207			
Decrease	187			
Stop	193			
Null	404			
Detailed model				
Panel estimation	553 observations (384 firms)	Of which:		
		1999–2001: 114 observations	2001–2003: 188 observations	2003–2005: 251 observations
		Firms	Universities	Public research centres
Start	Regional	21	33	22
	National	5	3	4
	Abroad	12	6	5
Increase	Regional	44	78	59
	National	20	21	15
	Abroad	45	31	31
Decrease	Regional	57	67	55
	National	18	28	16
	Abroad	42	15	24
Stop	Regional	28	30	20
	National	2	5	3
	Abroad	17	7	4

Notes: The category 'increase' includes 18 observations for which the share of outsourced R&D in total R&D remained equal between two years. 'Regional' refers to Brussels–Capital Region – Flemish Region – Walloon Region. 'National' refers to a location outside the firm's proper region.

Table 8.2 Mean values of the independent and the continuous control variables for observations included in the analysis (full model – see Table 8.1) and those not included

	Panel observations	Population (excl. panel) observations	Sig.*
Change in R&D employment intensity (R&D employment/average employment)	–0.02 (1,161)	–0.03 (936)	>0.10
Size (average employment)	320.89 (1,156)	171.69 (4,400)	<0.01
Export (share of turnover outside the home country)	0.43 (925)	0.53 (1,441)	<0.01
Research share (share of research in total R&D budget)	0.41 (871)	0.38 (1,095)	<0.10
R&D intensity	0.20 (1,161)	0.17 (2,276)	<0.01
Outsourced R&D (share of outsourced R&D budget in total – internal + outsourced – R&D budget)	0.13 (1,161)	0.16 (1,867)	<0.01

Notes: The number of observations is in brackets. The number of observations for some variables in the panel can be lower than 1,161 because the analysis in section 4 is based on a general-to-specific approach in which non-significant variables are gradually dropped. The symbol * denotes the significance of the difference between the averages of the two groups of observations based on a t-test.

Comparison of the distribution of firm ownership in the selected group of 1,161 observations (independent firms: 31.2%; firms belonging to a Belgian group: 31.3%; firms belonging to an international group: 37.6%) with the one for observations not included (respectively 39.5%; 29.1% and 31.4%) indicates that observations for firms belonging to a group (especially international) are over-represented in the analysis. In terms of industrial activity, a comparison of the analysed group with the group of observations not taken into account reveals no systematic changes for the industry dummies information-intensive, scale-intensive, science-based, resource-intensive or labour-intensive. However, a big discrepancy exists between specialised suppliers in services and manufacturing. In the manufacturing sector, firms belonging to this category were largely over-represented in the group of observations taken into account for the analysis (18.9% of observations in the analysed group versus 13.8% for those not analysed). The opposite was true in services (9.3% versus 14.9%).

A comparison of the 553 observations containing information on the type and geographical location of the contractor with the 1,161 observations included in the basic model revealed no significant bias in terms of the control variables presented in Table 8.2.

8.4 ANALYSIS

8.4.1 Descriptive Statistics on Firm Decisions

On average, over the period 1999–2005, a firm's outsourced R&D budget amounts to 13.1 % (standard deviation 0.23) of the total (internal and outsourced) R&D budget. Decisions to start, increase, decrease, or stop outsourcing imply absolute changes in the share of outsourced R&D in total R&D of respectively +0.28 (standard deviation equals 0.28); +0.14 (0.19); –0.18 (0.20); and –0.28 (0.25). As such, the share of outsourced R&D involved in starting and stopping outsourcing R&D seems of equal importance, whereas the change in the share in the total R&D budget is larger on average in case of a decrease than in case of an increase in outsourcing.

Table 8.3 highlights the frequencies of the different outsourcing decisions by head office, by size class (large refers to firms with 250 or more employees; medium-sized to 50–249; and small to less than 50), and by industrial activity. In terms of ultimate control of the firm it can be noted that firms belonging to a group are relatively more frequently involved in sustained outsourcing and less in starting and especially stopping such activities. Also, less than one third of these firms were not at all engaged in outsourcing R&D in two consecutive time periods (1999–2001; 2001–2003; 2003–2005). Independent firms relatively most often take the decision to stop the outsourcing of R&D and are less likely to be engaged in outsourcing of R&D.

The propensity to increase or decrease existing R&D outsourcing increases with firm size. The opposite is true for decisions to start or stop outsourcing R&D. These findings are in line with those presented by Veugelers and Cassiman (1999) indicating that large resourceful firms have a higher propensity to combine internal and outsourced R&D than their counterparts.

Finally, concerning industrial activity (Pavitt taxonomy) a clear distinction can be made between manufacturing and services firms. Close to 40% of firms in the manufacturing sector are involved in sustained (increased or decreased) outsourcing of R&D (with specialised suppliers at the lower end) compared to half as many in the services sector.

Table 8.3 Frequency of decisions to outsource R&D broken down by head office; size and industrial activity – in row percentages

	Start	In-crease	De-crease	Stop	Null	Obser-vations
Head office						
Independent firm	16%	10%	11%	20%	43%	362
Belgian group	13%	23%	19%	14%	31%	363
International group	15%	19%	18%	15%	32%	436
Size						
Large	15%	31%	23%	11%	20%	239
Medium-sized	13%	17%	18%	16%	36%	363
Small	16%	13%	12%	19%	39%	559
Industrial activity						
Labour-intensive	15%	17%	20%	20%	28%	142
Resource-intensive	12%	27%	17%	13%	31%	150
Science-based	13%	21%	22%	14%	30%	146
Scale-intensive	19%	20%	22%	16%	23%	208
Specialised suppliers	16%	17%	13%	20%	35%	220
Information-intensive services	13%	12%	10%	10%	56%	187
Specialised suppliers services	13%	9%	9%	27%	42%	108

Notes: The first five columns show row percentages, the last column shows the number of observations (total number of observations is 1,161).

For information-intensive services and specialised supplier services this is the case only for about one fifth of the firms. The findings corroborate Laursen and Salter (2006) who found knowledge-intensive small firms less likely to be open because they are often based on the exploitation of a new idea and face a greater danger of leakage of their ideas. Therefore, they limit the nature and scope of their external interaction.

Table 8.4 presents the correlations between the different types and geographical locations of the R&D contractor. The results presented refer to the sample of 553 observations included in the detailed model (see Table

8.1). Before investigating the correlation between outsourcing decisions with certain types of contractors at certain geographical locations, a snapshot is given of the number of contractors involved in each of these decisions (start; increase; decrease; or stop) respectively by type and by geographical location of the contractor. An analysis at the level of the detailed model (see Table 8.1 – 553 observations) for combined location and type of contractor revealed no significant differences (at 5% level) in terms of the four possible firm decisions related to R&D outsourcing between the observations included in the analysis and those pertaining to the basic model (Table 8.1 – 1,161 observations).

Firms starting to outsource do so in 43% of cases to another firm, in 49% of cases to a university and in 35% to a research centre. For increased outsourcing this amounts to 63%, 69%, and 68% respectively. For decreased outsourcing this amounts to 71%, 73%, and 55% respectively. For stopped outsourcing it is 58%, 51%, and 34% respectively. These data reveal that started and sustained outsourcing is relatively more likely to involve universities. Also, increased outsourcing occurs relatively more frequently in case of outsourcing to a research centre, whereas stopping most often involves other firms. This can be explained by the fact that associations with contract research firms or other technical service providers are generally well-delineated contractual agreements used to fulfil very specific needs in a relatively short period of time. Outsourcing to universities in many cases is part of a firm's long-term strategy to manage its pipeline of innovation.

In terms of geographical distribution, starting to outsource clearly most often includes regional contractors (77%), followed by foreign (24%) and national (15%) ones. Increased (decreased) outsourcing involves 93% (94%) regional contractors, followed by 65% (56%) foreign and 41% (39%) national. The corresponding percentages for firms deciding to stop outsourcing are 82%, 37%, and 12%.

The data presented above show that management decisions in terms of starting, increasing, decreasing, or stopping the outsourcing of R&D occur at the same time with different types of contractors (confirming the findings by Howells et al., 2004) and at different locations. On average, firms starting to outsource R&D do so with 1.27 different types of contractors at 1.16 different locations. For increased and decreased outsourcing this amounts to 2.01 and 2.00 types respectively at 2.00 and 1.89 locations. In case a firm decides to stop R&D outsourcing, on average, 1.43 different types of contractors at 1.31 different locations are involved. The information presented in Table 8.3 reveals important correlations between certain types of contractors at certain locations.

Table 8.4 Firms' decisions on R&D outsourcing – correlation between type and geographical location of the contractor

		Regional – regional	Regional – national	Regional – abroad	National-regional	National – national	National – abroad	Abroad – regional	Abroad – national	Abroad – abroad
Start	Firm – firm	n.a.			red.	n.a.		red.	red.	n.a.
	Firm – university									++
	Firm –- centre									
	University – university	n.a.			red.	n.a.		red.	red.	n.a.
	University – centre		+			++	+			
	Centre – centre	n.a.		++	red.	n.a.		red.	red.	n.a.
Increase	Firm – firm	n.a.	+++	+++	red.	n.a.	++	red.	red.	n.a.
	Firm – university			+		++		++		
	Firm – centre									
	University – university	n.a.	+++	+++	red.	n.a.	+++	red.	red.	n.a.
	University – centre			+++		+++	+++	+	+++	+++
	Centre – centre	n.a.	++		red.	n.a.	+++	red.	red.	n.a.

Table 8.4 (continued)

		Regional – regional	Regional – national	Regional – abroad	National-regional	National – national	National – abroad	Abroad – regional	Abroad – national	Abroad – abroad
Decrease	Firm – firm	n.a.	++	++	red.	n.a.	++	red.	red.	n.a.
	Firm – university					+			+	+
	Firm –- centre			+++		+		+		
	University – university	n.a.	++	+++	red.	n.a.	+++	red.	red.	n.a.
	University – centre				+++	+++		+++	+++	+++
	Centre – centre	n.a.		+	red.	n.a.		red.	red.	n.a.
Stop	Firm – firm	n.a.			red.	n.a.	+++	red.	red.	n.a.
	Firm – university									
	Firm – centre									
	University – university	n.a.	++		red.	n.a.	++	red.	red.	n.a.
	University – centre			++	++	+++	+++			
	Centre – centre	n.a.		++	red.	n.a.	++	red.	red.	n.a.

Notes: Correlations for the detailed model (see Table 8.1) with 553 observations. Correlation is statistically significant at 10% (+/-) , 5% (++/--), 1% (+++/---). n.a.: not applicable because no identification possible of contractor(s) involved. Red.: redundant with information presented in another cell.

In Table 8.4 a distinction is made between firms' decisions to start; increase; decrease; or stop R&D outsourcing. The second column gives all combinations of possible types of contractors and the first row does the same for possible locations of the contractor. The first location refers to the location of the first type of contractor mentioned in column 2, whereas the second refers to the location of the second type of contractor mentioned in column 2. For example, the cell in the second row in the third column indicates the correlation between a firm's decision to start outsourcing R&D to a regional firm on the one hand, and to a regional university on the other.

For firms deciding to start outsourcing R&D, a positive correlation is found between regional (respectively national) universities and national research centres; between national universities and world research centres, and between regional and foreign research centres. Also, for contractors located abroad, a positive correlation is found between decisions to start outsourcing to firms on the one hand and to universities on the other hand. These correlations suggest that firms deciding to start outsourcing R&D carefully choose a mix of partners based on directly applicable research (including other firms' expertise or the expertise of research centres).

For sustained outsourcing, a larger number of more significant correlations are found. This is not surprising given that the average number of contractors in these activities is higher (see before). Compared to starting outsourcing R&D, firms' decisions to increase or decrease R&D outsourcing involve a combination of regional, national, and foreign-based firms. A similar remark can be made for universities. Together with the involvement of research centres at different locations, this contributes to a broad mixture of contractors. An important difference between the case in which a firm decides to increase R&D outsourcing compared to deciding to decrease R&D outsourcing seems to be the absence of correlation between outsourcing to firms and outsourcing to research centres (at different combinations of geographical levels) in the case of increased outsourcing. This could point to a more prominent involvement of research centres at the point when the decision to start outsourcing is taken.

Stopping outsourcing R&D with national firms is strongly correlated with stopping with foreign firms. The same is true for research centres and universities (for which this relation also holds between the regional and the national level). Also, a high correlation exists between stopping outsourcing with universities on the one hand and with research centres on the other.

8.4.2 Estimation

The model presented in Section 8.3.1 is estimated using pooled ordinary least squares (OLS). The reason for performing pooled OLS is two-fold. First, all variables are considered in first differences, except for the dummy variables (year dummies; industry dummies; and group affiliation). As a specification in first differences is considered, time invariant firm-specific effects (fixed effects) are differenced out. Second, a fixed effects estimation is troublesome given the small number of firms for which observations over more than one year are available.

Estimation results are reported in Table 8.5. As the value of the dependent variable ranges from –1 to +1 it was transformed to apply a logit specification. Column (i) shows the results of the basic model of firms' decisions on R&D outsourcing and column (ii) reports the results of the detailed specification, in which the decision with respect to outsourcing R&D is broken down according to type of contractor and location (that is, both interacted). The two models presented are the final result of a general-to-specific approach, in which the most general unrestricted model is gradually simplified by dropping non-significant variables, as long as specification tests are satisfied (see, for example, Charemza and Deadman, 1992). One of the advantages of a general-to-specific approach is that the number of observations can increase when non-significant variables for which missing values occur can be dropped. Moreover, the quality of the specification often improves considerably by dropping irrelevant variables. The non-significant variables that remain in the specific models could not be dropped due to the significance of Ramsey's RESET test (a test of mis-specification). All models include unreported year dummies that are not statistically significant.

Table 8.5 R&D employment intensity and outsourcing

Dependent variable = Change in internal R&D employment intensity	Basic model (i)	Full model with interaction° (ii)
Pavitt: Labour-intensive	–0.06 (0.01)***	–0.10 (0.00)***
Pavitt: Science-based	0.11 (0.01)***	0.11 (0.15)
Pavitt: Information-intensive services	0.05 (0.21)	
Pavitt: Resource-intensive		–0.05 (0.08)*
Exports (share of turnover outside the nation)		–0.07 (0.70)
Share of research budget in total R&D		0.04 (0.40)
R&D intensity beginning of period	–0.61 (0.00)***	–0.60 (0.00)***
Share of outsourced R&D in total R&D budget	0.06 (0.12)	
Change in outsourced R&D – start	–0.17 (0.02)**	
Other firm within the region		–0.38 (0.10)*
Research centre within the nation		–0.77 (0.06)*
Change in outsourced R&D – increase	–0.31 (0.00)***	
Other firm abroad		1.32 (0.04)**
University abroad		1.92 (0.10)*
Research centre within the region		–0.80 (0.07)*
Change in outsourced R&D – decrease		
University within the nation		2.26 (0.02)**

Table 8.5 (continued)

Dependent variable = Change in internal R&D employment intensity	Basic model (i)	Full model with interaction° (ii)
Change in outsourced R&D – stop		
Other firm within the nation		–29.32 (0.02)**
University within the region		0.35 (0.05)**
Adjusted R^2	0.18	0.19
Obs.	1,161	553

Notes: Standard errors are heteroscedasticity-consistent; p-value in brackets; symbols *, **, *** denote significance at 5%, 1% and 0.1%. Symbol ° refers to the interaction between type of contractor and geographical location.

In the basic model (column (i)), the dummies denoting the industry to which a firm belongs indicate that resource-intensive firms witnessed a significantly lower increase in internal R&D employment intensity relative to the reference group (specialised suppliers) whereas science-based firms had significantly higher growth in internal R&D employment intensity, again compared to the reference group. Also, internal R&D employment intensity at the beginning of the period under consideration is statistically significant. The negative sign indicates that firms with an initially high internal R&D employment intensity are – all else being equal – more likely to witness a larger decrease (smaller increase) in R&D employment intensity in the following two-year period than their counterparts. The other control variables do not turn out to be statistically significant.

The decision of firms to start or to increase R&D outsourcing turns out to have a statistically-significant negative impact on internal R&D employment intensity. No significant impact on internal R&D employment is found for firms deciding to decrease or to stop R&D outsourcing.

Refining the outsourcing variables, using a breakdown of both the type of partner and its location (column (ii)) does not change the conclusions with respect to control variables, except for the statistical significance (10%) of the negative sign of resource-intensive firms. However, the estimation clearly shows that overall decisions to start, increase, decrease or stop the outsourcing of R&D conceal important differences between type and geographical location of the R&D contractor as to the impact on internal R&D employment intensity. This could be related to the different motives underlying the corporate decision to outsource R&D according to the type and geographical location of the R&D contractor and is in line with the correlations found in Table 8.4.

The decision to start outsourcing R&D to firms within the proper region and to research centres in another region has a significant negative impact – though only at the 10% level – on internal R&D employment intensity.

The coefficients of increased outsourcing towards other firms and universities outside the country suggest a positive relationship with firms' own R&D intensity. On the other hand, increased R&D outsourcing towards research centres in the proper region is negatively related to the internal R&D employment intensity.

In contrast with the lack of statistical evidence of an impact in the basic model of firms' decisions to decrease or stop R&D outsourcing, taking into account the type and geographical location of the contractor reveals that decreasing R&D outsourcing towards universities in another Belgian region and stopping outsourcing to a university in the proper region have a statistically-significant negative impact on the internal R&D employment intensity of firms. The opposite is true in the case of firms' decisions to cease to outsource R&D to other firms in another Belgian region. This decision has a significant and highly positive impact on internal R&D employment intensity suggesting that when the outsourcing of R&D to these contractors stops there is an increase in internal R&D employment intensity.

8.5 CONCLUSIONS

This chapter explored the effects of R&D outsourcing on R&D labour composition. Its aim was to assess whether firms' R&D outsourcing strategies changed in internal R&D employment intensity. A distinction was made between firms' decisions to start; increase; decrease; or stop R&D outsourcing. Differences in the type and location of the R&D contractor have also been considered.

In terms of R&D outsourcing activities, the analysis revealed that larger firms and firms belonging to a group have a higher propensity to be engaged in R&D outsourcing in a more permanent way. Smaller firms have a relatively higher propensity to start or to stop R&D outsourcing. Also, management decisions in terms of starting, sustaining or stopping R&D outsourcing occur at the same time with different types of contractors at different locations. Firms' decisions to start or to stop R&D outsourcing more frequently involve regional contractors, whereas sustained outsourcing involves national and international contractors relatively more frequently. Finally, the outcomes of the analysis pointed to a careful choice of a mix of different types of contractors at different geographical locations. In this respect, decisions to start or to stop R&D outsourcing frequently involve both universities and research centres. Sustained outsourcing is more often a mixture of both public and private contractors.

In terms of the effects of decisions related to the outsourcing of R&D, started or increased outsourcing – at least in the short run – has a significant negative effect on the internal R&D employment intensity, pointing to a change in labour composition. No significant impact is found in the case where a firm decides to decrease or stop R&D outsourcing. This indicates that decreases in internal R&D employment intensity related to started and increased R&D outsourcing are not turned around at the time when outsourcing stops or decreases. However, these decisions conceal marked differences between the types and geographical locations of the R&D contractor. Increased outsourcing towards other firms and universities outside the country affects internal R&D employment intensity in a significant and positive way. The opposite is true in the case of outsourcing towards research centres in the proper region.

These findings are not without policy relevance. The finding that starting or increasing R&D outsourcing negatively affects internal R&D employment intensity could stir fears of a hollowing-out of the local knowledge base. However, this fear seems unjustified because these negative findings seem to relate mainly to the outsourcing of – more applied – R&D to local research centres and it turns out that increased R&D outsourcing actually does have positive effects for other firms and universities outside the country, suggesting the existence of positive international spillovers.

Given the number of (data) limitations affecting the present analysis, the results presented in this chapter should be viewed with some caution. First, the analysis is based on a panel of (quasi-) permanent R&D-active firms in Belgium for the period 1999–2005. It cannot be ruled out that the analytical outcomes are somewhat biased towards large firms, research-intensive firms active in the specialised supplier manufacturing industry, and firms belonging to an international group.

Second, no data are available on the determinants of outsourcing and concerns of appropriability which are very likely to interact with the decision to start, increase, decrease, or stop outsourcing and its impact on internal R&D employment.

Finally, due to a lack of information, the type of R&D activity outsourced (for example, basic or applied research, product or process development) or the continuance of outsourcing to the same entity could not be taken into account in the analysis.

PART FOUR

The role of human resources in open innovation practices

9. The impact of occupation and education of R&D personnel on knowledge exchange

9.1 INTRODUCTION

Due to the complexity and knowledge intensity of developing new products and processes, the R&D-active company has to look elsewhere to complement its internal R&D efforts (Cassiman and Veugelers, 2006). To this end a company is able to engage in collaborative agreements that can have a formal or informal character (Bönte and Keilbach, 2005). At least two issues can be addressed by such an agreement. First, a collaboration agreement can spell out the exact arrangement between the partners involved as regards the costs and organisational engagements. Second, the agreement might include appropriability issues if the aim of collaboration is to devise a new product or set up a new process (Gulati and Singh, 1998; Helper et al., 2000). Hence the IP management of companies becomes a vital issue. This is especially the case in the early stages of formal R&D collaborations (Bader, 2006).

Knowledge stocks and flows are recognised as a crucial element of value creation in firms (Kang et al., 2007). This chapter adds to the knowledge-based view of firms, which sees them as repositories of competences and knowledge (Kogut and Zander, 1996; Grant, 1996) by devoting more attention to the embodiment of these competences with respect to the ability to cooperate and to share knowledge. Knowledge-generation often results from new combinations of existing knowledge (Cohen and Levinthal, 1990). This is not restricted to the boundaries of the firm as R&D increasingly necessitates the use of external sources (Chesbrough, 2003a; Coombs et al., 2003; Howells et al., 2003).

This chapter contributes to the literature on absorptive capacity. Absorptive capacity was coined by Cohen and Levinthal (1990) to indicate the need for firms to identify, assimilate and commercialise external knowledge. They point to the fact that absorptive capacity is path-dependent and cumulative, and is built through investments in its labour force. Usually, R&D investments are seen as an indicator for firms to invest in absorptive capacity. But the receptivity of the labour force to the R&D investments remains understudied (see, however, Roper et al. 2008).

This chapter intends to bridge this gap by looking more closely at the internal knowledge-base of firms. The focus is on the relation between the function and qualification of R&D personnel and external knowledge relations in terms of the development and exchange of knowledge in formal research cooperation. When investigating these relationships, attention is paid to the use of intellectual property and product innovation.

9.2 LITERATURE REVIEW AND HYPOTHESIS DEVELOPMENT

The necessity of firms to rely on external knowledge has brought the notion of absorptive capacity centre-stage. That is why the debate has attracted considerable attention from scholars (Lane and Lubatkin, 1998; Lane et al, 2006; Nieto and Quevedo, 2005; Todorova and Durisin, 2007; Zahra and George, 2002).

Human resources are a crucial element in the R&D activities of firms (Jain and Murray, 1984; Allen and Katz, 1992). This chapter deals with the role of two key dimensions in gaining from knowledge spillovers related to human resources – the function or skills and the formal qualification or educational level of the labour force – when they engage in formal collaborative agreements on behalf of the firm. Skills thus become a qualitative aspect of absorptive capacity because a firm's employees embody tacit knowledge which is considered to be a main source of spillovers (Kaiser, 2002).

Various authors have pointed to the fact that competitive pressures and increased complexity force firms to look for external knowledge and collaborative actions to innovate (Powell et al., 1996; Chesbrough, 2003a; Coombs et al., 2003; Howells et al., 2003). Collaborative agreements are a means of gaining access to external knowledge (Colombo and Gerrone, 1996; Lazaric and Marengo, 2000).

9.2.1 Occupation of R&D Personnel and Knowledge Exchange

The literature on open innovation conceptualises firms as organisations in which, continuously and simultaneously, 'make' or 'buy' decisions are being taken by individuals (Chesbrough, 2003a). In this way, the employees of the firm are making the boundaries permeable in order to attract the necessary knowledge and technology; while at the same time preventing unintended spillovers by scrutinising alternatives to commercialisation and other appropriability mechanisms.

Jacobides and Billinger (2006) and Rothaermel et al. (2006) have systematically analysed boundary construction. They introduced the concept of organisational absorptive capacity by acknowledging the roles of internal functions and competences which serve as antennas aimed at detecting external knowledge. They posit that these internal functions are a prerequisite to successfully interact and outsource.

Lazonick (2005) builds on the resource-based view of the firm to demonstrate how innovation depends on the development and accumulation of specialised internal capabilities. To stimulate the development of internal capabilities the firm needs organisational integration: a set of relations that creates incentives for employees who participate in hierarchical and functional divisions of labour to apply their skills and efforts to the innovation process (Helper et al., 2000). To absorb knowledge from the external environment, firms need organisational integration in which employees function as interfaces with the environment. These employees have to possess the skills to screen, interpret and assimilate the knowledge and transfer it through internal communication and diffusion on the shop floor. Lam (2000) finds that the narrower the interfaces with the external environment, the less knowledge and the fewer ideas are absorbed. The less internal employees learn about external ideas, the smaller the chance that they will succeed in their innovative efforts.

Hence, absorptive capacity is more than stimulating R&D investment in the vein of Cohen and Levinthal (1990). Absorptive capacity has to be enlarged by acknowledging its interface role towards the external environment while acting as internal evaluator as to the value of the external knowledge. From these considerations the first hypothesis reads that the higher the share of researchers and R&D managers, the higher the probability of cooperating to create new knowledge or exchange knowledge developed in the firm.

9.2.2 Education of R&D Personnel and Knowledge Exchange

Next to skills that can be developed in the workplace, skills are also obtained through education which results in formal qualifications in terms of degrees or diplomas (OECD, 1995). The relation between R&D and education has been established in the past (Scherer and Huh, 1992) and reiterated in recent research (Roach and Sauermann, 2010). Highly-qualified employees are associated with higher R&D investment levels. Education and training are, moreover, crucial to innovation (Lundvall and Nielsen, 1999; Lam, 2005).

Similar to skill levels, educational levels facilitate the detection and management of relevant external knowledge flows (OECD, 2008a), which is a key ingredient in absorptive capacity (Roach and Sauermann, 2010).

In the case of doctorate degrees in particular, the body of research is considerable (Stuart and Ding, 2006; Bercowitz and Feldman, 2008). This category of R&D personnel, at least of the most recent graduates, possesses knowledge that borders on the frontier of science and technology. They are especially receptive to so-called upstream research such as universities and public research centres that provide basic research and, also, their work is related to the patent activity of the firm (Sauermann and Cohen, 2008).

Roach and Sauermann (2010) raised the question as to whether corporate scientists (R&D personnel with a PhD) have a lower inclination to engage in collaborative relations than academic scientists. They found that corporate scientists are more inclined to be engaged in research involving downstream R&D than academic scientists. They also value the salaries and resources of research more than academic scientists.

The level of education is also thought relevant in the concept of absorptive capacity; where a positive relation between the two is assumed (Azagra-Caro et al., 2006; Roper et al., 2008). Based on the previous discussion, the second hypothesis can be formulated. Hence the second hypothesis posits that the higher the share of highly qualified employees, the higher the probability of cooperating to create new knowledge or exchange knowledge developed in the firm.

9.2.3 Using Intellectual Property in Knowledge Exchange

Intellectual property (IP) encompasses all protection mechanisms that enhance the willingness of firms to undertake risky R&D activities through cooperation (Gulati and Singh, 1998). By protecting the innovation, firms hope to capture the benefits attached to it (Chesbrough, 2003a). This might be through preventing the innovation from being imitated and functioning as a unique selling point for a period of time; or through licensing the innovation to other firms and creating additional revenue through this channel (Arora et al., 2001).

The returns from R&D activities can be negatively influenced by intellectual property because it reduces competitive elements, augments market prices and prevents a smooth technology transfer of the most recent technologies. Intellectual property is also highly industry-dependent (Granstrand, 2003).

The use of IP mechanisms, such as cross licensing, also serves to impact on: industrial standards; gaining access to technical portfolios of other firms; increasing innovation speed and freedom of operation (Lichtenthaler,

2005). From these elements a third set of hypotheses can be derived. First, the expectation of registering or obtaining patents stimulates the probability of developing knowledge in formal cooperation agreements; and second, the expectation of gaining revenues from inventions stimulates the exchange of knowledge to external parties.

All variables in the preceding paragraphs, together with the control variable, are described in Table A9.1 in the Appendix.

9.3 DATA DESCRIPTION AND VARIABLES

9.3.1 Data and Sample

The analysis makes use of the biannual OECD R&D business survey organised in Belgium in 2008. This survey starts from a population of 5,270 firms of which 2,083 stated that in 2006 and 2007 they performed R&D activities occasionally (302) or in a permanent (1,781) fashion. Not all of these firms engage in external knowledge relations. Deleting the missing values reduces the total observations for which we have information to a sample of 645. Table 9.1 gives an overview of the information on all variables for the entire sample and the reduced sample we are about to use for the analysis.

Table 9.1 Descriptive statistics: mean and standard deviation in the sample of R&D-active firms and the sample without missing values (N=645)

Variable	Obs.	Mean	St. dev.	Obs.	Mean	St. dev.	Difference in mean, unequal variances (p-value)
Dependent variables							
COOP	737	0.385	0.487	645	0.391	0.488	0.84
EXCH	736	0.363	0.481	645	0.366	0.482	0.90
Independent variables							
Occupation of R&D personnel							
FRESMAN	712	0.555	0.354	645	0.566	0.348	0.57
FTECHNIC	711	0.356	0.337	645	0.350	0.332	0.73
FSUPPORT	711	0.088	0.179	645	0.084	0.172	0.68

Table 9.1 (continued)

Variable	Obs.	Mean	St. dev.	Obs.	Mean	St. dev.	Difference in mean, unequal variances (p-value)
Education of R&D personnel							
QPHD	678	0.113	0.213	645	0.116	0.216	0.79
QMASTER	678	0.507	0.347	645	0.504	0.346	0.87
QBACHEL	678	0.225	0.272	645	0.224	0.273	0.98
QOTHER	678	0.155	0.257	645	0.155	0.258	0.99
Output variables							
INNOPROD	765	0.827	0.378	645	0.839	0.368	0.57
IPACQ	777	0.147	0.354	645	0.155	0.362	0.66
IPREG	775	0.268	0.443	645	0.273	0.446	0.85
IPEARN	774	0.124	0.330	645	0.136	0.344	0.49
Control variables							
LEMP	2,002	3.749	1.727	645	3.761	1.811	0.89
GP	1,186	0.557	0.497	645	0.550	0.498	0.76
IDCP	2,064	0.224	0.417	645	0.222	0.416	0.91
IDCS	2,064	0.017	0.129	645	0.019	0.135	0.79
IDFP	2,064	0.068	0.252	645	0.078	0.268	0.44
IDPE	2,064	0.223	0.417	645	0.217	0.413	0.74
IDSB	2,064	0.092	0.289	645	0.112	0.315	0.16
IDKI	2,064	0.305	0.461	645	0.310	0.463	0.82
IDOS	2,064	0.070	0.256	645	0.043	0.204	0.006***
RDOUT	2,050	0.584	0.376	645	0.482	0.500	0.000***
RDINT	2,002	0.202	0.493	645	0.231	0.301	0.045**
RSHR	743	0.551	0.362	645	0.555	0.356	0.83

Note: The symbols **, *** denote significance at 5%, and 0.1%.

Table 9.1 indicates that only in the case of three control variables do the two samples differ markedly. In the case of the industry dummy on other services, this category of firms is under-represented in the reduced sample since the share in the full sample is 7% whereas this is 4.3% in the case of the reduced sample.

9.3.2 Variables

Three groups of variables are discussed: dependent, independent and control variables. The dependent variables cover the external knowledge relations in which firms engage: formal collaborative agreements (COOP) and the exchange of knowledge that was generated in the framework of such an agreement (EXCH).

The independent variables can be grouped into three aspects that have an impact on the dependent variables. First, the OECD (2002a) acknowledges different functions that employees classified as R&D personnel can have: researchers and R&D managers (FRESMAN); technicians (FTECHNIC) and other supporting staff (FSUPPORT). Researchers and R&D managers are involved in the conception and generation of new knowledge, products, processes, methods and systems. Also the management of the R&D projects falls under this heading. Technicians and equivalent staff perform tasks that demand technical knowledge and experience. These employees are involved in the application of the new knowledge and are usually under the supervision of R&D managers. Other supporting staff captures the secretarial, clerical employees and craftsmen that participate in R&D projects or are closely related to these projects.

Second, the OECD (2002a) classifies R&D personnel using formal qualifications. Four academic qualifications are used: doctorate holders (QPHD); holders of basic university degrees at master's (QMASTER) and bachelor's (QBACHEL) level; and other qualifications below university level (QOTHER). Abstraction is made of the field of science in which these degrees have been obtained.

Both the functions and qualifications of R&D personnel have been measured in terms of full-time equivalents.

The third group of independent variables are related to the output of the R&D activities. Under normal circumstances the most obvious output are innovative products that the firms aim to bring to market (INNOPROD). Other types of output are related more to enhancing the knowledge used in the company. In this case firms can acquire patents, licences or other intellectual property rights (IPACQ). Firms can also register patents (IPREG), or sell patents, licences and other intellectual property rights to third parties and earn money from them (IPEARN). The latter activity is found to be popular in the context of open innovation (Chesbrough, 2003a) and requires properly functioning technology markets (Arora et al., 2001).

Six control variables are used in the analysis because they, too, have potential impacts on the external knowledge relations that firms maintain. First, the size of the firm is often used. The size is measured by the natural log of the total employment of the firm. The effect is quite unclear, since on the one hand smaller firms are assumed to lack the human resources to engage in external knowledge relations, yet, on the other, the need for additional information is making itself increasingly felt in the case of small firms.

Second, the firm that belongs to a group might rely on an internal division of labour that is aimed at source-relevant knowledge in host countries and apply this knowledge in their home countries. Also, subsidiaries may be more inclined to look for external information that helps them enter or integrate in local markets.

Third, a set of industry dummies accounts for the fact that R&D activities are known to differ according to the industry in which firms operate. Likewise, aspects such as patenting are also known to differ across industries (Griliches, 1990). This chapter relies on the industry classification by Marsili and Verspagen (2002). Drawing on Pavitt's (1984) taxonomy, they use empirical data (such as patents, R&D statistics, scientific inputs, innovation surveys) to develop a typology of five regimes. These regimes are based on a set of criteria: the nature of the knowledge base; technological opportunity conditions; and technological entry barriers. Marsili and Verspagen (2002), like Pavitt (1984) consider manufacturing only, which is inadequate in the Belgian context for it is highly specialised in terms of services. Therefore two additional regimes are added, the knowledge-intensive services and other services, in order to cover the services industry.

R&D intensity, an indicator that reflects the technological needs of the firm, is an important element in the context of open innovation. Cassiman and Veugelers (2006) have already pointed to the complementarity between internal R&D efforts and the use that can be made of external knowledge relations. Cohen and Levinthal (1990) further highlighted the use of R&D intensity in the development of absorptive capacity by internal employees in screening and using external knowledge flows. Following Caloghirou et al. (2004), R&D intensity is measured in terms of R&D personnel.

The fifth control variable, on R&D outsourcing, captures whether firms engage in subcontracting R&D to third parties. This activity also indicates whether a firm is open to external knowledge flows, but buying R&D results obviously requires less intensive involvement of the internal R&D personnel than the other external knowledge relations that have been implied by a formal collaborative agreement.

Finally, a sixth control variable relates to the ratio of fundamental and applied research to development. This distinction is less relevant in recent years. It might be the case, however, that external knowledge relations are aimed more at the development of a specific product or targeted at a well-defined result. All of these variables are concisely defined in Table A9.1 in the Appendix.

9.4 ANALYSIS

The undertakings of collaboration and the exchange of knowledge devised in the framework of the technological collaboration agreement are not independent of each other; the actions are, indeed, correlated. The binary dependent variables, COOP and EXCH, appear to have four mutually exclusive cases: (COOP, EXCH) = (0,0), (1,0), (0,1), and (1,1) respectively.

Because the data are cross-sectional we use a non-recursive bivariate probit model and both equations have the same specification (Greene, 2003).

Table 9.2 Bivariate probit regression

Dependent variable	COOP	EXCH
Intercept	−1.325	−1.418
	(0.463)***	(0.480)***
Independent variables		
Occupation of R&D personnel		
FRESMAN	−0.159	−0.727
	(0.354)	(0.375)*
FTECHNIC	−0.394	−0.908
	(0.359)	(0.379)**
Education of R&D personnel		
QPHD	0.672	0.882
	(0.318)**	(0.350)**
QMASTER	0.281	0.586
	(0.240)	(0.270)**
QBACHEL	−0.151	0.681
	(0.278)	(0.291)**
Output variables		
INNOPROD	0.331	0.174
	(0.157)**	(0.151)
IPACQ	0.290	0.127
	(0.148)**	(0.151)
IPREG	0.274	0.206
	(0.128)**	(0.134)
IPEARN	0.050	0.379
	(0.173)	(0.164)**

Table 9.2 (continued)

Dependent variable	COOP	EXCH
Control variables		
LEMP	0.182	0.129
	(0.044)***	(0.044)***
GP	–0.129	0.136
	(0.132)	(0.131)
IDCP	–0.573	-0.334
	(0.279)**	(0.276)
IDCS	0.276	0.186
	(0.479)	(0.471)
IDFP	–0.131	0.202
	(0.307)	(0.300)
IDPE	–0.339	-0.159
	(0.275)	(0.273)
IDSB	–0.212	0.305
	(0.299	(0.294)
IDKI	–0.366	0.032
	(0.284)	(0.275)
RDOUT	0.459	0.309
	(0.110)***	(0.115)***
RDINT	0.885	0.937
	(0.249)***	(0.243)***
RSHR	–0.098	0.100
	(0.153)	(0.152)
Number of observations	645	
Rho	0.870	
	(0.026)***	
Wald test of full model	186.08***	

Notes: The robust standard errors are denoted between brackets. The support R&D staff (FSUPPORT), the other qualifications (QOTHER) and the industry dummy other services (IDOS) are used as reference categories to avoid singularity. The symbols *, ** and *** denote significance at the 10%, 5% and 1% levels.

Technological collaboration agreements that favour collaboration are influenced by a relatively high share of doctorate holders in the company (as opposed to other qualifications). This indicates that the knowledge transferred from the outside demands the highest level of education. The fact that firms also acquire patents and licenses and, at the same time, register patents implies that the internal personnel has to be independent and autonomous. Collaborative activities are motivated by willingness to generate product innovations (INNOPROD).

Exchanging knowledge that was developed internally requires distinct functions of R&D personnel. In this case, researchers and R&D managers (FRESMAN) and technical personnel (FTECHNIC) are less involved in this type of knowledge exchange. R&D personnel with supporting roles are relatively more involved. This does not imply, as can be seen from the qualifications of the R&D personnel, that personnel with lower educational levels are implied. As expected, the exchange of knowledge to other firms and organisations is positively influenced by the prospects of earning money from the internally developed knowledge (IPEARN).

Based on these results, the analysis is at odds with the first hypothesis. In the case of cooperation there is no effect; whereas in the case of knowledge exchange the first hypothesis ought to be rejected. Hypothesis two, on the other hand, can be accepted: higher shares of qualified employees give rise to higher probabilities of engaging in cooperation and knowledge exchange. The first part of hypothesis three, on acquiring patents for use or registration of new patents, is also acceptable in the case of cooperation with other organisations and is found to be positively motivated by the prospect of creating product innovations. The second part of hypothesis three, on gaining revenues from selling internally developed knowledge, is also corroborated.

The control variables demonstrate that large firms that outsource their R&D and have a high R&D intensity in terms of personnel have a higher probability of entering into collaborative agreements aimed at cooperation and aimed at exchange of knowledge.

The bivariate probit regression predicts the outcomes of the dependent variables as follows. The probability of engaging in both cooperation and exchange is 28.5%; exclusively in cooperation is 9.0%; exclusively in exchange is 6.5%; and neither in cooperation nor exchange is 56.0%.

Making use of marginal effects in Table 9.3, the arguments developed above can be interpreted with greater precision. Looking at the function or occupational skills of R&D personnel, the impact appears where exchange of internally developed knowledge is concerned. It seems that these are tasks requiring attention from the R&D support staff since no new knowledge generation is involved. As we have already observed, this does not mean that a lower level of education is needed, though firms that are exclusively involved in knowledge exchange benefit from a relatively higher share of R&D personnel with a bachelor's degree. When both cooperation and knowledge exchange are involved, the mix of education level plays a role, since a higher share of R&D personnel with a doctorate degree has a positive impact.

Table 9.3 Marginal effects after the bivariate probit model

Variable	COOP = 1 and EXCH = 1		COOP = 1 and EXCH = 0		COOP = 0 and EXCH = 1		COOP = 0 and EXCH = 0		Mean value
	dy/dx	z-value	dy/dx	z-value	dy/dx	z-value	dy/dx	z-value	
FRESMAN	−0.163	−1.38	0.103	1.79*	−0.106	−2.23**	0.167	1.22	0.566
FTECHNIC	−0.234	−1.97**	0.085	1.41	−0.102	−2.05**	0.251	1.84*	0.350
QPHD	0.273	2.62***	−0.018	−0.29	0.054	1.02	−0.308	−2.58***	0.116
QMASTER	0.155	1.93*	−0.049	−1.07	0.062	1.59	−0.168	−1.83*	0.504
QBACHEL	0.106	1.18	−0.163	−3.06***	0.147	3.31***	−0.089	−0.87	0.224
INNOPROD	0.082	1.91*	0.038	1.72*	−0.020	−0.82	−0.100	−1.87*	0.839
IPACQ	0.070	1.39	0.043	1.31	−0.022	−1.15	−0.091	−1.66*	0.155
IPREG	0.084	1.93*	0.021	0.87	−0.007	−0.36	−0.099	−2.07*	0.273
IPEARN	0.072	1.24	−0.053	−2.65***	0.073	2.21**	−0.092	−1.48	0.136
LEMP	0.053	3.82***	0.016	2.09**	−0.005	−0.84	−0.064	−3.95***	3.761
GP	0.006	0.16	−0.056	−2.32**	0.044	2.48**	0.005	0.11	0.550
IDCP	−0.144	−1.98**	−0.058	−1.62	0.026	0.59	0.176	1.89*	0.222
IDCS	0.082	0.49	0.026	0.28	−0.011	−0.19	−0.097	−0.54	0.019
IDFP	0.006	0.06	−0.055	−1.59	0.071	1.07	−0.022	−0.20	0.078
IDPE	−0.082	−1.04	−0.042	−1.07	0.024	0.56	0.099	1.03	0.217
IDSB	−0.001	−0.01	−0.077	−3.02***	0.118	1.74*	−0.040	−0.36	0.112
IDKI	−0.057	−0.68	−0.078	−2.07**	0.069	1.47	0.066	0.65	0.310
RDOUT	0.130	3.68***	0.043	2.19**	−0.016	−1.03	−0.157	−3.88***	0.482
RDINT	0.317	4.13***	0.019	0.41	0.030	0.81	−0.366	−4.10***	0.231
RSHR	0.004	0.08	−0.041	−1.54	0.033	1.54	0.004	0.07	0.555

Notes: The symbols *, ** and *** denote significance at the 10%, 5% and 1% levels.

Firms showing cooperation and knowledge exchange are lightly motivated by the prospect of registering patents. If the firm exchanges knowledge only, the motive is clearly the creation of additional revenue (this is not the case if the firm engages in cooperation). The control variables corroborate our earlier findings. Large firms that are already outsourcing R&D and are R&D intensive in terms of R&D personnel have a higher probability of engaging in cooperation and knowledge exchange at the same time.

9.5 CONCLUSIONS AND LIMITATIONS

This chapter set out to look more closely at a few key characteristics of R&D personnel in capturing external knowledge in the framework of technological collaborative agreements. Two such characteristics came under consideration: the function and the qualification of R&D personnel. This means that people were the central issue in this study.

In policy circles, R&D activities are closely related to monetary flows (for example, the European 3% target). This is also the case in business environments when concepts such as absorptive capacity are formulated exclusively in terms of R&D investments. Behind these financial streams there are real, flesh-and-blood people; people with their own motivations and aspirations. And their characteristics are vital if firms want to source external knowledge or if they want to commercialise the knowledge that they have developed internally.

Based on the previous analysis, the main conclusion is that the level of educational attainment is quite important if firms are to engage in research cooperation and exchange of knowledge. This finding does not imply that on-the-job training is not useful, but that the skills necessary for external knowledge relations are more to be found in people who already have a high degree as a threshold. Moreover, as pointed out by Lundvall and Nielsen (1999), it enabled firms to introduce more product innovations.

In the case of firms pursuing knowledge exchange of internally developed knowledge, the support function of R&D personnel becomes crucial, and this might be performed by R&D personnel with a bachelor's degree.

The research is not without limitations, which point to interesting avenues for further research. First, the research does not account for the type of R&D cooperation or the type of partners with which knowledge is exchanged. From the literature it became apparent that this aspect was important (Roach and Sauermann, 2010), whether corporate R&D personnel are more inclined to enter into downstream relations than firms are to motivate upstream relationships with universities. As R&D becomes more complex, this relationship might not be neglected.

Second, the research focused on formal collaborative agreements. This might be important considering the emphasis on dealing with appropriability issues; but these issues are even trickier in the case of informal collaborative actions and relations. In this case, social constructs such as trust become a key determinant.

A third limitation is that the research is focused on having at least one collaborative agreement; whereas the number of agreements is not without relevance. Gulati and Singh (1998) find that maintaining larger numbers of collaborative agreements might have negative effects due to coordination problems. At the same time, Lazaric and Marengo (2000) state that larger numbers of such agreements give rise to scale effects due to the cumulativeness of experiences over time. Since the sample in our research was cross-sectional, a longitudinal analysis might shed more light on this aspect.

Despite these limitations, the research provided insights into the determinants of engaging in technological collaborative agreements by demonstrating that people matter. Their characteristics, in terms of function and qualification, have an impact on the probability of being able to engage in external knowledge relations. If firms are to rely on external relations, the human resources departments of firms have to focus on hiring R&D personnel with the appropriate profiles.

APPENDIX

Table A9.1 Description of the variables

Variable	Description
Dependent variables	
Cooperation (COOP)	COOP = 1 if the firm develops technical knowledge in the framework of a collaborative agreement; 0 otherwise
Exchange (EXCH)	EXCH = 1 if the firm exchanges technical knowledge developed from R&D in the firm in the framework of a collaborative agreement; 0 otherwise
Independent variables	
Occupation of R&D personnel	
Researchers and R&D managers (FRESMAN)	The share of researchers and R&D managers in all R&D personnel
Technicians (FTECHNIC)	The share of technicians in all R&D personnel
Support functions (FSUPPORT)	The share of supporting R&D staff in all R&D personnel
Education of R&D personnel	
PhD degrees (QPHD)	The share of PhD degrees in all R&D personnel
Master's degrees (QMASTER)	The share of master's degrees in all R&D personnel
Bachelor's degrees (QBACHEL)	The share of bachelor's degrees in all R&D personnel
Other qualifications (QOTHER)	The share of other qualifications in all R&D personnel
Output variables	
Product innovation (INNOPROD)	INNOPROD = 1 if the firm develops a new product; 0 otherwise
IP acquisition (IPACQ)	IPACQ = 1 if the firm acquired patents, licenses or other intellectual property
IP registration (IPREG)	IPREG = 1 if the firm registered patents, licenses or other intellectual property
IP earnings (IPEARN)	IPEARN = 1 if the firm sold patents, licenses or other intellectual property to third parties

Table A9.1 (continued)

Variable	Description
Control variables	
Log of employment (LEMP)	Natural log of employment in 2007
Group membership (GP)	GP = 1 if the firm is member of a group; 0 otherwise
Industry dummy continuous process (IDCP)	CP = 1 if the firm is classified as active in 'continuous process' (a variety of production activities such as metallurgical process industries, metals and building materials, and chemical process industries, textiles and paper, food and tobacco); and 0 otherwise
Industry dummy complex systems (IDCS)	CS = 1 if the firm is classified as active in 'complex systems' (that is, combination of mechanical, electrical and transportation technologies. It includes motor vehicles and other transport equipment); and 0 otherwise
Industry dummy fundamental process (IDFP)	FP = 1 if the firm is classified as active in 'fundamental process' (that is, associated with chemistry-based technologies and is mainly oriented towards process innovation); and 0 otherwise
Industry dummy product engineering (IDPE)	PE = 1 if the firm is classified as active in 'product engineering' (that is, uses mechanical engineering technologies, within areas such as machinery, equipment and instruments); and 0 otherwise
Industry dummy science-based (IDSB)	SB = 1 if the firm is classified as 'science-based' (that is, pharmaceuticals, computers and other information-processing equipment, electronics, and telecommunication equipment); and 0 otherwise
Industry dummy knowledge-intensive services (IDKI)	KI = 1 if the firm is classified as active in 'knowledge-intensive services' (that is, which includes telecommunications, computer services and technical business services); and 0 otherwise
Industry dummy other services (IDOS)	OS = 1 if the firm is classified as 'other services' (that is, activities in other services than KI); and 0 otherwise
R&D outsourcing (RDOUT)	RDOUT = 1 if the firm outsourced R&D activities; 0 otherwise
R&D intensity (RDINT)	The share of all R&D personnel in total employment
Research share (RSHR)	The share of research (fundamental and applied) in total R&D expenditures

10. Research experts' role in knowledge development and exchange in formal research cooperation

10.1 INTRODUCTION

In addition to labour and capital, knowledge has become an important production factor. Knowledge can be codified or tacit. If codified, knowledge can be traded on a specialised market, provided that it is standardised and fairly priced. Tacit knowledge remains embodied in the minds of people, and is part and parcel of their skills. These skills are either the result of their formal educational training or have been acquired en route during their job.

The literature on 'distributed' or 'open' innovation (Chesbrough, 2003a; Coombs et al., 2003) emphasises the need for enterprises to co-depend on external knowledge in order to perform research and innovate. One way to do so is by means of research cooperation. Firms are, therefore, engaging in formal research collaboration agreements. Such a research cooperation may improve learning efficiency in absorbing external knowledge, thereby increasing the impact on the innovative performance of incoming spillovers. It may, at the same time, provide a mechanism to reduce outgoing spillovers, by helping to internalise them. In order to allow cooperative agreements, a sufficient degree of knowledge appropriability is required to reduce free-rider behaviour by outsiders. Second, research cooperation may give access to tacit knowledge and know-how, which does not spill over and cannot easily be contracted through market transactions. Third, research partnerships allow exploitation of economies of scale and scope in both research and development, thereby reducing innovation costs and, in the same vein, help share risks.

The focus of our research is on knowledge development – in terms of the generation or creation of new knowledge – and the exchange of internally-developed knowledge within formal research cooperation agreements. A central feature is the particular role played by the qualification and occupation of the firm's internal R&D personnel. To our knowledge this aspect of research has not received much attention. This can be partly explained by the fact that the topic of absorptive capacity in firms is usually concentrated on the monetary dimension of R&D intensity.

The novelty of the chapter lies in its combination of three elements. First, it reviews the importance of research experts or managers in R&D projects in relation to external linkages in the form of research cooperation. Second, it makes within research cooperation a distinction between the development of knowledge and the exchange of knowledge developed in-house. Third, the empirical analysis relates the appropriability and commercialisation of knowledge, in terms of patenting and product innovation, to research cooperation (and as such also indirectly to the efforts of R&D personnel).

Section 10.3 provides a brief literature overview on research cooperation and the importance of the qualification and occupation of internal R&D personnel and develops the research questions. The database and descriptive information on research cooperation and R&D personnel are presented in Section 10.4. Section 10.5 provides an empirical analysis on the relationship between knowledge exchange and development in research cooperation and R&D personnel qualification and occupation; and Section 6 presents the conclusions of the chapter.

10.2 LITERATURE REVIEW AND HYPOTHESES

Due to the public good nature of new knowledge and the related problem of appropriability (see Cohen and Levinthal, 1989), firms' incentives to invest in R&D tend to be distorted. This causes firms to 'under-invest' in R&D because they cannot entirely internalise the social returns of their private efforts in the presence of R&D spillovers. In addition to government intervention (mainly by means of tax incentives and direct subsidies) to stimulate engagement in R&D, ex ante R&D cooperation is a means of internalising R&D spillovers, reducing wasteful duplication of R&D efforts and distributing risk and fixed costs among participants (see Jacquemin, 1988).

Apart from the element of market failure, the increasing complexity, budgets and risks related to innovation are important drivers for companies to use external knowledge (for example, through research cooperation) to complement their in-house innovative activities. This influences the innovation strategies of firms, and innovation becomes increasingly 'distributed' over various partners (von Hippel, 1988) and 'open' in the sense that firms adapt their business model in favour of R&D activities and technical changes that take place outside the firm (Chesbrough, 2003a), and has important implications in terms of dealing with external knowledge relations for innovation management (Chesbrough et al., 2006). From a distributed or open innovation perspective not only does knowledge development as part of research cooperation contribute to the internal R&D efforts, but the exchange (in a cooperation agreement) of the knowledge developed on the basis of internal R&D is important and these should be seen in relation to each other (see also, for example, Cohen and Levinthal, 1990; Arora and Gambardella, 1994; Veugelers and Cassiman, 1999; Chiesa, 2001; Coombs et al., 2003).

As such, knowledge development and exchange in research cooperation form part of the active and effective management of valuable knowledge which is considered a core competency in strengthening the competitive advantage of an organisation (see, for example, Wong and Aspinwall, 2004).

Evolutionary economics stress the impact of changing firm boundaries on the dynamic capabilities of the firm, and hence on its ability to integrate, build and reconfigure internal and external competencies to address changing environments (Teece et al., 1997). Cassiman and Veugelers (2006) found that firms that combine internal and external R&D strategies for innovation show superior innovative performance. It also turns out that large, resourceful firms have a higher propensity to combine both strategies (Veugelers and Cassiman, 1999).

It is widely accepted that access to external technology, in combination with the presence of internal R&D needed to absorb external knowledge (Cohen and Levinthal, 1990), allows firms to improve the performance of internal R&D efforts (Cassiman and Veugelers, 2006). In addition, internal R&D enables firms to translate external knowledge into innovation opportunities and outsourced R&D facilitates spillovers from outside the network to the firm's specific knowledge base (Audretsch et al., 1996).The most important component reflecting internal R&D is internal R&D employment. Whereas IT solutions can efficiently facilitate codified knowledge access and utilisation, tacit knowledge-sharing and transformation is decided by people-to-people interactions (Gold et al., 2001). Despite its importance, most of the economic literature focuses on R&D expenditure and disregards R&D employment characteristics when dealing with external knowledge relations within formal research cooperation agreements.

From the perspective of organisational absorptive capacity (Jacobides and Billinger, 2006) and the resource-based view that emphasises the need to innovate on the development and accumulation of specialised internal capabilities, a first set of hypotheses can be formulated as follows: hypothesis 1a posits that the higher the share of a firm's internal R&D

experts/managers, the higher the propensity to develop knowledge within a cooperation agreement.

Hypothesis 1b posits that the higher the share of a firm's internal R&D experts/managers, the higher the propensity to exchange knowledge based on internal R&D activities within a cooperation agreement.

Based on the insights from the literature, we expect a positive relationship between research employment and the development and exchange of in-house developed knowledge within research cooperation agreements, since firms with high levels of absorptive capacity are found to be more open to external R&D because the expert staff enables them to access wider networks (Rosenberg, 1990). At the same time, knowledge-intensive small firms are found less likely to be involved in this kind of activity because they often rely on the exploitation of new ideas. Given the danger they face from leakage of their ideas, they limit the nature and scope of their external interaction (Laursen and Salter, 2006). Also, the development and in particular the exchange of in-house developed knowledge in research cooperation cannot be seen entirely independently from the issues of intellectual property and knowledge commercialisation. In this respect, the knowledge cycle is found to present an evolution from knowledge creation and collection, knowledge sharing and distribution to final knowledge application and transformation (Wang and Ariguzo, 2004). Furthermore, increasing challenges and uncertainties in the business environment drive organisations to pay greater attention to the productivity of knowledge assets. Good knowledge management can accelerate the information flow and enhance integration of knowledge, innovation and creativity in the organisation, or, in other words, enhance knowledge exploitation. To achieve this ideal, a well-developed knowledge management system should not only take advantage of the IT solutions but also of cultural change, rewards and an execution team (Malhotra, 2003), and pay considerable attention to knowledge exchange by means of ex-post R&D cooperation through patents and licensing (Katz and Ordover, 1990).

These views are shared by an extensive literature on firm strategies and their relation to technology and/or innovation strategies that has developed over the past decade. The literature shows that internal competences, sourcing technology, leadership or follower strategies, managing assets complementary to innovation, and protection of intellectual property turn out to be important. Increasingly, the management of external relationships with universities, start-up companies etc. is part of the 'new industrial ecology' (Coombs and Georghiou, 2002).

As we have pointed out, little empirical evidence exists on the relationship between R&D employment and knowledge development and exchange in research cooperation. However, since this topic is in line with studies investigating the impact of technological change on the composition of employment we can rely on evidence in this broader context. Kahn (2000) reveals that technological standardisation reduces the importance of firm-specific knowledge, as standardisation reduces the specificity of internal skills and induces convergence of firm-specific skills to general skills over time. Pianta (2004) studied – at the macro level – the impact of technological innovation on the quality of jobs by looking at the relative importance of high-skilled and low-skilled jobs and found the diffusion of technologies to have a strong skill-bias effect in favour of high-skilled jobs. Tarabusi (1997) also found a trend towards skill upgrading, but revealed that the effects on qualifications depend on the importance of the nature of technology and the context in which technological changes are embedded. The latter finding seems supportive of the management literature highlighting the importance of the organisational context for changes in the skill composition of employment, and leads to our second set of research hypotheses. Hypothesis 2a posits that the higher the share of highly skilled internal R&D experts/managers, the higher the propensity to develop knowledge within a cooperation agreement. And hypothesis 2b posits that the higher the share of highly skilled internal R&D experts/managers, the higher the propensity to exchange knowledge based on internal R&D activities within a cooperation agreement.

Again, a priori, we expect a positive relationship between educational levels and knowledge development and exchange in research cooperation because educational levels are assumed to facilitate the detection and management of relevant external knowledge flows (OECD, 2008a). Doctorate holders in particular, who possess knowledge bordering on the frontier of science and technology, are supposed to be receptive to upstream research performed at universities or public research centres providing basic (and applied) research and its management (Sauermann and Cohen, 2008).

10.3 DATA DESCRIPTION

The analysis is based on an inventory of private enterprises that perform R&D on a regular basis in Belgium. The data are provided by the OECD biannual business R&D survey organised in 2008 and contains detailed information on R&D personnel for the year 2007 and information on research cooperation agreements for the period 2006–2007. The target population of R&D-active firms in Belgium in the period 2006–2007 consists of 1,947 private enterprises. Due to both unit and item non-response in the R&D survey, only 551 firms revealed detailed information on R&D (personnel data, cooperation agreements, and protection mechanisms).

The variables of interest are knowledge development and exchange in research cooperation agreements, and the presence and qualification of research experts/managers. Based on the Frascati Manual (OECD, 2002a), total R&D personnel includes all persons employed directly on R&D, as well as those providing direct services, such as R&D managers, administrators and clerical staff. Researchers (experts and managers) are professionals engaged in the conception or creation of new knowledge, products, processes, methods and systems, as well as in the management of these projects. Technicians and equivalent staff are persons whose main tasks require technical knowledge and experience in one or more fields of engineering, physical and life sciences, or social sciences and humanities. They participate in R&D by performing scientific and technical tasks involving the application of concepts and operational methods, normally under the supervision of researchers. Equivalent staff perform the corresponding R&D tasks under the supervision of researchers in social sciences and humanities. Other supporting staff includes skilled and unskilled craftsmen, and secretarial and clerical staff participating in R&D projects or directly associated with such projects. For the level of qualification (education), a distinction can be made between: PhD holders; holders of a master's diploma; holders of a bachelor's diploma; and holders of a diploma with other qualifications (OECD, 2002a).

Table 10.1 describes the data-set by means of the variables considered in the analysis, their mean values and standard deviation.

Table 10.1 Explanatory variables for knowledge development and exchange in research cooperation and the way they are measured (N=551)

Variable	Description	Mean	St. dev.
Dependent variables			
COOP_DEV	COOP_DEV = 1 if the firm developed knowledge used for internal R&D in the framework of a research cooperation in the period 2006–2007; 0 otherwise	0.42	0.49
COOP_EXCH	COOP_EXCH = 1 if the firm exchanged knowledge based on internal R&D in the framework of a research cooperation in the period 2006–2007; 0 otherwise	0.39	0.49
Independent variables – research experts/managers			
REM_INT	Share of R&D experts/managers in total firm employment	0.17	0.25
REM_RDSHR	Share of R&D experts/managers in firm R&D employment	0.64	0.30
REM_PhDSHR	Share of PhD holders in R&D experts/managers	0.20	0.33
REM_MSRSHR	Share of master's in R&D experts/managers	0.62	0.39
REM_BCHSHR	Share of bachelor's in R&D experts/managers	0.13	0.27
REM_OTHSHR	Share of other diplomas in R&D experts/managers	0.05	0.18
Independent variables – other R&D characteristics			
R&D_PERS	Total R&D personnel in full-time equivalents	20.3	64.0
R&D_LAB	R&D_LAB = 1 if the firm has a separate research lab	0.66	0.48
RESEARCH	Share of research expenditures in internal R&D expenditures	0.55	0.35
R&D_INT	Share of R&D employment (in full-time equivalents) in average firm employment	0.25	0.30
R&D_OUT	Share of R&D outsourcing budget in total (internal and outsourced) firm R&D budgets	0.11	0.18

Table 10.1 (continued)

Variable	Description	Mean	St. dev.
Independent variables – knowledge protection			
INNOPROD	INNOPROD = 1 if the firm had product innovations in the period 2006–2007; 0 otherwise	0.85	0.36
PATACQ	PATACQ = 1 if the firm purchased patents, licences or other intellectual property rights in the period 2006–2007; 0 otherwise	0.16	0.36
PATREG	PATREG = 1 if the firm launched patent registration demands in the period 2006–2007; 0 otherwise	0.28	0.45
PATEARN	PATEARN = 1 if the firm sold patents, licences or other intellectual property rights in the period 2006–2007; 0 otherwise	0.14	0.35
Control variables			
SECTOR_CP	Industry dummy for 'continuous process' (a variety of production activities such as metallurgical process industries, metals and building materials, and chemical process industries, textiles and paper, food and tobacco)	0.21	0.41
SECTOR_CS	Industry dummy for 'complex systems' (that is, combination of mechanical, electrical and transportation technologies. It includes motor vehicles and other transport equipment)	0.02	0.15
SECTOR_FP	Industry dummy for 'fundamental process' (that is, associated with chemistry-based technologies mainly oriented towards process innovation);	0.08	0.27
SECTOR_PE	Industry dummy for 'product engineering' (that is, uses mechanical engineering technologies, in areas as machinery, equipment and instruments	0.21	0.41

Table 10.1 (continued)

Variable	Description	Mean	St. dev.
Control variables			
SECTOR_SB	Industry dummy for 'science-based' (that is, pharmaceuticals, computers and other information processing equipment, electronics, and telecommunication equipment)	0.11	0.32
SECTOR_KI	Industry dummy for 'knowledge-intensive services' (that is, which includes telecommunications, computer services and technical business services)	0.32	0.47
SECTOR_OS	Industry dummy for 'other services' (that is, activities in other services than KI)	0.05	0.20
SIZE	Average firm employment expressed in full-time equivalents	219.0	585.9
GROUP	Binary variable = 1 if the firm belongs to a group	0.57	0.50

Notes: Sectors are derived from the Marsili–Verspagen (2002) classification. Reference year 2007 unless otherwise mentioned.

Five groups of variables are considered: dependent variables (related to knowledge diffusion in cooperation); independent variables (related to research experts/managers); other R&D-related variables; knowledge-protection variables; and general firm characteristics as control variables.

For the first group, the dependent variables related to the development and exchange of knowledge in research cooperation, it can be noted that within research cooperation, over two-fifths of the R&D-active firms develop knowledge useful for internal R&D and close to two-fifths of the R&D-active firms exchange knowledge based on internal R&D activities.

Concerning the independent variables of main interest, that is, those related to the presence and qualification of research experts/managers, it turns out that close to two-thirds of the R&D personnel are experts or managers, of which one-fifth are PhD holder . In terms of share in the total firm employment, this category of R&D personnel represents 17%.

For the broader R&D characteristics, on average, one-quarter of firm employment is engaged in R&D activities representing over 20 full time

R&D personnel equivalents and close to two-thirds of the R&D-active firms have a separate R&D lab.

Regarding knowledge production, 85% of the firms are engaged in product innovation and one-sixth of the firms traded patents, licences or other intellectual property rights in the period 2006–2007. Almost one-third of the R&D-active firms in the sample launched a patent registration demand in the period under consideration.

In terms of the control variables, an R&D-active firm has on average 219 employees, in one-third of the cases belonging to the knowledge-intensive services, in over half of the cases to part of a group, devoting 55% of its R&D budgets to research, and has a budget for R&D outsourcing that represents over 10% of the total R&D budget (that is, internal and outsourced R&D budgets).

Table 10.2 Comparison of firms included and excluded in the estimation – mean values and significance of the difference between means

	Firms with full data (N=551)	R&D-active firms with missing data (N=1,396)
SIZE	219.0	184.8
R&D_PERS	20.3*	13.1
R&D_INT	0.25*	0.20
R&D_OUT	0.11	0.11
SECTOR_CP	0.21	0.23
SECTOR_CS	0.02	0.02
SECTOR_FP	0.08	0.07
SECTOR_PE	0.21	0.23
SECTOR_SB	0.11	0.08
SECTOR_KI	0.32	0.29
SECTOR_OS	0.05	0.08

Notes: Significant difference between group means (at 5% level); a chi-square test including all sectors at the same time revealed no significant differences (at 5% level) in terms of sector distribution between firms in the selected sample and those excluded from the analysis.

As can be seen from the standard deviations reported in Table 10.1, large differences exist between firms. Therefore, it can be misleading to put too much emphasis on the profile of the 'average' R&D-active firm. Moreover, the analysis of the R&D-active population is faced with a large number of firms for which information is missing for one or more variables. Therefore, in Table 10.2 we briefly address the representativeness of the selected sample (limited to the variables for which we have information for the entire population of R&D-active firms). From the Table it turns out that the firms in the sample have a higher internal R&D employment and R&D employment intensity. General firm characteristics in terms of size and sector of activity and R&D outsourcing intensity are not biased in the selected sample.

10.4 EMPIRICAL RESEARCH

A high (0.65) significantly positive correlation is found between knowledge development and knowledge exchange in research cooperation. Therefore, we make use of a bivariate probit model to investigate the relation between the relative importance of the presence and qualifications of research experts/managers and firm decisions (i) to develop technological knowledge within a cooperation agreement which influences the firm's internal R&D activities or (ii) to exchange knowledge that has been developed in internal R&D activities in formal research cooperation. The model can be expressed as follows.

$$COOP\_DEV = X.\beta1 + REM\_INT.\beta2 + REM\_RDSHR.\beta3 + REM\_QUAL.\beta4 + \varepsilon1 \quad (10.1)$$

$$COOP\_EXCH = X.\beta5 + REM\_INT.\beta6 + REM\_RDSHR.\beta7 + REM\_QUAL.\beta8 + \varepsilon2 \quad (10.2)$$

Where X denotes the control variables (general firm characteristics), the research-related set of explanatory variables and the knowledge protection variables are as presented in Equation 10.1. The variable QUAL refers to the levels of qualification.

In order to consider the robustness of the model we first estimate a ('base') model including the overall R&D personnel intensity disregarding the presence and qualification of the research experts/managers. The latter variables are included in the 'detailed model'. Following Veugelers (1997), along with the research expert/managers-related variables, we include the variable 'R&D-LAB' which provides information on the way R&D activities are organised within the firm (in a formal lab versus more informally internally distributed).

Table 10.3 Bivariate probit model relating the presence and qualification of research experts/managers to knowledge exchange and development in research cooperation

	Base model		Full model	
Dependent variable	COOP_DEV	COOP_EXCH	COOP_ DEV	COOP_EXCH
Constant	–1.10	–1.28	–1.56	–1.63
	(–3.11)***	(–3.58)***	(–3.08)***	(–2.85)***
SECTOR_CP	–0.69	–0.47	–0.65	–0.44
	(–2.39)*	(–1.64)	(–2.19)*	(–1.49)
SECTOR_CS	0.04	0.04	0.07	0.13
	(0.08)	(0.08)	(0.15)	(0.27)
SECTOR_FP	–0.04	0.13	–0.12	0.11
	(–0.12)	(0.41)	(–0.36)	(0.34)
SECTOR_PE	–0.47	–0.31	–0.45	–0.24
	(–1.67)°	(–1.08)	(–1.55)	(–0.83)
SECTOR_SB	–0.38	0.12	–0.39	0.15
	(–1.24)	(0.39)	(–1.26)	(0.49)
SECTOR_KI	–0.38	–0.08	–0.39	0.03
	(–1.33)	(–0.29)	(–1.33)	(0.10)
lnSIZE	0.18	0.14	0.19	0.10
	(3.94)***	(3.11)***	(3.88)***	(2.20)*
GROUP	–0.18	0.02	–0.19	0.01
	(–1.28)	(0.12)	(–1.38)	(0.05)
INNOPROD	0.32	0.16	0.36	0.11
	(1.90)°	(0.97)	(2.05)*	(0.65)
PATACQ	0.38	0.21	0.38	0.18
	(2.45)*	(1.28)	(2.39)*	(1.13)
PATREG	0.32	0.19	0.26	0.13
	(2.35)*	(1.38)	(1.92)°	(0.96)
PATEARN	–0.04	0.29	–0.08	0.24
	(–0.23)	(1.75)°	(–0.46)	(1.47)
RESEARCH	–0.07	0.07	–0.09	0.28
	(–0.44)	(0.45)	(–0.54)	(0.17)
R&D_OUT	0.95	0.53	0.94	0.50
	(2.87)**	(1.74)°	(2.86)**	(1.59)
R&D_INT	0.93	1.07		
	(3.49)***	(4.15)***		

Table 10.3 (continued)

	Base model		Full model	
Dependent variable	COOP_DEV	COOP_EXCH	COOP_ DEV	COOP_EXCH
REM_INT			1.25	1.15
			(3.75)***	(3.55)***
REM_RDSHR			–0.16	–0.43
			(–0.73)	(–1.91)°
REM_PhDSHR			0.73	0.78
			(1.91)°	(1.75)°
REM_MSRSHR			0.52	0.64
			(1.49)	(1.51)
REM_BCHSHR			–0.09	0.51
			(–0.23)	(1.06)
R&D_LAB			0.12	0.37
			(0.94)	(2.85)**
RHO	0.85 ***		0.85 ***	
Wald chi²	125.32 ***		141.84 ***	
Obs.	551		551	

Notes: Reference categories: SECTOR_OS (other services). The z-values appear between brackets (two-tailed tests); and the symbols °, *, **, *** denote the significance at 10%, 5%, 1% and 0.1% level.

In terms of general firm characteristics, in the base model, firms active in 'continuous processes' are less likely to be involved in knowledge development in research cooperation. To a lesser extent (10% significance level) this is also the case for firms classified in product engineering. Concerning firm size, there is a highly significant positive both with the development and exchange in knowledge in research cooperation (in line with the findings by Veugelers and Cassiman, 1999). Of course, R&D cooperation is not a goal in itself but rather one measure among many for attaining innovation objectives, for example by establishing relationships with third parties to help enter a company in a new area of technology. From this perspective, it is not surprising to find a positive relation between product innovation, patent acquisition and registration for knowledge development in research cooperation. In terms of knowledge exchange, a positive relation can be identified with patent earnings. As for research characteristics, R&D outsourcing is positively related to knowledge development and to a lesser extent also to knowledge exchange. This is supported by the fact that research cooperation is only one means of improving innovation in a distributed or open innovation model. No differences exist between research versus development activities. Finally, in the base model, R&D intensity (R&D personnel as a share of total firm employment) is highly significantly related in a positive way to both knowledge development and knowledge exchange in research cooperation.

These findings and robustness for general and broader R&D and knowledge protection related variables form a solid background for the interpretation of the variables related to research experts/managers. Based on the outcomes in Table 10.3 and related to the research questions, it turns out that the share of research managers/experts in total firm employment positively influences the propensity to develop and to exchange knowledge in research cooperation. This supports the resource-based and the organisational absorptive capacity views emphasising the need for a sufficient internal R&D capacity to assimilate and commercialise external knowledge. However, when considering the share of research experts/managers in the total R&D personnel, there is no significant association with the development of knowledge in research cooperation and the relation with knowledge exchange in research cooperation becomes significantly negative. In other words, the higher the share of research managers/experts in the R&D personnel, the lower the propensity of knowledge exchange based on research cooperation. This could be related to the highly significant positive relation between knowledge exchange in cooperation and a more formal organisation of R&D activities in R&D laboratories.

When we look at the requirements in terms of formal qualification of research experts/managers, it turns out that the involvement of PhD holders in research increases the propensity of being engaged both in knowledge exchange and development in research cooperation (although only at 10% significance level).

10.5 DISCUSSION AND CONCLUSIONS

This chapter started from the observation that the theoretical and empirical literature on external knowledge relations mainly focuses on research expenditures. However, knowledge development and exchange is generally accepted to be highly related to human resources. To address part of this gap, the empirical analysis presented here investigates the relation between the presence and relative importance of research experts/managers and their formal qualifications and the propensity to engage in knowledge development and exchange in research cooperation.

The analysis is based on the OECD bi-annual business R&D survey for Belgium for the years 2006–2007 and includes firms active in R&D on a (quasi-) permanent basis. The analysis proved somewhat biased towards firms with a higher number of R&D personnel and with a higher share of R&D personnel in the overall firm employment.

Two research questions have been put forward. The first one relates the presence and relative importance of research experts/managers to knowledge development and exchange in research cooperation. In line with the resource-based views and the literature on organisational absorptive capacity, the share of research managers/experts in total firm employment influences the propensity to exchange and to develop knowledge in research cooperation positively. However, when considering the share of research experts/managers in the total R&D personnel, there is no significant association with the development of knowledge in research cooperation and the relation with the exchange of knowledge in research cooperation, becomes significantly negative. In other words, the higher the share of research managers/experts in the R&D personnel, the lower the propensity to engage in exchange of knowledge based on research cooperation. This could be related to larger opportunities for commercialisation of knowledge developed within a cooperation agreement. Also, it seems to be supported by the highly significant positive relation between knowledge exchange in cooperation and the formal organisation of R&D activities in an R&D laboratory.

The second research question investigated the relation between the formal qualification level of research experts/managers and knowledge exchange and development in research cooperation. The involvement of PhD holders in research turned out to increase the propensity of being engaged both in knowledge development and exchange in research cooperation agreements. This could also be related to a higher probability of commercialising research results. However, differences in networking between PhD-holders and other education levels for research experts/managers could also help explain this finding. The latter remains unclear and further investigation is needed.

One way to further refine the analysis is to differentiate cooperation according to type of partner. Moreover, it would be interesting to look at the effects of research cooperation since it is not an end in itself but rather one measure among many for attaining innovation objectives. For this, a longitudinal analysis would be more appropriate. Also, a further distinction could be made between research experts and research managers.

Finally, from a policy perspective the results show that knowledge development and exchange in research cooperation can be stimulated by public authorities by influencing the qualifications and occupations of R&D

personnel. From this perspective it will be interesting to investigate the effects of the recently introduced tax-credit measure in Belgium for R&D personnel holding a master's or PhD diploma in a scientific domain. By restricting tax credits to highly qualified R&D workers there could be a change in the composition of the R&D personnel and indirectly, and in the longer term, in the propensity for knowledge development and exchange in research cooperation.

General conclusions

In the past decade, innovation has been described as becoming more 'open' than before, with firms engaging in external knowledge relations with other organisations. By the turn of the century, the world in which innovation took root had changed, in that it offered an abundance of highly skilled workers, good access to financial means for firms and a large share in the knowledge base of innovation systems. The discussion can be framed largely in terms of the consequences for business enterprises of open innovation. However, for policy-makers, the question is one of whether their policy instruments have been out-dated by the world of open innovation.

Based on the empirical work presented in this book, there are a few lessons to be learned with regard to the management and governance of open innovation. The main messages are presented in line with the four research areas identified for further development at the beginning of this book, both from a theoretical and empirical perspective: firm heterogeneity in innovation performance; the role of government in shaping open innovation practices; the implications for internal research and innovation R&D outsourcing activities; and the role of internal R&D personnel in a firm's decision to engage in open innovation practices.

First, with regard to firm heterogeneity in the management of open innovation, systematic differences have been identified between the behaviours of poor and strong innovative performers. Hence, the relevance to management science of an understanding of the behaviour at the extremes, that is, firms that perform exceptionally well, or poorly.

The internalisation of external knowledge generated through incoming knowledge spillovers and research cooperation has, clearly, become a major purpose of the research activities of innovative firms. The empirical analyses revealed that the impact of knowledge-related variables on research intensity differs sharply between firms. This appeared to be particularly the case where reliance on strategic appropriability is concerned. Moreover, innovative firms are found to devote an increasing part of their R&D effort to the absorption of external knowledge and it is particularly among the strongest innovative performers that we find firms prepared to sell excess knowledge in disembodied form. Incoming knowledge spillovers, research cooperation, appropriability, workers' qualifications, export orientation, as well as the general technological environment, matter.

A breakdown of the various types of innovating firms into weak and strong performers in terms of research intensity and innovative turnover revealed that the effects of all knowledge-related variables on innovative turnover are indirect. Incoming knowledge spillovers are more important, in terms of statistical significance, in explaining research intensity in the case of strong innovators. Also, research collaboration is most significant for strong innovators. In sharp contrast to this, strategic appropriation seems to be highly relevant to weak innovators, but loses significance for strong innovators. Although appropriability does, on average, continue to form an incentive for undertaking research, it is relatively less important, it seems, in the case of the strongest R&D performers. This would suggest that it is here that we find the majority of firms prepared to market excess knowledge in disembodied form. The greater impact of research intensity on innovative turnover among strong product innovators would tend to confirm that they have the most efficient R&D processes.

The fiction of a 'representative firm' may, therefore, be highly misleading in view of the heterogeneity in innovative behaviour. This is highly relevant since open innovators are likely to be found among the most dynamic firms in particular.

To summarise, there seems to be a typology of firms when we consider the joint relative impact of incoming knowledge spillovers and research cooperation on research intensity. This impact tends to increase quantile-by-quantile in relation to that of appropriability, and, if anything, the impact of research intensity on innovative turnover becomes stronger in the higher quantiles. This suggests that dynamic innovators have a more open and efficient innovation process. This has important consequences both for understanding heterogeneity in business models on innovation and for innovation policy that focuses all too much on the entire population of firms, although recent policy measures have been devised to enhance new technology-based firms (academic and corporate spin-offs, independent hi-tech start-ups).

Second, attention was paid to the determinants for firm-level decisions to engage in research cooperation and the effects on innovation performance, particular attention being devoted to a mixture of public funding mechanisms. The main lessons are that the influence of funding on innovative turnover has turned out to be exclusively indirect (through research intensity) and that more attention will have to be paid to networking behaviour in terms of the instalment and persistence of research collaborations.

A limited number of studies have addressed the issue of whether public funding for innovation succeeds in influencing cooperative practices. Moreover, a common weakness of these respective studies is that they tend to concentrate on the impact of a single measure of funding, relating to total or national funding. However, since the various funding authorities apply partly different project selection criteria, it is of interest to discover whether there are differences in impact between different types of funding. Therefore, attention was paid to different types of funding, to the type of partner and to geographical location. More specifically, two types of research cooperation were considered (cooperation with business and cooperation with public research organisations) and three levels of public funding by means of R&D subsidies (regional government funding, federal government funding and EU funding, as well as different combinations) have been accounted for.

The impact of public funding on innovative turnover is indirect, through research intensity. Research intensity is not only the key explanatory variable of innovative turnover; its impact also tends to increase quantile-by-quantile. When considered in conjunction with the results on research intensity, this implicitly suggests that the greater openness of the innovation process of highly research-intensive firms allows them to achieve significant gains in product-innovative efficiency. Helping to achieve such gains in innovative efficiency is one of the main justifications for encouraging research cooperation through public funding.

Funding policies play their part in helping to open up the research process. In the relationship between source of funding and research cooperation, both the type (industry–science versus inter-firm) and geographical scope (domestic versus international) of cooperation matter. The indirect effect is strongest for weak innovators, indicating the highest need for public intervention to open up the research process for these innovators. In terms of innovative turnover, funding showed only an indirect influence through research intensity. The impact of the latter turned out to increase (quantile-by-quantile) with the strength of the innovator.

From a policy-making point of view, when dealing with (questions on) behavioural additionality it is important to understand that effects may (or may not) be intentional on the part of either the policy-maker or the recipient of funding. Also, it is of interest to assess both the effects of funding on the instalment and the persistence of the effects, since behavioural additionality effects may be expected to last beyond the period of R&D and become integrated in the firms' general capabilities. Innovation policies founded in the systems perspective place a heavy emphasis on the formation and promotion of the resulting networks.

In order to more adequately assess the additionality of policy measures for R&D, the necessity to take into account the policy mix as a whole (the complementarity of the different programs) and assess the performance of the innovation system in an international perspective can be highlighted.

In the past, policies aimed at boosting innovation started from the theoretical arguments legitimising knowledge spillovers between organisations and stressing systemic and market failures. By and large, many of these arguments still hold in times of open innovation. Innovating firms like to benefit optimally from their innovations and rank the issue of protective mechanisms on intellectual property and information asymmetry, in a world in which information is freely flowing, high on the political agenda. Many of these insights were derived from the study of the national and regional innovation systems. But open innovation practices in the field affect these policy insights. Policy-makers have to be aware of the changed ways in which innovation is managed at firm level.

The third topic refers to the impact of firm-level decisions for R&D outsourcing on internal R&D activities. The main messages of the analyses are the complementarity of in-house R&D activities and R&D outsourcing, and the differences in terms of impacts on internal R&D activities according to the type and location of the outsourcing partner. The latter is closely related to strategic decisions at firm level in terms of starting, increasing, or stopping R&D outsourcing activities.

The most influential, formal explanations for the outsourcing of economic activities by firms have been offered by transaction cost economics and contract theory. According to these, outsourcing occurs under conditions of low transaction costs. However, the tacit components of technological knowledge tend to make it specific and the uncertainty of the outcome of the R&D process and the information differentials between buyer and supplier (subcontractor) will create monitoring problems. A complementary explanation for outsourcing focuses on the capabilities of the firm. According to this, the firm will outsource non-core activities in which it is not particularly specialised, but which may still be important as support. Technology outsourcing should, therefore, relate to relatively standardised technological knowledge which is non-core to the firm's activities.

The importance of creating prior knowledge to effectively absorb the knowledge obtained through research partnerships has been highlighted. In-house R&D and R&D outsourcing have, therefore, become major complements to one another.

In-house R&D has been found to have a strong positive impact on R&D outsourcing intensity, emphasising the importance of knowledge-absorptive effort and, therefore, the complementary nature of in-house and external R&D. Strategic appropriability has a stronger positive impact on R&D

outsourcing intensity than patenting, and this, mainly, through complexity of design (contrary to lead-time advantage on competitors, which is often too brief to affect R&D partnership behaviour). These findings suggest that the required appropriation occurs in a more informal manner than through deliberate actions of legal protection or secrecy. Formal cooperation also has a significant positive impact on R&D outsourcing intensity and this is mainly due to cooperation with public research institutions, such as universities and research centres. This suggests that the firms in question tend to spend relatively more on external R&D.

Empirical evidence on the effects of the outsourcing of R&D on the composition of skills in the total labour force of the firm, and more precisely the relative importance of R&D workers (that is, the internal R&D employment intensity) remains a blank spot in the literature. Therefore, the effects of R&D outsourcing on R&D labour composition were explored. The aim was to assess whether firms' R&D outsourcing strategies changed the internal R&D employment intensity. A distinction was made between firms' decisions to start, increase, decrease or stop R&D outsourcing. Differences in the type and location of the R&D contractor have also been considered. The analysis revealed that larger firms and firms belonging to a group have a higher propensity to be engaged in R&D outsourcing in a more permanent way. Smaller firms have a relatively higher propensity to start or to stop R&D outsourcing. Also, management decisions in terms of starting, sustaining or stopping R&D outsourcing occur at the same time with different types of contractors at different locations. Within the setting of the small open Belgian economy, firms' decisions to start or to stop R&D outsourcing more frequently involve regional contractors, whereas sustained outsourcing involves national and international contractors relatively more frequently. Finally, the outcomes of the analysis pointed to a careful choice for a mix of different types of contractors at different geographical locations. In this respect, decisions to start or stop R&D outsourcing frequently involve both universities and research centres. Sustained outsourcing is more often a mixture of both public and private contractors.

These findings are not without policy relevance. The finding that starting or increasing R&D outsourcing negatively affects internal R&D employment intensity could stir fears of a hollowing-out of the local knowledge base. However, this fear seems unjustified in that these negative findings seem to relate mainly to the outsourcing of – more applied – R&D to local research centres, and it turns out that increased R&D outsourcing actually does have positive effects for other firms and universities outside the country, suggesting the existence of positive international spillovers.

A final set of conclusions is formulated with regard to the role of human resources in open innovation practices. The main outcomes of the studies are that both the qualification and occupation of the internal R&D personnel and the availability and qualification of the R&D experts and managers are important determinants in firm-level decisions relating to open innovation practices.

Both in business analysis and policy-making circles, R&D activities have been closely related to monetary flows (for example, the European 3% target and the definition in business environments of concepts such as absorptive capacity). Behind these financial streams there are real flesh-and-blood people; people with their own motivations and aspirations. And their characteristics are vital if firms want to source external knowledge or if they want to commercialise the knowledge they have developed internally. There is a gap in the theoretical and empirical literature on external knowledge relations with regard to human resources for R&D.

The novelty of approach in this work lies in the combination of the importance of research experts or managers in R&D projects in relation to external linkages in the form of research cooperation. A distinction has been made within research cooperation between the development of knowledge and the exchange of knowledge developed in-house. At the same time, the appropriability and commercialisation of knowledge in terms of patenting and product innovation to research cooperation (and as such, indirectly, to the efforts of R&D personnel) has been accounted for.

The two key dimensions in the process of gaining from knowledge spillovers related to human resources – the function or skills and formal qualifications or educational level of the labour force – turned out to be important for firms when they engage in formal collaborative agreements on behalf of the firm. Skills, therefore, become a qualitative aspect of absorptive capacity because a firm's employees embody tacit knowledge which is considered to be a main source of spillovers.

The level of educational attainment turned out to be of importance if firms engage in research cooperation and exchange of knowledge. This finding does not imply that on-the-job training is not useful, but that the skills necessary for external knowledge relations are to be found more in people who already have a high-level degree as a threshold. However, in the case of firms pursuing the knowledge-exchange of internally developed knowledge, the support function of R&D personnel becomes crucial and this might be performed by R&D personnel having lower formal degree qualifications. If firms are to rely on external relations, their human resources departments have to focus on hiring R&D personnel with the appropriate profiles. Besides human resources, in terms of absorptive capacity, it is found that patent registration and innovative outputs also influence the propensity to engage in knowledge-development in R&D

collaboration, but not in terms of engagement of knowledge-exchange in R&D collaboration agreements.

In line with the resource-based views and the literature on organisational absorptive capacity, the share of research managers/experts in total firm employment has been found to influence the propensity to exchange and to develop knowledge in research cooperation positively. Also, a positive relation is detected between the formal qualification level of research experts/managers and knowledge exchange and development in research cooperation. In particular, the involvement of PhD holders in research can be positively associated with the propensity to be engaged in both knowledge-development and exchange in research cooperation agreements.

These findings are not without policy relevance, since they imply that the results show that public authorities can stimulate knowledge-development and exchange in research cooperation by influencing the qualifications and occupations of R&D personnel.

References

Abramovsky, L., Kremp, E., Lopez, A., Schmidt, T. and Simpson, H. (2005), 'Understanding co-operative R&D activities: evidence from four European countries', Working Paper 05-23, Institute for Fiscal Studies.

Acha, V. and Cusmano, L. (2005), 'Governance and co-ordination of distributed innovation processes: patterns of R&D co-operation in the upstream petroleum industry', *Economics of Innovation and New Technology*, **14**, 1–21.

Afuah, A. (1998), *Innovation Management: Strategies, Implementation and Profits*. Oxford University Press, New York.

Allen, T.J. and Katz, R. (1992), 'Age, education and the technical ladder', *IEEE Transactions on Engineering Management*, **39**, 237–245.

Antras, P. (2005), 'Incomplete contracts and the product cycle', *American Economic Review*, **95**, 1054–1073.

Arora, A. and Gambardella, A. (1994a), 'The changing technology of technological change: general and abstract knowledge and the division of innovative labour', *Research Policy*, **23**, 523–532.

Arora, A. and Gambardella, A. (1994b), 'Evaluating technological information and utilizing it: scientific knowledge, technological capability and external linkages in biotechnology', *Journal of Economic Behaviour and Organisation*, **24**, 91–114.

Arora, A., Fosfuri, A. and Gambardella, A. (2001), *Markets for Technology. The Economics of Innovation and Corporate Strategy*. MIT Press, Cambridge (Mass.).

Arrow, K. (1962), 'Economic welfare and the allocation of resources for invention', in: Nelson, R. (ed.), *The Rate and Direction of Inventive Activity*. Princeton University Press, Princeton, pp. 609–626.

Arundel, A. (2001), 'The relative effectiveness of patents and secrecy for appropriation', *Research Policy*, **30**, 611–624.

Aschhoff, B. and Sofka, W. (2008), 'Successful patterns of scientific knowledge sourcing – mix and match', ZEW Discussion Paper 08–033.

Aschhoff, B., Fier, A. and Fier, H. (2006), 'Detecting behavioural additionality – an empirical study on the impact of public R&D funding on firm's cooperative behaviour in Germany', ZEW Discussion Paper 06–037.

Atallah, G. (2002), 'Vertical R&D spillovers, cooperation, market structure and innovation', *Economics of Innovation and New Technology*, **11**, 179–209.

Audretsch, D. and Feldman, M. (1986), 'R&D spillovers and the geography of innovation and production', *American Economic Review*, **86**, 630–640.

Audretsch, D., Menkveld, A. and Thurik, R. (1996), 'The decision between internal and external R&D', *Journal of Institutional and Theoretical Economics*, **152**, 519–530.

Autio, E., Kanninen, S. and Gustafsson, R. (2008), 'First- and second-order additionality and learning outcomes in collaborative R&D programs', *Research Policy*, **37**, 59–76.

Azagra-Caro, J.M., Archontakis, F., Gutierrez-Gracia, A. and Fernandez-de-Lucio, I. (2006), 'Faculty support for the objectives of university–industry relations versus degree of R&D cooperation: the importance of regional absorptive capacity', *Research Policy*, **35**, 37–55.

Bacchiocchi, E. and Montobbio, F. (2009), 'Knowledge diffusion from university and public research. A comparison between US, Japan and Europe using patent citations', *Journal of Technology Transfer*, **34**, 169–181.

Bader, M.A. (2006), *Intellectual Property Management in R&D Collaborations: the Case of the Service Industry*. Physica-Verlag, Heidelberg.

Barney, J. (1999), 'How a firm's capabilities affect boundary decisions', *Sloan Management Review*, **40**, 137–145.

Bayona, C., Garcia-Marco, T. and Huerta, E. (2001), 'Firms' motivations for cooperative R&D: an empirical analysis of Spanish firms', *Research Policy*, **30**, 1289–1307.

Becker, W. and Dietz, J. (2004), 'R&D cooperation and innovation activities of firms – evidence for the German manufacturing industry', *Research Policy*, **33**, 209–223.

Belderbos, R., Carree, M., Diederen, B., Lokshin, B. and Veugelers, R. (2004), 'Heterogeneity in R&D cooperation strategies', *International Journal of Industrial Organization*, **22**, 1237–1263.

Beneito, P. (2006), 'The innovative performance of in-house and contracted R&D in terms of patents and utility models', *Research Policy*, **35**, 502–517.

Bercowitz, J. and Feldman, M. (2008), 'Academic entrepreneurs: organizational change at the individual level', *Organization Science*, **19**, 69–89.

Bessant, J. and Rush, H. (1995), 'Building bridges for innovation: the role of consultants in technology transfer', *Research Policy*, **24**, 97–114.

Blanes, J.V. and Busom, I. (2004), 'Who participates in R&D subsidy programs? The case of Spanish manufacturing firms', *Research Policy*, **33**, 1459–1476.

Boden, R., Cox, D., Georghiou, L., and Barker, K. (2001), 'Administrative reform of United Kingdom government research establishments: case studies of new organisational forms', in: Cox, D., Gummett, P. and Barker, K. (eds), *Government Laboratories. Transition and Transformation*. IOS Press, Amsterdam. pp. 77–96.

Bönte, W. (2003), 'R&D and productivity: internal vs. external R&D – evidence from West German manufacturing industries', *Economics of Innovation and New Technology*, **12**, 343–360.

Bönte, W. and Keilbach, M. (2005), 'Concubinage or marriage? Informal and formal cooperations for innovation', *International Journal of Industrial Organization*, **23**, 279–302.

Bozeman, B. (1994), 'The cooperative technology paradigm: an evaluation of U.S. government laboratories' technology transfer activities', *Policy Studies Journal*, **22**, 322–327.

Bruce, M. and Morris, B. (1998), 'In-house, outsourced or a mixed approach to design', in: Bruce, M. and Jevnaker, B. (eds), *Management of Design Alliances: Sustaining Competitive Advantage*. Wiley, Chichester. pp. 39–69.

Bruneel, J., Spithoven, A., Maesen, A. and Clarysse, B. (2007), 'External linkages of high tech start-ups in Flanders: networking and proximity', in: Bruneel, J., Clarysse, B., Maesen, A., Moray, N. and Spithoven, A. (eds), *An Analysis of Industry–Science Relations and Knowledge Networks of High Tech Start-ups in Flanders*. Flanders District of Creativity, Gent. pp. 44–95.

Brusoni, S., Prencipe, A. and Pavitt, K. (2001), 'Knowledge specialization, organizational coupling, and boundaries of the firm: why do firms know more than they make?', *Administrative Science Quarterly*, **46**, 597–621.

Buchinsky, M. (1995), 'Estimating the asymptotic covariance matrix for quantile regression models: a Monte Carlo study', *Journal of Econometrics*, **68**, 303–338.

Buchinsky, M. (1998), 'Recent advances in quantile regression models: a practical guide for empirical research, *Journal of Human Resources*, **33**, 88–126.

Buisseret, T., Cameron, H. and Georghiou, L. (1995), 'What difference does it make? Additionality in the public support of R&D in large firms', *International Journal of Technology Management*, **10**, 587–600.

Busom, I. (2000), 'An empirical evaluation of the effects of R&D subsidies', *Economics of Innovation and New Technology*, **9**, 111–148.

Busom, I. and Fernandez-Ribas, A. (2004), 'Firm strategies in R&D: cooperation and participation in R&D programs', Working Paper 04-05, Deparment d'Economia Applicada, Universitat Autonoma de Barcelona.

Busom, I. and Fernández-Ribas, A. (2008), 'The impact of firm participation in R&D programmes on R&D partnerships', *Research Policy*, **37**, 240–257.

Caloghirou, Y., Ioannides, S. and Vonortas, N. (2003), 'Research joint ventures', *Journal of Economic Surveys*, **17**, 541–570.

Caloghirou, Y., Kastelli, I. and Tsakanikas, A. (2004), 'Internal capabilities and external knowledge sources: complements or substitutes for innovative performance?', *Technovation*, **24**, 29–39.

Capron, H. and Cincera, M. (2005), 'The commitment of Belgian teams to European S&T cooperative agreements', in: Spithoven, A. and Teirlinck, P. (eds), *Beyond Borders. Internationalisation of R&D and Policy Implications for Small Open Economies*. Elsevier, Amsterdam.

Capron, H., Cincera, M. and Dumont, M. (2000), 'The institutional profile', in: Capron, H. and Meeusen, W. (eds), *The National Innovation System of Belgium*. Physica-Verlag, Heidelberg, pp. 43–69.

Carmichael, J. (1981), 'The effects of mission-oriented public R&D spending on private industry', *Journal of Finance*, **36**, 617–627.

Cassiman, B. and Veugelers, R. (2002), 'R&D cooperation and spillovers: some empirical evidence from Belgium', *American Economic Review*, **92**, 1169–1184.

Cassiman, B. and Veugelers, R. (2006), 'In search of complementarity in innovation strategy: internal R&D and external knowledge acquisition', *Management Science*, **52**, 68–82.

Cassiman, B., Di Guardo, C. and Valentini, G. (2005), 'Organizing for innovation: R&D projects, activities and partners', IESE Business School Working Paper No. 597.

Cassiman, B., Perez-Castrillo, D. and Veugelers, R. (2002), 'Endogenizing know-how flows through the nature of R&D investments', *International Journal of Industrial Organization*, **20**, 775–799.

Cesaretto, S. and Mangano, S. (1992), 'Technological profiles and economic performance in the Italian manufacturing sector', *Economics of Innovation and New Technologies*, **2**, 237–256.

Cesaroni, F. (2004), 'Technological outsourcing and product diversification: do markets for technology affect firms' strategies?', *Research Policy*, **33**, 1547–1564.

CFS/STAT (2001), 'Méthodologie pour l'estimation de l'effort de R&D des entreprises'. Rapport interne SSTC, Brussels.

CFS/STAT (2007), www.belspo.be – accessed 9 December 2008.

Charemza, W.W. and Deadman, D.F. (1997), *New Directions in Econometric Practice: General to Specific Modelling, Cointegration and Vector Autoregression* (Second Edition). Edward Elgar, Cheltenham.

Chesbrough, H. (2003a). *Open Innovation. The New Imperative for Creating and Profiting from Technology*. Harvard Business School Press, Boston.

Chesbrough, H. (2003b), 'The logic of open innovation: managing intellectual property', *California Management Review*, **45**, 33–58.

Chesbrough, H. (2003c), 'The governance and performance of Xerox's technology spin-off companies', *Research Policy*, **32**, 403–421.

Chesbrough, H. (2006a), 'Open innovation: a new paradigm for understanding industrial innovation', in: Chesbrough, H., Vanhaverbeke, W. and West, J. (eds), *Open Innovation: Researching a New Paradigm*. Oxford University Press, Oxford, pp. 1–12.

Chesbrough, H. (2006b), *Open Business Models. How to Thrive in the New Innovation Landscape*. Harvard Business School Press, Boston.

Chesbrough, H. and Crowther, A.K. (2006), 'Beyond high tech: early adopters of open innovation in other industries', *R&D Management*, **36**, 229–236.

Chesbrough, H. and Rosenbloom, R.S. (2002), 'The role of the business model in capturing value from innovation: evidence from xerox corporation's technology spin-off companies', *Industrial and Corporate Change*, **11**, 529–555.

Chesbrough, H. and Schwartz, K. (2007), 'Innovating business models with co-development partnerships', *Research Technology Management*, January–February, 55–59.

Chesbrough, H., Vanhaverbeke, W. and West, J. (2006), *Open Innovation: Researching a New Paradigm*. Oxford University Press, Oxford.

Chiesa, V. (2001), *R&D Strategy and Organisation*. Imperial College Press, London.

Chiesa, V., Manzini, R. and Pizzurno, E. (2004), 'The externalisation of R&D activities and the growing market of product development services', *R&D Management*, **34**, 65–75.

Clarysse, B., Wright, M. and Mustar, P. (2009), 'Behavioural additionality of R&D subsidies: a learning perspective', *Research Policy*, **38**, 1517–1533.

Coase, R.H. (1937), 'The nature of the firm', *Economica*, **4**, 386–405.

Cohen, W. and Levinthal, D. (1989), 'Innovation and learning: the two faces of R&D', *Economic Journal*, **99**, 569–596.

Cohen, W. and Levinthal, D. (1990), 'Absorptive capacity: a new perspective on learning and innovation', *Administrative Science Quarterly*, **35**, 128–152.

Cohen, W.M. (1995), 'Empirical studies of innovative activity', in Stoneman, P. (ed.), *Handbook of The Economics of Innovation and Technological Change*. Blackwell, Oxford. pp. 182–264.

Cohen, W.M. and Klepper, S. (1996), 'A reprise of size and R&D', *Economic Journal*, **106**, 925–951.

Colombo, M. and Gerrone, P. (1996), 'Technological cooperative agreements and firm's R&D intensity. A note on causality relations', *Research Policy*, **25**, 923–932.

Coombs, R. (1996), 'Core competencies and the strategic management of R&D', *R&D Management*, **25**, 345–355.

Coombs, R. and Georghiou, L. (2002), 'A new "industrial ecology"', *Science*, **296** (5567), 471.

Coombs, R., Harvey, M. and Tether, B. (2003), 'Analysing distributed processes of provision and innovation', *Industrial and Corporate Change*, **12**, 1125–1155.

Crépon, B., Duguet, E. and Mairesse, J. (1998), 'Research, innovation and productivity analysis at the firm level', *Economics of Innovation and New Technology*, **7**, 115–158.

Cusmano L., Luisa, M. and Morrison, A. (2008), 'Innovation and the geographical and functional dimensions of outsourcing: an empirical investigation based on Italian firm level data', Papers in Evolutionary Economic Geography, n°6, Utrecht University, Section of Economic Geography.

Czarnitzki, D., Ebersberger, B. and Fier, A. (2004), 'The relationship between R&D collaboration, subsidies and patenting activity: empirical evidence from Finland and Germany', ZEW Discussion Paper 04–37.

Dachs, B., Ebersberger, B. and Pyka, A. (2004), 'Why do firms co-operate for innovation? A comparison of Austrian and Finnish CIS 3 results', Volkswirtschaftliche Diskussionsreihe 255, Institut für Volkswirtschaftslehre der Universität Augsburg.

Dahlander, L. and Gann, D. (2007), 'How open is innovation?', Paper presented at the DRUID Summer Conference on 'Appropriability, Proximity, Routines and Innovation', Denmark, Copenhagen, 1–34.

Dasgupta, P. and David, P. (1994). 'Towards a new economics of science', *Research Policy*, **23**, 487–521.

D'Aspremont, C. and Jacquemin, A. (1988), 'Cooperative and noncooperative R&D in duopoly with spillovers', *American Economic Review*, **78**, 1133–1137.

David, P.A., Hall, B.H. and Toole, A.A. (2000), 'Is public R&D a complement or substitute for private R&D? A review of the econometric evidence', *Research Policy*, **29**, 497–529.

Davidson, R. and MacKinnon, J.G. (1993), *Estimation and Inference in Econometrics*. Oxford University Press, New York.

Davidson, W. and McFetridge, D. (1984), 'International technology transactions and the theory of the firm', *Journal of Industrial Economics*, **32**, 253–264.

Defazio, D., Lockett, A. and Wright, M. (2009), 'Funding incentives, collaborative dynamics and scientific productivity: evidence from the EU framework program', *Research Policy*, **38**, 293–305.

Delmar, F., Davidson, P. and Gartner, W. (2003), 'Arriving at the high-growth firm', *Journal of Business Venturing*, **18**, 189–216.

Dhondt-Peltrault, E. and Pfister, E. (2007), 'R&D cooperation versus R&D subcontracting: empirical evidence from French survey data', BETA Working Paper, 2007–17, Strasbourg.

Dodgson, M., Gann, D. and Salter, A. (2008), *The Management of Technological Innovation. Strategy and Practice* (Second edition). Oxford University Press, Oxford.

Dosi, G., Malerba, F., Ramello, G.B. and Silva, F. (2006), 'Information, appropriability, and the generation of innovative knowledge four decades after Arrow and Nelson: an introduction', *Industrial and Corporate Change*, **15**, 891–901.

Dyer, J. and Singh, H. (1998), 'The relational view: cooperative strategy and sources of interorganizational competitive advantage', *Academy of Management Review*, **23**, 660–679.

Ebersberger, B., Marsili, O., Reichstein, T. and Salter, A. (2006), 'Into thin air: using a quantile regression approach to explore the relationship between R&D and innovation', Working Paper, Tanaka Business School, Imperial College London.

Edler, J., Meyer-Krahmer, F. and Reger, G. (2002), 'Changes in the strategic management of technology: results of a global benchmarking study', *R&D Management*, **32**, 149–164.

European Commission (2000), 'Mid-term review of the 5th Framework Programme for research and technological development'. Commission Staff Working Paper SEC (2000)1780, 1–12.

European Commission (2001), 'The regions and the new economy. Guidelines for innovative actions under the ERFD in 2000–2006'. Communication from the Commission COM(2001)60final, 1–27.

European Commission (2003), 'Third European report on science & technology indicators, towards a knowledge-based economy', European Commission, Brussels.

European Commission (2005), 'Responsible partnering. Joining forces in a world of open innovation'. European Commission, Brussels.

European Commission (2008), 'A more research-intensive and integrated European Research area', European Commission, Luxemburg.

Falk, R. (2007), 'Measuring the effects of public support schemes on firms' innovation activities – survey evidence from Austria', *Research Policy*, **36**, 665–679.

Fischer, M. and Varga, A. (2002), 'Technological innovation and interfirm co-operation: an exploratory analysis using survey data from manufacturing firms in the metropolitan region of Vienna', *International Journal of Technology Management*, **24**, 724–742.

Fontana, R., Aldo, G. and Matt, M. (2004), 'Firm size and openness: the driving force of university–industry collaboration', EARIE Conference, Berlin, 2–5 September 2004.

Frantzen, D. (2000), 'R&D, human capital and international technology spillovers: a cross-country analysis', *Scandinavian Journal of Economics*, **102**, 57–75.

Fritsch, M. and Lukas, R. (2001), 'Who co-operates on R&D?', *Research Policy*, **30**, 297–312.

Gann, D. (2005), 'Open innovation: The new imperative for creating and profiting from technology', *Research Policy*, **34**, 122–123.

Gassmann, O. (2006), 'Opening up the innovation process: towards an agenda', *R&D Management*, **36**, 223–228.

Gassmann, O. and Enkel, E. (2004), 'Towards a theory of open innovation: three core process archetypes', Paper presented at the R&D Management Conference, Lisbon, Portugal, July 5–6, 1–18.

Georghiou, L. (2002), 'Additionality and impact of R&D subsidies', *IWT Studies*, **40**, 57–64.

Georghiou, L., Clarysse, B., Steurs, G., Bilsen, V. and Larosse, J. (2004), 'Making the difference – the evaluation of behavioural additionality of R&D subsidies', *IWT Studies*, **48**, 1–75.

Gereffi, G., Humphrey, J. and Sturgeon, T. (2005), 'The governance of global value chains', *Review of International Political Economy*, **12**, 78–104.

Gerfin, M. and Lechner, M. (2002), 'A microeconometric evaluation of the active labour market policy in Switzerland', *Economic Journal*, **112**, 854–893.

Gilley, K. and Rasheed, A. (2000), 'Making more by doing less: an analysis of outsourcing and its effects on firm performance', *Journal of Management*, **26**, 763–790.

Gold, A., Malhotra, A. and Segars, A. (2001), 'Knowledge management: an organizational capabilities perspective', *Journal of Management Information Systems*, **18**, 185–214.

Gonzalez, X., Jaumandreu, J. and Pázo, C. (2005), 'Barriers to innovation and subsidy effectiveness', *RAND Journal of Economics*, **36**, 930–950.

Granstrand, O. (ed.) (2003), *Economics, Law and Intellectual Property*. Kluwer Academic Publishers, Dordrecht.

Granstrand, O., Bohlin, E., Oskarson, C. and Sjöberg, N. (1992), 'External technology acquisition in large multitechnology corporations', *R&D Management*, **22**, 111–113.

Grant, R.M. (1996), 'Toward a knowledge-based theory of the firm', *Strategic Management Journal*, **17**, 109–122.

Greene, W.H. (1996), 'Marginal effects in the bivariate probit model', Working Paper EC-96-11, June.

Greene, W.H. (2003), *Econometric Analysis*. Fifth Edition. Prentice Hall, New Jersey.

Griffith, R., Huergo, E., Mairesse, J. and Peeters, B. (2006), 'Innovation and productivity across four European countries', *Oxford Review of Economic Policy*, **22**, 482–498.

Griffith, R., Redding, S. and Van Reenen, J. (2004), 'Mapping the two faces of R&D: productivity growth in a panel of OECD industries', *Review of Economics and Statistics*, **86**, 883–895.

Griliches, Z. (1990), 'Patent statistics as economic indicators: a survey', *Journal of Economic Literature*, **28**, 1661–1707.

Guellec, D. and Van Pottelsberghe de la Potterie, B. (2003), 'The impact of public R&D expenditure on business R&D', *Economics of Innovations and New Technology*, **12**, 225–244.

Gulati, R. and Singh, H. (1998), 'The architecture of cooperation: managing coordination costs and appropriation concerns in strategic alliances', *Administrative Science Quarterly*, **43**, 781–814.

Hagedoorn, J. (1993), 'Understanding the rationale of strategic technology partnering: interorganizational modes of cooperation and sectoral differences', *Strategic Management Journal*, **14**, 371–385.

Hagedoorn, J., Link, A. and Vonortas, N. (2000), 'Research partnerships', *Research Policy*, **29**, 567–586

Hall, B. and Van Reenen, J. (2000), 'How effective are fiscal incentives for R&D? A review of the evidence', *Research Policy*, **29**, 449–469.

Hall, B.H. (2002), 'The financing of research and development', *Oxford Review of Economic Policy*, **29**, 35–51.

Hansen, P. (1982), 'Large sample properties of generalized method of moments estimation', *Econometrica*, **50**, 1029–1054.

Heckman, J. (1979), 'Sample selection bias as a specification error', *Econometrica*, **47**, 153–161.

Helfat, C. (2006), 'Open innovation: the new imperative for creating and profiting from technology', *Academy of Management Perspectives*, **20**, 86–88.

Helper, S., MacDuffie, J.P. and Sabel, C. (2000), 'Pragmatic collaborations: advancing knowledge while controlling opportunism', *Industrial and Corporate Change*, **9**, 443–488.

Hofstede, G. (1991), *Cultures and Organizations: Software of the Mind*. McGraw-Hill, Berkshire.

Holmstrom, B. and Milgrom, P. (1994), 'The firm as an incentive system', *American Economic Review*, **84**, 972–991.

Howells, J. (1999), 'Research and technology oursourcing', *Technology Analysis and Strategic Management*, **11**, 17–29.

Howells, J. (2006), 'Intermediation and the role of intermediaries in innovation', *Research Policy*, **35**, 715–728.

Howells, J., Gagliardi, D. and Malik, K. (2008), 'The growth and management of R&D outsourcing: evidence from UK pharmaceuticals', *R&D Management*, **38**, 205–219.

Howells, J., James, A. and Malik, K. (2003), 'The sourcing of technological knowledge: distributed innovation process and dynamic change', *R&D Management*, **33**, 395–409.

Howells, J., James, A. and Malik, K. (2004), 'Sourcing external technological knowledge: a decision support framework for firms', *International Journal of Technology Management*, **27**, 143–155.

Hussinger, K. (2006), 'Is silence golden? Patents versus secrecy at the firm level', *Economics of Innovation and New Technology*, **15**, 735–752.

Huston, L. and Sakkab, N. (2006), 'Connect and develop: inside Procter & Gamble's new model for innovation', *Harvard Business Review*, **48,** 58–66.

Huston, L. and Sakkab, N. (2007), 'Implementing open innovation', *Research Technology Management*, **50**, 21–25.

Jacobides, M. and Billinger, S. (2006), 'Designing the boundaries of the firm: from "make, buy or ally" to the dynamic benefits of vertical architecture', *Organization Science*, **17**, 249–261.

Jacquemin, A. (1988), 'Cooperative agreements in R&D and European antitrust policy', *European Economic Review*, **32**, 551–560.

Jaffe, A.B. (1986), 'Technological opportunity and spillovers of R&D: evidence from firms' patents, profits and market value', *American Economic Review*, **76**, 984–1001.

Jaffe, A.B. (2002), 'Building programme evaluation into design of public research-support programmes', *Oxford Review of Economic Policy*, **18**, 22–34.

Jain, H. and Murray, V. (1984), 'Why the human resources management function fails', *California Management Review*, **26**, 95–110.

Jaumandreu, J. (2003), 'Does innovation spur employment? A firm-level analysis using Spanish CIS data'. Working Paper European Project 'Innovation and Employment in European Firms: Microeconometric Evidence', 1–14.

Kahn, S. (2000), 'The bottom-line impact of nonstandard jobs on companies' profitability and productivity', in: Carré, F., Ferber, M.A., Golden, L. and Herzenberg, S.A. (eds), *Nonstandard Work: The Nature and Challenges of Changing Employment Relations*. University of Illinois, Urbana-Champaign.

Kaiser, U. (2002), 'Measuring knowledge spillovers in manufacturing and services: an empirical assessment of alternative approaches', *Research Policy*, **31**, 125–144.

Kamien, M., Muller, E. and Zang, I. (1992), 'Research joint ventures and R&D cartels', *American Economic Review*, **82**, 1293–1292.

Kamien, M. and Zang, I. (2000), 'Meet me halfway: research joint ventures and absorptive capacity', *International Journal of Industrial Organization*, **18**, 995–1012.

Kang, S.-C., Morris, S.S. and Snell, S.A. (2007), 'Relatinal archetypes, organizational learning, and value creation: extending the human resource architecture', *Academy of Management Review*, **32**, 236–256.

Katsoulakos, Y. and Ulph, D. (1998), 'Endogenous spillovers and the performance of research joint ventures', *Journal of Industrial Economics*, **46**, 333–354.

Katz, M. and Ordover, J. (1990), 'R&D cooperation and competition', Brooking Papers on Economic Activity: Microeconomics, pp. 137–203.

Kesteloot, K. and Veugelers, R. (1995), 'Stable R&D cooperation with spillovers', *Journal of Economics of Management Strategy*, **4**, 651–672.

Klette, T.J., Moen, J. and Griliches, Z. (2000), 'Do subsidies to commercial R&D reduce market failures? Microeconometric evaluation studies', *Research Policy*, **29**, 471–495.

Klevorick, A., Levin, R., Nelson, R. and Winter, S. (1995), 'On the sources and significance of interindustry differences in technological opportunities', *Research Policy*, **24**, 185–205.

Kline, D.J. and Rosenberg, N. (1986), 'An overview of innovation', in: Landau, R. and Rosenberg, N. (eds), *The Positive Sum Strategy. Harnessing Technology for Economic Growth*. National Academy Press, Washington, pp. 275–305.

Klomp, L. and Van Leeuwen, G. (2001), 'Linking innovation and firm performance: a new approach', *International Journal of the Economics of Business*, **3**, 343–364.

Kogut, B. and Zander, U. (1992), 'Knowledge of the firm, combinative capabilities, and the replication of technology', *Organization Science*, **3**, 383–397.

Kogut, B. and Zander, U. (1993), 'Knowledge of the firm and the evolutionary theory of the multinational corporation', *Journal of International Business Studies*, **24**, 625–645.

Kogut, B. and Zander, U. (1996), 'What firms do? Coordination, identity, and learning', *Organization Science*, **7**, 502–518.

Kristensen, F. (1999), 'Towards a taxonomy and theory of the interdependence between learning regimes and sectoral patterns of innovation and competition: an empirical analysis of an elaborated Pavitt taxonomy applying Danish data', Paper presented at the DRUID 1999 winter conference, Seeland, Denmark.

Lach, S. (2002), 'Do R&D subsidies stimulate or displace private R&D? Evidence from Israel', *Journal of Industrial Economics*, **50**, 369–390.

Lam, A. (2000), 'Tacit knowledge, organisational learning and societal institutions: an integrated framework', *Organization Studies*, **21**, 487–513.

Lam, A. (2005), 'Work roles and careers of R&D scientists in network organizations', *Industrial Relations*, **44**, 242–275.

Lane, P., Koka, P. and Pathak, S. (2006), 'The reification of absorptive capacity: a critical review and rejuvenation of the construct', *Academy of Management Review*, **31**, 833–863.

Lane, P.J. and Lubatkin, M. (1998), 'Relative absorptive capacity and inter-organizational learning', *Strategic Management Journal*, **19**, 461–477.

Laursen, K. and Salter, A. (2004), 'Searching high and low: what types of firms use universities as a source of innovation?', *Research Policy*, **33**, 1201–1215.

Laursen, K. and Salter, A. (2006), 'Open for innovation: The role of openness in explaining innovation performance among UK manufacturing firms', *Strategic Management Journal*, **27**, 131–150.

Laursen, K., Paananen, M. and Salter, A. (2007), 'Profiting from openness: exploring the relation between profits and openness among innovative firms', Paper presented at the DRUID conference, Copenhagen, Denmark, 18–20 June.

Lazaric, N. and Marengo, L. (2000), 'Towards a characterization of assets and knowledge created in technological agreements: some evidence from the automobile robotics sector', *Industrial and Corporate Change*, **9**, 53–86.

Lazonick, W. (2005), 'The innovative firm', in: Fagerberg, J., Mowery, D. and Nelson, R. (eds). *Understanding innovation.* Oxford University Press, New York, pp. 29–55.

Lechner, M. (1999), 'Earnings and employment effects of continuous off-the-job training in East Germany after reunification', *Journal of Business and Economics Statistics*, **17**, 74–90.

Lechner, M. (2001), 'Identification and estimation of causal effects of multiple treatments under the conditional independence assumption', in: Lechner, M. and Pfeiffer, F. (eds), *Econometric Evaluation of Active Labour Market Policies*. Physica-Verlag, Heidelberg, pp. 43–58.

Lichtenthaler, U. (2005), 'External commercialisation of knowledge: review and research agenda', *International Journal of Management Reviews*, **7**, 231–255.

Liebeskind, J.P. (1997), 'Keeping organizational secrets: protective institutional mechanisms and their costs', *Industrial and Corporate Change*, **6**, 623–663.

Lööf, H. and Heshmati, A. (2002), 'Knowledge capital and performance heterogeneity: a firm level innovation study', *International Journal of Production Economics*, **76**, 61–85.

Lööf, H. and Heshmati, A. (2006), 'Knowledge capital and heterogeneity in firm performance: a sensitivity analysis', *Economics of Innovation and New Technology*, **15**, 317–344.

Love J.H. and Roper, S. (2001), 'Outsourcing in the innovation process: locational and strategic determinants', *Papers in Regional Science*, **80**, 317–336.

Lundvall, B. (1992), *National Systems of Innovation. Towards a Theory of Innovation and Interactive Learning*. Pinter Publishers, London.

Lundvall, B.-Å. and Nielsen, P. (1999), 'Competition and transformation in the learning economy - the Danish case', *Revue d'Economie Industrielle*, **88**, 67–90.

Luukkonen, T. (2001), 'Networking impacts of the EU Framework Programme', in: OECD, Innovative Networks: Co-operation in National Innovation Systems. OECD, Paris, pp. 193–208.

Malhotra, Y. (2003), 'Knowledge assets of a nation: knowledge systems for development', Research paper prepared for the invited keynote presentation at ad hoc group of expert meeting for knowledge systems development, United Nations.

Mansfield, E. and Romeo, A. (1980), 'Technology transfer to overseas subsidiaries by U.S.-based firms', *Quarterly Journal of Economics*, **95**, 737–750.

Marsili, O. and Verspagen, B. (2002), 'Technology and the dynamics of industrial structures: an empirical mapping of Dutch manufacturing', *Industrial and Corporate Change*, **11**, 791–815.

Martin, S. (2002), 'Spillovers, appropriability, and R&D', *Journal of Economics*, **75**, 1–32.

Martin, S. and Scott, J. (2000), 'The nature of innovation market failure and the design of public support for private innovation', *Research Policy*, **29**, 437–447.

McGaughey, S.L. (2002), 'Strategic interventions in intellectual asset flows', *Academy of Management Review*, **27**, 248–274.

Mehta, S. and Peters, L. (2007), 'Outsourcing a core competency', *Research and Technology Management*, **50**, 26 34.

Milgrom, P. and Roberts, J. (1987), 'Informational asymmetries, strategic behavior, and industrial organization', *American Economic Review*, **77**, 184–193.

Miotti, L. and Sachwald, F. (2003), 'Cooperative R&D: why and with whom? An integrated framework analysis', *Research Policy*, **32**, 1481–1499.

Mohnen, P. and Hoareau, C. (2003), 'What type of enterprise forges close links with universities and government labs? Evidence from CIS2', *Managerial and Decision Economics*, **24**, 133–145.

Mol, M.J. (2005), 'Does being R&D intensive still discourage outsourcing? Evidence from Dutch manufacturing', *Research Policy*, **34**, 571–582.

Monjon, S. and Waelbroeck, P. (2003), 'Assessing spillovers from universities to firms: evidence from French firm-level data', *International Journal of Industrial Organization*, **21**, 1255–1270.

Muldur, U., Corvers, F., Delanghe, H., Dratwa, J., Heimberger, D., Sloan, B. and Vanslembrouck, S. (2006), *A New Deal for an Effective European Research Policy: The Design and Impacts of the 7th Framework Programme*. Springer, Dordrecht.

Nahm, J.H. (2001), 'Nonparametric quantile regression analysis of the R&D-turnover relationship for Korean firms', *Empirical Economics*, **26**, 259–270.

Narula, R. (2001), 'Choosing between internal and non-internal R&D activities: some technological and economic factors', *Technology Analysis and Strategic Management*, **13**, 365–387.

Nauwelaers, C. (2001), 'The S&T institutional profile', in: Cincera, M. and Clarysse, B. (eds), *Belgian Report on Science, Technology and Innovation*. Federal Office for Scientific, Technical and Cultural Affairs, Brussels. pp. 129–256.

Nauwelaers, C. (2004), 'De instellingen en het beleid inzake wetenschap, technologie en innovatie in België, Belgian report on science, technology and innovation', Federal Office for Scientific, Technical and Cultural Affairs, Brussels.

Nauwelaers, C. (2009), 'Policy mixes for R&D in Europe', a study commissioned by the European Commission, Directorate General for Research. European Commission, Brussels.

Negassi, S. (2004), 'R&D co-operation and innovation a microeconometric study on French firms', *Research Policy*, **33**, 365–384.

Nelson, R.R. (ed.) (1993), *National Innovation Systems: A Comparative Analysis*. Oxford University Press, New York.

Nieto, M. and Quevedo, P. (2005), 'Absorptive capacity, technological opportunity, knowledge spillovers, and innovative effort', *Technovation*, **25**, 1141–1157.

Nooteboom, B. (2001), 'From evolution to language and learning', in: Foster, J. and Metcalfe, J.S. (eds), *Frontiers of Evolutionary Economics. Competition, Self-Organization and Innovation Policy*. Edward Elgar, Cheltenham, pp. 41–69.

OECD (1995), *Manual on the Measurement of Human Resources Devoted to S&T*. OECD, Paris.

OECD (1997), *Oslo Manual. The Measurement of Scientific and Technological Activities. Proposed Guidelines for Collecting and Interpreting Technological Innovation Data*. Second Edition. OECD, Paris.

OECD (2001), *Innovative Networks: Co-operation in National Innovation Systems*. OECD, Paris.

OECD (2002a), *Frascati Manual. The Measurement of Scientific and Technological Activities: Proposed Standard Practice for Surveys on Research and Experimental Development*. OECD, Paris.

OECD (2002b), *Benchmarking Industry–Science Relations*. OECD, Paris.

OECD (2005), *Oslo Manual. Guidelines for Collecting and Interpreting Innovation Data. Third Edition*. OECD, Paris.

OECD (2007), *Main Science and Technology Indicators*. OECD, Paris.

OECD (2008a), *OECD Science, Technology and Industry Outlook*. OECD, Paris.

OECD (2008b), *The Internationalisation of Business R&D – Evidence, Impacts and Implications*. OECD, Paris.

OECD (2008c), *Globalisation and Open Innovation*. OECD, Paris.

OMC/CREST (2007), *Peer Review of the Belgian Policy Mix*. OMC/CREST, Paris.

O'Regan, N., Ghobadian, A. and Gallear, D. (2006), 'In search of the drivers of high growth in manufacturing SMEs', *Technovation*, **26**, 30–41.

Parisi, M.L., Schiantarelli, F. and Sembenelli, A. (2006), 'Productivity, innovation and R&D: micro evidence for Italy', *European Economic Review*, **50**, 2037–2061.

Pavitt, K. (1984), 'Sectoral patterns of technical change: towards a taxonomy and a theory', *Research Policy*, **13**, 343–373.

Perkmann, M. and Walsh, K. (2007), 'University–industry relationships and open innovation: towards a research agenda', *International Journal of Management Reviews*, **9**, 259–280.

Pianta, M. (2004), 'The impact of innovation on jobs, skills and wages', *Economia e Lavoro*, **1**, 7–26.

Piga, C. and Vivarelli, M. (2004), 'Internal and external R&D: a sample selection approach', *Oxford Bulletin of Economics and Statistics*, **66**, 457–482.

Pisano, G. (1990), 'The R&D boundaries of the firm: an empirical analysis', *Administrative Science Quarterly*, **35**, 153–176.

Piva, M., Santarelli, E. and Vivarelli, M. (2003), 'The skill bias effect of technological and organisational change: evidence and policy implications', IZA Discussion Papers 934, Institute for the Study of Labor (IZA).

Polt, W. (2001), 'The role of governments in networking', in: OECD (ed.), *Innovative Networks: Co-operation in National Innovation Systems*. OECD, Paris, pp. 307–320.

Polt, W. and Psarra, F. (2006), 'Behavioural additionality of the EU's 5th Framework Programme', in: OECD (ed.), *Government R&D Funding and Company Behaviour*. OECD, Paris.

Powell, W.W., Koput, K.W. and Smith-Doerr, L. (1996), 'Interorganizational cooperation and the locus of innovation: networks of learning in biotechnology', *Administrative Science Quarterly*, **41**, 116–145.

Quinn, J.B. (2000), 'Outsourcing innovation. The new engine of growth', *MIT Sloan Management Review*, **41**, 13–28.

Redding, S. (1996), 'The low-skill, low-quality trap: strategic complementarities between human capital and R&D, *Economic Journal*, **106**, 458–470.

Roach, M. and Sauermann, H. (2010), 'A taste for science? PhD scientists' academic orientation and self-election into research careers in industry', *Research Policy*, **39**, 422–434.

Roijakkers, N., van Ekert, M. and Hagedoorn, J. (1994), 'The spatial dimension of R&D networks', Report to the European Commission, STI-NET Project. MERIT, Maastricht.

Röller, L., Tombak, M. and Siebert, R. (1997), 'Why firms form research joint ventures: theory and evidence', CEPR Discussion Paper 1654.

Romer, P. (1990), 'Endogenous technological change', *Journal of Political Economy*, **98**, 71–102.

Roper, S., Du, J. and Love, J.H. (2008), 'Modelling the innovation chain', *Research Policy*, **37**, 961–977.

Rosenbaum, P. and Rubin, D. (1983), 'The central role of the propensity score in observational studies for causal effects', *Biometrika*, **70**, 41–55.

Rosenberg, N. (1990), 'Why do firms do basic research (with their own money)?', *Research Policy*, **19**, 165–174.

Rothaermel, F.T., Hitt, M.A. and Jobe, L.A. (2006), 'Balancing vertical integration and strategic outsourcing: effects on product portfolio, product success, and firm performance', *Strategic Management Journal*, **27**, 1033–1056.

Rothwell, R. (1992), 'Successful industrial innovation: critical factors in the 1990s', *R&D Management*, **22**, 221–239.

Rothwell, R. and Zegveld, W. (1985), *Reindustrialization and Technology*. Longman Group Ltd, Essex.

Rubin, D. (1977), 'Assignment to treatment group on the basis of covariate', *Journal of Educational Statistics*, **2**, 1–26.

Sauermann, H. and Cohen, W. (2008), 'What makes them tick? Employee motives and industrial innovation', NBER Working Paper no. 14443.

Scherer, F.M. and Huh, K. (1992), 'Top managers' education and R&D investment', *Research Policy*, **21**, 507–511.

Schibany, A. and Schartinger, D. (2001), 'Interactions between universities and enterprises in Austria: an empirical analysis at the micro and sector levels', in: OECD (ed.), *Innovative Networks: Co-operation in National Innovation Systems*. OECD, Paris, pp. 235–252

Schilling, M.A. (2008), *Strategic Management of Technological Innovation* (Second edition). McGraw-Hill, Boston.

Schmookler, J. (1966), *Invention and Economic Growth*. Harvard University Press, Cambridge (Mass.).

Schwartz, S. (1999), 'A theory of cultural values and some implications for work', *Applied Psychology: An International Review*, **48**, 23–47.

Sena, V. (2004), 'The return of the prince of Denmark: a survey on recent developments in the economics of innovation', *Economic Journal*, **114**, 312–332.

Shapiro, C. and Willig, R. (1990), 'On the antitrust treatment of production joint ventures', *Journal of Economic Perspectives*, **4**, 113–130.

Soete, L. (2007), *Eindrapport - Expertgroep voor de Doorlichting van het Vlaams Innovatie-Instrumentarium*. UNU-MERIT, Maastricht.

Sorenson, O., Rivkin J.W. and Fleming, L. (2006), 'Complexity, networks and knowledge flow', *Research Policy*, **35**, 994–1017.

Spence, M. (1984), 'Cost reduction, competition and industry performance', *Econometrica*, **52**, 101–121.

Spithoven, A. (2005), 'Internationalisation of innovation in a small open economy. Exploring the role of ownership structure on performances and R&D collaboration', in: Spithoven, A. and Teirlinck, P. (eds), *Beyond Borders. Internationalisation of R&D and Policy Implications for Small Open Economies*. Elsevier, Amsterdam, pp. 259–295.

Spithoven, A. (2007), *Kennisproductie in België. Analyse van de O&O-Activiteiten in de Non-Profitsector*. Federaal Wetenschapsbeleid, Brussel.

Spithoven, A. and Teirlinck, P. (2005), 'De appreciatie van netwerkvorming in innovatiestrategieën', *Tijdschrift voor Economie en Management*, **51**, 269–307.

Spithoven, A. and Teirlinck, P. (2006), 'The geography of collaboration in times of open innovation: the position of small open European economies', Paper presented at the conference, 'Shaping EU Regional Policy', Regional Studies Association, Leuven, Belgium.

Spithoven, A. and Teirlinck, P. (2010), 'External R&D: exploring the functions and qualifications of R&D personnel', *International Journal of Innovation Management*, **14**, 967–987.

Spithoven, A., Frantzen, D. and Clarysse, B. (2010), 'Heterogeneous effects of knowledge exchanges on product innovation: differences between lagging and dynamic product innovators', *Journal of Product Innovation Management*, **27**, 363–382.

Storper, M. (1997), *The Regional World. Territorial Development in a Global Economy*. The Guilford Press, New York.

Stuart, T. and Ding, W. (2006), 'When do scientists become entrepreneurs? The social structural antecedents of commercial activity in the life sciences', *American Journal of Sociology*, **112**, 97–144.

Swan, K. and Allred, B. (2003), 'A product and process model of the technology-sourcing decision', *Journal for Product and Innovation Management*, **20**, 485–497.

Tarabusi, C. (1997), 'Technology and employment: the role of organisational change and learning', *STI Review*, **20**, OECD, Paris.

Teece, D.J. (1986), 'Profiting from technological innovation: implications for integration, collaboration, licensing and public policy', *Research Policy*, **15**, 285–305.

Teece, D.J., Pisano, G. and Shuen, A. (1997), 'Dynamic capabilities and strategic management', *Strategic Management Journal*, **18**, 509–533.

Teirlinck, P., Dumont, M. and Spithoven, A. (2010), 'Corporate decision-making in R&D outsourcing and the impact on internal R&D employment intensity', *Industrial and Corporate Change*, **19**, 1741–1768.

Teirlinck, P. and Spithoven, A. (2008), 'The spatial organization of innovation: open innovation, external knowledge relations and urban structure', *Regional Studies*, **42**, 689–704.

Tether, B.S. (2002), 'Who co-operates for innovation, and why? An empirical analysis', *Research Policy*, **31**, 947–967.

Tether, B.S. and Tajar, A. (2008), 'Beyond industry–university links: sourcing knowledge for innovation from consultants, private research organisations and the public science-base', *Research Policy*, **37**, 1079–1095.

Tidd, J., Bessant, J. and Pavitt, K. (1997), *Integrating Technological, Market and Organisational Change*. John Wiley and Sons Ltd, Chichester.

Tiwana, A. and Keil, M. (2007), 'Does peripheral knowledge complement control? An empricial test in technology outsourcing alliances', *Strategic Management Journal*, **28**, 623–634.

Todorova, G. and Durisin, B. (2007), 'Absorptive capacity: valuing a reconceptualization', *Academy of Management Review*, **32**, 774–786.

Trefler, D. (2005), 'Offshoring: threats and opportunities', in: Collins, S. and Brainard, L. (eds), *Offshoring White-Collar Work – The Issues and the Implications*. Brookings Institution Press, New York.

Trott, P. (1998), *Innovation Management and New Product Development*. Financial Times Management, London.

Tushman, M. and Anderson, Ph. (1997), *Managing Strategic Innovation and Change: A Collection of Readings*. Oxford University Press, Oxford.

Van Leeuwen, G. and Klomp, L. (2006), 'On the contribution of innovation to multi-factor productivity growth', *Economics of Innovation and New Technology*, **15**, 367–390.

Verbeek, M. (2000), *A Guide to Modern Econometrics*. John Wiley and Sons Ltd, Chichester.

Veugelers, R. (1997), 'Internal R&D expenditures and external technology sourcing', *Research Policy*, **26**, 303–315.

Veugelers, R. and Cassiman, B. (1999), 'Make and buy in innovation strategies: evidence from Belgian manufacturing firms', *Research Policy*, **28**, 63–80.

Veugelers, R. and Cassiman, B. (2005), 'R&D cooperation between firms and universities. Some empirical evidence from Belgian manufacturing', *International Journal of Industrial Organization*, **23**, 355–379.
von Hippel, E. (1988), *The Sources of Innovation*. Oxford University Press, New York.
Walker, G. and Weber, D. (1984), 'A transaction cost approach to make-buy decisions', *Administrative Science Quarterly*, **29**, 373–391.
Wallsten, S. (2000), 'The effects of government – industry R&D programs on private R&D: the case of the SBIR program', *RAND Journal of Economics*, **31**, 82–100.
Wang, S. and Ariguzo, G. (2004), 'Knowledge management through the development of information schema', *Information and Management*, **41**, 445–456.
West, J., Vanhaverbeke, W. and Chesbrough, H. (2006), 'Open innovation: a research agenda', in: Chesbrough, H., Vanhaverbeke, W. and West, J. (eds), *Open Innovation. Researching a New Paradigm*. Oxford University Press, Oxford, pp. 285–307.
Wheelwright, S.C. and Clark, K.B. (1992), *Revolutionizing Product Development. Quantum Leaps in Speed, Efficiency, and Quality*. The Free Press, New York.
Williamson, O. (1975), *Markets and Hierarchies: Analysis and Antitrust Implications*. The Free Press, New York.
Williamson, O. (1981), 'The economics of organization: the transaction cost approach', *American Journal of Sociology*, **87**, 548–577.
Williamson, O.E. (1985), *The Economic Institutions of Capitalism*. The Free Press, New York.
Wilson, R. (1977), 'The effects of technological environment and product rivalry on R&D efforts and licensing of inventions', *Review of Economics and Statistics*, **59**, 171–178.
Wong, K. and Aspinwall, E. (2004), 'Knowledge management implementation frameworks: a review', *Knowledge and Process Management*, **11**, 92–104.
Zaheer, A., McEvily, B. and Perrone, V. (1998), 'Does trust matter? Exploring the effects of interorganizational and interpersonal trust on performance', *Organization Science*, **9**, 141–159.
Zahra, L.G. and George, G. (2002), 'Absorptive capacity: a review, reconceptualization and extension', *Academy of Management Review*, **27**, 185–203.

Index